Modeling in Computational Biology and Biomedicine

Frédéric Cazals • Pierre Kornprobst
Editors

Modeling in Computational Biology and Biomedicine

A Multidisciplinary Endeavor

Springer

Editors
Frédéric Cazals
Inria Sophia Antipolis Méditerranée
Algorithms-Biology-Structure
Sophia Antipolis
France

Pierre Kornprobst
Inria Sophia Antipolis Méditerranée
Neuromathcomp project team
Sophia Antipolis
France

Foreword by:
Olivier Faugeras
Inria Sophia Antipolis Méditerranée
Neuromathcomp project-team
Sophia Antipolis
France

Joël Janin
Institut de Biochimie et BiologieMoléculaire
Université Paris-Sud
Orsay
France

ISBN 978-3-642-31207-6 ISBN 978-3-642-31208-3 (eBook)
DOI 10.1007/978-3-642-31208-3
Springer Heidelberg New York Dordrecht London

Library of Congress Control Number: 2012949649

Printed on acid-free paper

Springer is part of Springer Science+Business Media (www.springer.com)

Foreword

This book is about the use and usefulness of mathematical and computational models in biology and biomedicine. Using mathematical models in the natural sciences, in particular in physics, is not new (it can be traced back to ancient Greek scholars and even further in the past) and has proved to be exceptionally successful. The use of mathematical models in the life sciences, in particular biology and biomedicine, is more recent and its success or lack of success is still heartily debated.

In 1960, the physicist and Nobel Prize winner Eugene Wigner wrote a famous article [7] entitled "The unreasonable effectiveness of mathematics in the natural sciences" in which he stated that "the enormous usefulness of mathematics in the natural sciences is something bordering on the mysterious" and that "there is no rational explanation for it." Without entering the discussion about the existence or not of a rational explanation for this effectiveness of mathematics (see, e.g., [3]), it is sufficient for this preface to say that he was of course thinking mostly of physics, and I want to briefly examine how this statement applies to life sciences.

First I would like to quote Israel Moiseevich Gelfand, a famous mathematician who also worked in biomathematics and molecular biology: "There is only one thing which is more unreasonable than the unreasonable effectiveness of mathematics in physics, and this is the unreasonable ineffectiveness of mathematics in biology." This citation can be found in an article by Lesk [4] discussing a subject closely related to this preface, where it is proposed to call the union of these two principles the Wigner-Gelfand principle.

In order to understand this dilemmna we examine the different meanings of mathematical effectiveness. Following Étienne Klein, it is convenient to distinguish three such meanings:

1. The ability to produce predictions or retrodictions. The model should be able to predict new experimental findings or to reproduce existing data. This is the usual acceptation of the word effectiveness.
2. The ability to provide "explanatory" structures that go beyond the strictly predictive ability. This was also pointed out by René Thom in [6] who stressed

the importance of not confusing the two ideas of prediction and explanation. Note that the ability to explain goes hand in hand with the ability to unify since to explain is to derive the diversity of phenomena from a small number of principles.

3. The ability to generate new ideas, concepts, or even original answers to old questions. This aspect has been particularly stressed by Alain Connes in his book written jointly with Jean-Pierre Changeux [2] (English translation of [1]).

I would be inclined to say that if physics can be proud of having developed theories that display all three abilities, called superb theories in Roger Penrose's classification [5], the life sciences and in particular biology and biomedicine are still struggling with the development of theories at level 1.

Does this mean that we will never be able to reach in biology and biomedicine the level of the theoretical elegance and effectiveness that has been achieved in physics and to prove wrong the Gelfand part of the Wigner-Gelfand principle?

I believe that the answer to the first part of the question is no and to the second, yes: we will eventually produce as powerful theories as it has been the case in physics and this book is a magnificent example of the kind of progress that has been made in the last few years in this direction.

Sophia Antipolis, France Olivier Faugeras
May 2012

References

1. J. Changeux and A. Connes, *Matière à pensée*, vol. 22, Paris: Odile Jacob, 1989.
2. J.-P. Changeux and A. Connes, *Conversations on Mind, Matter, and Mathematics*, Princeton N. J.: Princeton University Press, 1995. Edited and translated by M.B. DeBevoise.
3. N. Goodman, *Mathematics as natural science*, The Journal of symbolic logic, 55 (1990), pp. 182–193.
4. A. Lesk, *The unreasonable effectiveness of mathematics in molecular biology*, The Mathematical Intelligencer, 22 (2000), pp. 28–37.
5. R. Penrose and M. Gardner, *The emperor's new mind: concerning computers, minds, and the laws of physics*, Oxford University Press, USA, 1999.
6. R. Thom, *Prédire n'est pas expliquer*, Eshel, 1991.
7. E. Wigner, *The unreasonable effectiveness of mathematics in the natural sciences*, Communications in Pure and Applied Mathematics, 13 (1960), pp. 1–14.

Foreword

Computational biology was born in the late 1960s at the boundary of chemical physics with two emerging fields: molecular biology and computer science. Modeling the structure of proteins from their sequence was its main objective at the time [1]. That proved to be a hard problem, solved only in recent years, and only for small proteins, but meanwhile, computational biology has diversified to cover many aspects of life sciences and biomedicine, and bioinformatics has developed at a fast pace as part of genomic and post-genomic sciences. Computational biology and bioinformatics now form a meeting ground for mathematics and the study of life. These two old human activities blend together readily in Ancient Greece, but examples of productive encounters in modern times are not that many. Personally, I would pick D'Arcy Wentworth Thompson (1860–1948), a polymath who translated Aristotle and wrote in *On Growth and Form*: "when (the zoologist) meets with a simple geometrical construction, for instance in the honeycomb, he would fain refer it to psychical instinct, or to skill and ingenuity, rather than to the operation of physical forces or mathematical laws." My other favorite example is the "plausible model" of allosteric protein regulations that Monod, Wyman, and Changeux published in 1965 [3]. Monod's fascination with symmetry [2] is at the heart of that model, but like D'Arcy Thompson, Monod was an exception. The point of view that prevailed among biologists and mathematicians in the 1960s was still the one Claude Bernard had developed in his *Introduction à la Médecine Expérimentale* a century before: life is too complex and poorly understood to be formalized in mathematics. If anything, the large body of knowledge that has accumulated afterwards, has proved the complexity of life to be far beyond what Bernard could imagine, and we still understand very little. Nevertheless, we should consider Bernard's conclusion as mistaken. Mathematics deserves a central position in biology and in medical research as it does in other fields of science. Mathematicians view complexity as a challenge that stimulates their curiosity, and biologists need mathematics to make the best use of their data, and turn their intuitions into robust models.

The very complexity of the living matter implies that biologists reason on models rather than on the objects themselves. A good model is one that recapitulates both what we know of the system under study and the hypotheses we make about it. Initially, data and hypotheses are all mixed together and the model is subjective. In the next step, it must become explicit and take a form that can be checked in an objective manner. Obviously, that form will be different in different fields, but, in all cases, it will involve some sort of mathematics. The model of allostery proposed by Monod et al. [3] led to a particular expression of a partition function subject to symmetry constraints. When applied to hemoglobin, the function accurately described its response to the concentration of oxygen, pH, and other environmental factors. A few years later, the same response could be modeled at the atomic level, thanks to Perutz's crystal structure of hemoglobin. Modeling the biological function from a molecular structure is now common practice in protein, DNA, or RNA studies. The atomic coordinates of these macromolecules, stored in the Protein Data Bank, can be used in many ways to model their geometry by ribbons folded in space, by surfaces, sets of hard spheres, Voronoï polyhedra, etc. But the model can also take the form of a force field or an energy surface that describes how the structure responds as the macromolecule interacts with other components of the cell, like oxygen with hemoglobin. In other circumstances, the model can be a probability distribution, a matrix or a graph. Each representation makes assumptions and approximations which are also part of the model. They too must be made explicit, since they may render the model suitable for a given application, and quite wrong in another. Students and scientists active in life and health sciences do not necessarily have the mathematical background needed to build a formal model. If it involves elaborate geometry, functions, or graphs, only expert mathematicians may be in a position to make the best use of it. However, the biologists must be aware of what formalization can offer, and the mathematicians must understand the nature of the questions being asked. The book edited by F. Cazals and P. Kornprobst, which addresses both publics and covers modeling methods in several major aspects of biology and biomedicine from structural biology to neuroscience, will undoubtedly contribute to making the encounter fertile.

Orsay, France
May 2012

Joël Janin

References

1. M. Levitt M and S. Lifson. Refinement of protein conformations using a macromolecular energy minimization procedure. *J. Mol. Biol.*, 46:269–79, 1969.
2. J. Monod. On symmetry and function in biological systems. In Arne Engström and Bror Strandberg, editors, *Symmetry and function of biological systems at the macromolecular level, Nobel Symposium #11*, pages 15–27. Almqvist and Wiksell, 1968.
3. J. Monod, J. Wyman, and J-P. Changeux. On the nature of allosteric transitions: a plausible model. *J. Mol. Biol.*, 12:88–118, 1965.

Preface

Biology is concerned with living systems and their constituents at all scales, namely, molecule, cell, tissue, organ, individual, organism, and ecosystem. Studies which aim at describing, understanding, and monitoring these systems are grounded in the data provided by experiments and observations. While such activities had a prominent qualitative component, as illustrated by biological classifications and taxonomies, the advent of novel experimental techniques has triggered a shift to the quantitative side. Two examples will illustrate these dramatic changes. At one end of the scale, where crystallizing a protein used to be a tour de force, for which Kendrew and Perutz were awarded the Nobel Prize for Chemistry in 1962, high-throughput structural genomics projects nowadays allow crystallizing hundreds of new proteins and complexes per week. At the other end of the scale, sequencing a genome used to be another tour de force, for which Sanger was awarded a Nobel Prize (his second) in Chemistry in 1980. Nowadays, the genomes of whole populations are uncovered by meta-genomics projects, shedding light on a given ecosystem as a whole. Similar examples can be found at the intermediate scales as well.

This spectacular progress is due to a synergy between technological advances, allowing data acquisition on biological objects at complementary scales, and conceptual syntheses of ideas from biology, physics, chemistry, mathematics, statistics, and computer science. Such advances are motivated not only by outstanding scientific challenges but also by the potential value added by new protocols and techniques in biotechnologies and (bio-)medicine. A principal goal of this book is to illustrate that in modeling biological systems, deeper insights can be gained using more advanced mathematical and algorithmic developments that implicate a wide spectrum of techniques from applied mathematics and computer science. But if modeling in biological sciences is expanding rapidly, the specificities of the data dealt with and the different natures of the mother disciplines must be understood in order to develop a full synergy. Before discussing the contents of the book, we therefore briefly address selected properties of biological systems, as well as characteristics of mathematical and algorithmic modeling.

Specificities of Biological Systems

Biological Systems Integrate Multiple Scales

The fact that biological systems span multiple scales is obvious from the enumeration given above, which encompasses molecules to ecosystems. But in complex organisms, apprehending a complex function may also involve multi-scale and global considerations. One such example is the regulation of blood pressure in vertebrates. This regulation indeed involves molecules (e.g., the thyroid hormones affecting the calcium homeostatis), tissues (e.g., the cardiac cells accounting for electric phenomena), as well as whole organs (the heart of course, but also the kidneys which regulate electrolytes and the volume of the blood). Modeling such a complex phenomenon requires integrating across these scales, as well as coupling together models of the individual building blocks.

Biological Systems Are Governed by a Mix of Deterministic and Probabilistic Behaviors

In physics, the study of a perfect gas at the atomic level starts with the computation of its mean free path, while a macroscopic description is given by the law of perfect gases. The same holds for a number of biological systems: while their description at the finest scale requires probabilistic models, integrating across scales typically yields deterministic behaviors. This integration explains the (seemingly) deterministic phenotypes of living systems. But the probabilistic nature of computations at a fine scale may also affect coarser ones. An example is perceptual multi-stability which is observed in different sensory modalities such as vision. If the underlying neural mechanisms are still unclear (i.e., physiological correlates of perceptual switches, origins of transitions, or populations dynamics), it is clear that the influence of noise underpins multistability. Thus, deterministic and/or probabilistic models may be developed independently or jointly, depending on the phenomenon of interest.

The Complexity of Biological Systems Warrants Phenomenological Models

Biological systems have evolved under the dual mechanism of mutation and selection. They often involve redundant features, which may be seen as backups. In metabolic phenomena, for example, alternative, competing pathways related to a particular function often exist, and one may replace the other in case of failure, such as a mutation impairing a particular protein. This complexity touches on the very nature of biology itself, and it is unclear whether simple laws governing the behavior of complex systems will ever be discovered. It is in sharp contrast with physics, where Newton's or Coulomb's laws, just to mention two, have been instrumental in modeling gravitation and electrostatic interactions, respectively, and in triggering technological developments. This observation explains why modeling for biology comes in two guises. On the one hand, selected models are derived from first principles. On the other hand, phenomenological models are developed based on the investigation of correlations between parameters describing the system of interest, calling for methods in the realm of machine learning and inferential modeling.

The Variability of Biological Systems Calls for Statistical Assessments

A cell in a given tissue shares the phenotype of its mates concurring in the function contributed by the organ, but may have specific features as well, for example, if one of its genes has undergone a mutation. The hearts of two humans share the same design pattern, but each may also have specific features. These two examples illustrate two key features of biological models. First, generic models accommodating individual-specific variations are called for. Second, the parameters used to single out specific properties should be scrutinized under the lens of statistical assessments.

Modeling for Biology and Biomedicine

The fact that biological systems provide a mine of complex modeling problems should not come as a surprise. Yet any mathematical or algorithmic development for biological sciences requires reconciling somewhat different perspectives, and a number of specificities need to be accepted in order to appreciate the contributions discussed in this book.

System-Centric Versus Generic Developments

Biology is often a system-centric activity, with a focus on a cell, an organ, or a pathology, while mathematics and computer science aim at exhibiting general properties and algorithms which may be instantiated in a number of settings.

Ill-Posed Problems, Models, and Their Validation

While mathematics and computer science have traditionally been concerned with well-posed problems, biology essentially raises ill-posed problems. In fact, modeling in computational biology and biomedicine is equally, if not more, about designing models, than it is about solving well-posed problems. Also, since models are simplified representations of complex phenomena, validations are called for. In fact, any model should be confronted to experiments at some point, in order to be confirmed or falsified. Numerous such examples will be discussed in this book.

Multidisciplinary Models

Models for complex biological systems are often multidisciplinary as they involve mathematics, physics, biology, and computer science.

Biology is of course the starting point, as any modeling study requires circumscribing the properties of interest and the data involved, but it is also the yardstick to be used for assessment (knowledge from the literature, additional data, etc). In making these specifications, one needs to compromise between biological accuracy on the one hand and conceptual simplicity on the other, as the latter warrants uncluttered models retaining the essential features (leaving aside the amenability of such models to calculation and simulation). Mathematics allows specifying abstract properties of the system studied (e.g., geometry, topology, hierarchical organization, and dynamics) and allows solving systems of equations, performing statistics, etc. Physics and chemistry are used to endow this abstract model with selected properties, for example, mechanical or electrical. Biology provides the semantics of the model by embedding it in a biological context. Finally, computer science allows automating certain tasks, running simulations and performing analyses. The role of simulations in investigating complex systems cannot be overstated; such simulations can be seen as numerical experiments, from which properties can be inferred.

We note in passing that the systems simulated may be continuous or discrete, and the question of discretizing continuous biological processes while retaining essential properties is always a critical one.

Mathematical and Algorithmic Contributions

From a transverse, rather than applied, perspective, mathematics and computer science can contribute to biology in at least two ways. First, an existing methodological development may be improved in terms of robustness and efficiency. These aspects include mastering the numerics involved in floating-point calculations, improving the convergence properties of algorithms, for optimization in particular, and designing algorithms with enhanced asymptotic properties – which *scale better* as biologists might put it.

Second, and perhaps more importantly, concepts and algorithms from mathematics and computer science may lay the groundwork for more advanced and more accurate models, and several topics in mathematics and computer science are currently undergoing major developments. One of them is stochastic modeling, which is especially prominent in problems from neurosciences, such as interpreting spike trains. Another is inverse problem solving, for inverse problems are faced not only in physiology and neurosciences but also in image processing. In the former, one wishes to infer the inner structure of an organ, for example, the heart or the brain, from peripheral measurements. In the latter, one is concerned with the enhancement of a blurred image acquired by a microscope or by tomography. Yet another contribution is machine learning and statistical inference, as the investigation of correlations between the various parameters describing a system is a ubiquitous challenge.

Software and Programs as Telescopes

An algorithm is an effective method which needs to be implemented in software in order to become operational. This coding process is in general nontrivial, as it may be influenced by a number of factors such as the need to certify selected critical tasks, requirements on the numerical accuracy of the calculations carried out, constraints inherent to the size of the data processed, portability and parallelization issues, compatibility with programs upstream and downstream, etc. This complexity explains the emergence of highly specialized libraries which are often community-wide efforts, and which can be seen as research instruments equivalent to telescopes for astronomers.

Motivation to Write This Book

Computational biology and biomedicine is a vast field where intensive research is currently being carried out, with outstanding perspectives both in terms of the complexity of the scientific problems to be addressed and technological developments to be made. Taking up these challenges requires developing an enhanced synergy between biology and biomedicine on the one hand and applied mathematics and computer science on the other hand.

In line with this observation, the motivation to write this book has been to show that researchers trained in more quantitative and exact sciences can make major contributions in this emerging discipline, and those with roots in biology and biomedicine can benefit from a true leveraging power tailored to their specific needs.

The need to train actors with multidisciplinary skills explains the ever-increasing number of advanced-level classes at the master's level which have begun over the last few years. This book itself emerged from a series of lectures given within the scope of the *Computational Biology and Biomedicine (CBB)*[1] program at the Université Nice Sophia Antipolis, France. The goal of this Master is to provide advanced training at the crossroads of biology, biomedicine, applied mathematics and computer science. This book adopts the organization of these class units, namely, bioinformatics, biomedical signal and image analysis, and modeling in neuroscience.

This book deals with the problem of modeling certain biological phenomena, and we believe that is well suited for two communities. The first is the vast community

---

[1]MSc in Computational Biology and Biomedicine, Université Nice Sophia Antipolis, France, http://cbb.unice.fr

of applied mathematicians and computer scientists, whose interests should be captured by the added value generated by the application of advanced concepts and algorithms to challenging biological or medical problems. The book also highlights some unsolved and outstanding theoretical questions, with potentially high impact on these disciplines. The second is the equally vast community of biologists, and more generally life scientists, concerned by the three fields covered in this book: bioinformatics, biomedicine, and neuroscience. Whether scientists or engineers, they will find in this book a clear and self-contained account of concepts and techniques from mathematics and computer science, together with success stories on their favorite systems.

Book Organization

While the topics discussed span a wide spectrum, the presentations of the chapters adopt a unified plan. In short, each chapter, which has been written in tandem by computer scientists/applied mathematicians and biologists/biophysicists/medical doctors, presents the biological or medical context, follows up with mathematical or algorithmic developments triggered by a specific problem, and concludes with one or two *success stories*, namely, new insights gained thanks to these methodological developments.

At the end of each chapter, the reader will find a list of online resources grouped into three categories: tutorials, databases, and software. In assembling these resources, the goal has been to provide pointers to more specific topics, but also to expand the horizon and to list important programs related to the material presented.

To facilitate the reading experience, a list of acronyms is also supplied at the end of each chapter. Finally, the book ends with two thematic indexes corresponding to biology, medicine, physics, and biophysics on the one hand and to mathematics and computer science on the other hand. These indexes are also accompanied by a flat index amenable to alphabetical search. Again, this index shows the plurality of the contributions and the complementarity of the disciplines involved.

Let us now briefly characterize the content of each chapter.

Part I: Bioinformatics

Chapter 1: Modeling Macromolecular Complexes: A Journey Across Scales. Motivated by the question of unraveling the key features of the macromolecular complexes which account for biological functions, this chapter discusses phenomenological models for protein complexes, namely, atomic resolution models for binary complexes, and coarse-grain models for large assemblies. In both cases, accurate geometric and topological models allow uncovering subtle biological and biophysical signals. These approaches are validated through enhanced correlations between biophysical and structural properties of protein complexes in the Protein Data Bank.

Chapter 2: Modeling and Analysis of Gene Regulatory Networks. This chapter addresses the problem of modeling networks of interacting genes and macromolecules, a central topic in systems biology. Having recalled classical models based on systems of ordinary differential equations, whose nonlinear features are not easily dealt with, and which also pose parameter identification problems, the authors proceed with linearization schemes of such systems and with formal analysis based on finite-state transition graphs. An application to the elucidation of the reason why the bacteria *Pseudomonas aeruginosa* over-secretes mucus in lungs affected by cystic fibrosis is discussed.

Part II: Biomedical Signal and Image Analysis

Chapter 3: Noninvasive Cardiac Signal Analysis Using Data Decomposition Techniques. This chapter discusses techniques to analyze electrocardiogram signals, with applications to the prevention and the cure of cardiac arrhythmia. More precisely, the authors develop statistical techniques, based on principal component analysis and independent component analysis, to analyze signals incurring subtle variations and burdened with recording noise. Some of these techniques are currently used by cardiologists to make clinical decisions.

Chapter 4: Deconvolution and Denoising for Confocal Microscopy. Fluorescence light microscopes such as confocal microscopes have become powerful tools in life sciences for observing biological samples, in order to measure the spatial distribution of proteins or other molecules of interest. However, there are some inherent imaging limitations in confocal images such as blurring due to the diffraction limit of the optics and low signal levels. To overcome these limitations, the method developed in this chapter consists of improving the resolution by deconvolution. This illustrates the concept of regularized solutions which allow solving ill-posed problems (in the variational framework or in the Bayesian probabilistic frameworks as proposed here). This chapter develops general concepts that can be applied to any similar imaging problem, allowing one to go beyond current limitations in imaging systems.

Chapter 5: Statistical Shape Analysis of Surfaces in Medical Images Applied to the Tetralogy of Fallot Heart. This chapter develops tools to model the morphology and the physiology of the heart, when affected by the so-called tetralogy of Fallot, a severe congenital defect requiring open-heart surgery in infancy followed by subsequent operations. The methodology developed consists of modeling the heart using the mathematical formalism of differential forms and currents, from which correlations between the morphology of the heart and physiology properties can be inferred, thus guiding clinical decisions. This chapter emphasizes the need to develop generic models amenable to patient-specific tailoring.

Chapter 6: From Diffusion MRI to Brain Connectomics. Diffusion MRI (dMRI) is a unique modality of MRI which allows one to indirectly examine the microstructure and integrity of the cerebral white matter in vivo and *noninvasively*. In this chapter, the authors present an overview of the mathematical framework for dMRI. Two methodologies can be distinguished: The first is based on physics and aims at modeling the imaging process from a physical point of view. The second is based on computer science and aims at extending approaches from computer vision to dMRI images seen as images in higher dimensions. Thanks to these recent mathematical and modeling advances dMRI is becoming a key element in the study and diagnosis of important pathologies of the cerebral white matter, such as Alzheimer's and Parkinson's diseases, as well as in studying its physical structure in vivo.

Part III: Modeling in Neuroscience

Chapter 7: Single-Trial Analysis of Bioelectromagnetic Signals: The Quest for Hidden Information. This chapter deals with the analysis of multitrial electrophysiology datasets coming from neuroelectromagnetic recordings by electroencephalography and magnetoencephalography (EEG and MEG). In order to correctly capture the inter-trial variability, the authors develop techniques such as nonlinear dimensionality reduction and extended Matching Pursuit methods. These techniques to interpret multitrial signals are already largely applied in clinical or cognitive science research. The rapidly growing field of Brain Computer Interfaces is also driving research on the online interpretation of EEG signals, so that much progress on single-trial biosignal is expected in the future.

Chapter 8: Spike Train Statistics from Empirical Facts to Theory: The Case of the Retina. This chapter focuses on methods from statistical physics and probability theory allowing the analysis of spike trains in neural networks. Taking as an example the retina, the authors present recent contributions aiming at understanding how retina ganglion cells encode the information transmitted to the visual cortex via the optical nerve, by analyzing their spike train statistics. Understanding the neural code remains an open challenge and this chapter makes a stride in this direction, based on the joint advances of MEA recording devices, spike sorting algorithms, statistical methods, and efficient algorithms accommodating the huge amount of data inherent to neural activity.

Outlook

Each topic covered in this book has of course been the subject of different publications presenting in-depth treatments of the state of the art. But these highly specialized texts often exhibit a bias inherent to the discipline the authors originate from, and tend to target experts and researchers already in the field. The same holds on the mathematical and computer science side; a large number of books are available, each focusing on one specific set of techniques. But none strikes a balance between biological and methodological developments.

As should be clear from this Preface, this book instead touches upon a limited number of topics in the vast realm of computational biology and biomedicine, and for each of them seeks to balance the applied and the formal considerations. Along the way, it clearly shows that a panoply of complementary techniques are needed to apprehend the complex issues encountered in life sciences.

We believe that our initiative will be a successful complement to existing material, both from the biological and modeling sides, and we hope that this book will serve as a useful source of reference, inspiration, and motivation for both students and fellow researchers in applied mathematics and biology.

Sophia Antipolis, France
May 2012

Frédéric Cazals
Pierre Kornprobst

Contents

Contributors

Christian Bénar Institut des Neurosciences des Systèmes, UMR 1106 INSERM, Aix-Marseille Université, Faculté de Médecine La Timone, Marseille France

Gilles Bernot I3S – UMR 7271 CNRS/UNS, Algorithmes-Euclide-B, Sophia Antipolis, France

Laure Blanc-Féraud I3S – (CNRS/UNS) UMR 7271 CNRS/UNS and Inria, Algorithmes-Euclide-B, Sophia Antipolis, France

Frédéric Cazals Inria Sophia Antipolis Méditerranée, ABS project-team, Sophia Antipolis, France

Bruno Cessac Inria Sophia Antipolis Méditerranée, Neuromathcomp project-team, Sophia Antipolis, France,

Madalena Chaves Inria Sophia Antipolis Méditerranée, Biocore project-team, Sophia Antipolis, France

Maureen Clerc Inria Sophia Antipolis Méditerranée, Athena project-team, Sophia Antipolis, France

Jean-Paul Comet I3S – UMR 7271 CNRS/UNS, Algorithmes-Euclide-B, Sophia Antipolis, France

Pierre Comon GIPSA-Lab - UMR 5216, Grenoble, France

Frédéric Dayan SOBIOS SA, Valbonne Sophia Antipolis, France

Rachid Deriche Inria Sophia Antipolis Méditerranée, Athena project-team, Sophia Antipolis, France

Tom Dreyfus Inria Sophia Antipolis Méditerranée, ABS project-team, Sophia Antipolis, France

Gilbert Engler Inra, Sophia Antipolis, France

Aurobrata Ghosh Inria Sophia Antipolis Méditerranée, Athena project-team, Sophia Antipolis, France

Jean-Luc Gouzé Inria Sophia Antipolis Méditerranée, Biocore project-team, Sophia Antipolis, France

Pierre Kornprobst Inria Sophia Antipolis Méditerranée, Neuromathcomp project-team, Sophia Antipolis, France

Decebal Gabriel Latcu Cardiology Department, Princess Grace Hospital, Monaco, France

Tommaso Mansi Siemens Corporate Research, Image Analytics and Informatics, Princeton, NJ, USA

Kristin McLeod Inria Sophia Antipolis Méditerranée, Asclepios project-team, Sophia Antipolis, France

Olivier Meste I3S – UMR 7271 CNRS/UNS, Algorithmes-Euclide-B, Sophia Antipolis, France

Adrian G. Palacios CINV-Centro Interdisciplinario de Neurociencia de Valparaiso, Universidad de Valparaiso, Chile

Praveen Pankajakshan Institut Pasteur, Paris, France

Théodore Papadopoulo Inria Sophia Antipolis Méditerranée, Athena project-team, Sophia Antipolis, France

Xavier Pennec Inria Sophia Antipolis Méditerranée, Asclepios project-team, Sophia Antipolis, France

Giacomo Pongiglione Ospedale Pediatrico Bambino Gesù, Rome, Italy

Adrien Richard I3S – UMR 7271 CNRS/UNS, Algorithmes-Euclide-B, Sophia Antipolis, France

Charles H. Robert Laboratoire de Biochimie Théorique – UPR 9080 CNRS, Institut de Biologie Physico Chimique, Université Paris Diderot Sorbonne Paris Cité, Paris, France

Nadir Saoudi Cardiology Department, Princess Grace Hospital, Monaco, France

Maxime Sermesant Inria Sophia Antipolis Méditerranée, Asclepios project-team, Sophia Antipolis, France

Vicente Zarzoso I3S – UMR 7271 CNRS/UNS, Algorithmes-Euclide-B, Sophia Antipolis, France

Josiane Zerubia Inria Sophia Antipolis Méditerranée, Ariana project-team, Sophia Antipolis, France

Part I
Bioinformatics

Chapter 1
Modeling Macromolecular Complexes: A Journey Across Scales

Frédéric Cazals, Tom Dreyfus, and Charles H. Robert

1.1 Introduction

Proteins and nucleic acids thoroughly pervade biological organisms. Fully appreciating biological development and pathologies implies understanding the nature of these macromolecules in detail. But if macromolecules are the fundamental components of an organism, then Biology itself is based on their mutual interactions. For example, a receptor complex that spans the cell membrane may be activated by binding a hormone on the outside of a cell, and only then interact with a partner protein on its inner-membrane side in the cell to trigger a signaling cascade. The biological interactions we address in this chapter are direct and physical, involving formation of a transient or permanent complex. In the signaling system just mentioned, the binding of the hormone, and of the partner protein afterwards, create transient complexes. In contrast, when 50 proteins and numerous nucleic acids assemble to make a ribosome [64], which is responsible for physically translating the genetic code into new proteins, the resulting machine is more permanent. Another example of a long-lived structure is the nuclear pore complex (NPC), made up of about 450 proteins, which allows active transport of molecules from the nucleus to the cytoplasm in eukaryotic cells and vice-versa.

F. Cazals (✉) · T. Dreyfus
Inria Sophia Antipolis Méditerranée, ABS project-team, 2004 Route des Lucioles, 06902, Sophia Antipolis, France
e-mail: frederic.cazals@inria.fr; tom.dreyfus@inria.fr

C.H. Robert
Laboratoire de Biochimie Théorique – UPR 9080 CNRS, Institut de Biologie Physico Chimique, Université Paris Diderot Sorbonne Paris Cité, 13 rue Pierre et Marie Curie, 75005, Paris, France
e-mail: robert@ibpc.fr

F. Cazals and P. Kornprobst (eds.), *Modeling in Computational Biology and Biomedicine*, DOI 10.1007/978-3-642-31208-3_1,

1.1.1 Structure of Macromolecular Systems

Although structure determination is an essential part of the study of all macromolecules, we will mainly be concerned with proteins in this chapter, so we briefly describe the basic, hierarchical rules of protein structural organization. A protein is made of only a few atom types, principally H, C, N, O and S. These atoms are chemically bonded in different ways to form 20 distinct types of amino acids, each made of about 10–20 atoms. The amino acids are themselves chemically linked (via a peptide bond) in a topologically linear polypeptide chain: the sequence of amino acids in this chain is called the *primary* structure of the protein. In the present chapter we will not be concerned with either the forming or the breaking of such covalent bonds, which involves energy changes well beyond those involved in the formation of the majority of protein complexes.

Although the primary structure of a protein is often constant, the chain itself has many degrees of freedom, allowing atoms in different regions to interact via weaker, *non-covalent* forces. Such influences include van der Waals interactions, hydrogen bonding, the hydrophobic effect, etc. They tend to produce local folding of the chain, in particular *secondary* structural elements, of which examples are regularly repeating structures such as alpha helices or beta hairpins. At a higher level, the different secondary structural elements and remaining chain regions fold up still further into a globular domain, which is stabilized by so-called *tertiary* interactions. If a given protein involves multiple domains, their precise structural interaction is described by the *quaternary* structure. An essential fact for most proteins is that the polypeptide chain is thus folded into a particular, highly ordered three-dimensional shape, which is what we usually will refer to as the structure of the protein.[1]

Structural information obtained for macromolecular systems has proven essential in interpreting physical, biochemical, and functional data. An elite club of experimental techniques, dominated by X-ray crystallography and NMR, plays an enormous role in biological and pharmaceutical research by providing three-dimensional structures of macromolecules—that is, the x, y and z coordinates of each of the thousands of atoms in the molecule. This information is stored in a public repository, the Protein Databank (PDB, http://www.rcsb.org/pdb) [58]. Determination of the structures of individual proteins is now commonplace: the PDB currently contains over 75,000 entries, with hundreds added each month.

Two main experimental techniques are being used to solve the structures of macromolecular systems and populate the PDB. The first, X-ray crystallography, gained importance after Kendrew and Perutz used it to solve the structures of myoglobin and hemoglobin, a key achievement for which they were awarded the Nobel prize in chemistry in 1962. X-ray crystallography has proved to be particularly well adapted to biological structure determination, as it allows one to obtain atomic

[1]While single-chain proteins are common, one protein may include more than one polypeptide chain. Multiple chains are frequently covalently bonded via a disulfide bond formed between cysteine residues.

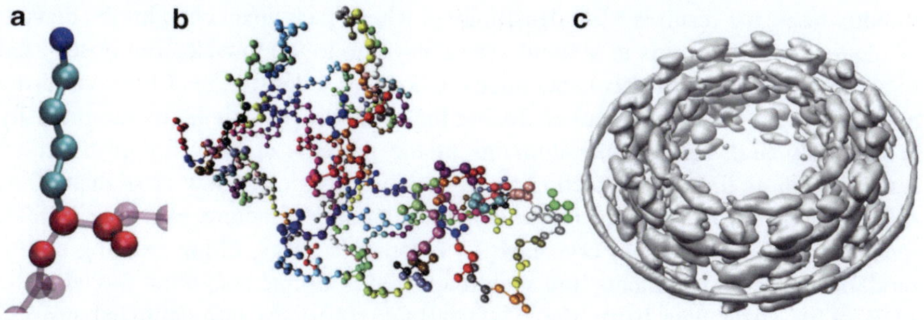

Fig. 1.1 Relevant objects in structural modeling vary in size from 1 Å to 100 nm. (**a**) The radius of an atom (nucleus and electron cloud) lies in the range 1–2 Å, depending on the element and chemical group to which it belongs. Each amino acid contributes four heavy atoms (*magenta*) to the backbone of the protein and has a variable side chain (*cyan*) and *blue*). Lysine is represented. (**b**) The backbone of lysozyme, a protein involving 162 amino-acids, colored by amino-acid type. The diameter of the protein is circa 40 Å, that is 4 nm. (**c**) With a size of circa 100 nm, the Nuclear Pore Complex is the largest protein assembly known to date in eukaryotic cells

coordinates for a wide variety of sizes of structures, from small molecules to large proteins. However, this technique has practical limits. Fewer than about 5 % of the structures in the PDB correspond to non-redundant protein-protein or protein-nucleic acid complexes [22]. An important factor here is the size of the complex. Single-domain proteins are generally in the range of 1–2 nanometers (nm) [44] (Fig. 1.1b). Although structures of virus particles having a high degree of symmetry have been solved, at 30 nm the ribosome is currently the largest asymmetric structure solved by X-ray crystallography. Multidomain proteins, oligomers and complexes can be much larger than this; the nuclear pore complex just mentioned measures 100 nm across [3] (Fig. 1.1c).

After X-ray crystallography, the importance of NMR spectroscopy to structural determination of proteins is steadily increasing. Originating with the work of K. Wüthrich, who was awarded the Nobel prize in Chemistry in 2002, successful resolutions for proteins up to a few hundred residues are now commonplace. Further, NMR data obtained for isolated proteins can be used in determining the structures of complexes when only minor conformational changes occur upon association, e.g., [49]. Protein-protein docking techniques [10] can be leveraged in such determinations in much the same way that X-ray crystallography exploits molecular replacement to resolve crystal structures of related proteins.

Another approach under active development is cryo-electron microscopy (Cry-oEM) [27]. Structures as large as whole cells and as small as individual proteins can be imaged with electrons, and with cryo techniques final resolutions on the order of 0.3 nm have been attained.

In single particle analysis, bombarding isolated samples with electrons yields images corresponding to different viewpoints, and these can be combined into a 3D model of the particle. In cryoEM tomography, a given sample is instead bombarded at incremental degrees of rotation, from which a 3D model can also be reconstructed.

In both cases, the result is a 3D density map, where each voxel encodes the density of matter. This density is in general very noisy due to the low electron doses used to avoid damaging biological specimens. Choosing a density level for contouring a surface (called the envelope) enclosing the model is non trivial, as the intensity is generally high for globular domains of the proteins but low for unstructured regions such as linkers connecting these domains. Typically, low (less than 10 Å, domains visible) to medium (around 5 Å, secondary structure elements visible) resolutions are achieved in cryo-EM. In favourable cases, fitting existing and/or modelled structural elements into such maps yields atomic resolution models. The PDB or its sister data bank, the EMDataBank ([40], see http://emdatabank.org, currently contains CryoEM models for over 1,000 biological complexes. These resources contain detailed structural information ranging from the Å scale, relevant to small molecules and individual amino acids, to hundreds of nanometers and beyond for large complexes.

1.1.2 Dynamics of Macromolecular Systems

Besides the 1,000-fold range of molecular dimensions, even more pronounced and challenging differences exist in the characteristic timescales describing biological processes. Biological timescales range from picoseconds for localized side-chain transitions in amino acids at a protein surface, up to hundreds of nanoseconds or microseconds for slow loop rearrangements, and from milliseconds to hours for folding reactions and global conformational changes [1]. Different experimental techniques provide dynamic information. The temperature factors in a crystal structure are obtained along with the atomic coordinates. Each reflects the spatial dispersion of the electron density around a given atomic position and thus the atom's mobility in the crystal, although static disorder and errors contribute as well. Conformational dynamics is also measured using methods such as time-resolved spectroscopy, which can provide exceptional detail on changes in structural features such as bond lengths, coupled with some method of rapidly initiating a change, such as laser temperature jump or photo-dissociation of a ligand. NMR, in addition to macromolecular structure resolution, furnishes dynamic information: in an external magnetic field, the interaction between two protons of a molecule influences their rates of transition between magnetic energy levels. Such effects are both time- and conformation-dependent, and can be exploited in different ways. For example, in partially orienting solutions, incomplete rotational averaging allows one to extract comprehensive information concerning conformational dynamics of the macromolecule [59].

Understanding such dynamics in detail often entails the use of numerical simulations. As mentioned above, for the processes we are considering here, no chemical bond making or breaking takes place, for which quantum mechanical descriptions would be necessary. The dynamics that occur in protein folding, conformational changes, and association principally involve changes in weaker, non-covalent interactions. These can be treated using conventional classical mechanics, and notably the integration of Newton's equations of motion, such as is used for all-atom molecular dynamics simulations. Such methods require a timestep on the order of the fastest nuclear vibrations, with a period on the order of 1 fs. Significant time and computational resources are called for in simulating microsecond (10^9 steps) processes, but dedicated hardware has allowed in silico folding of small proteins on the sub-msec scale to be approached [62].

The corpus of experimental and theoretical results demonstrate unequivocally that a macromolecule or complex in thermal equilibrium is best represented by an ensemble of interconverting conformations, in which an individual molecule at a given instant in time is represented by a single point in a high-dimensional conformational space. The nature of the energy surface corresponding to this space is the subject of intensive research, and provides a means of establishing the probability of occurrence of a given configuration of the system. Experimental observables reflect the intrinsically statistical quality of the phenomena, in which a large number of macromolecules, under thermal agitation, repeatedly visit and revisit favorable regions of the conformational space and only rarely venture into higher energy zones. The strength (affinity) of a specific interaction between two macromolecules is typically determined by the properties of three regions in the configurational space, which regroups the conformational spaces of the two isolated components and of the complex. Each region is associated with a basin in the energy surface associated with the stable conformation of the corresponding macromolecular species. On the other hand, the kinetics, or rate, of the association depends not only on the character of the stable regions of the space but also of the accessible paths linking them. The principal pathways describe conformational changes and different association geometries, and can be approached theoretically via transition pathway exploration [30].

1.1.3 Simulation, Geometry, and Insight

Two tempting but false conclusions from this brief exposition must be avoided. The first is that biological macromolecular systems are so complex that meaningful predictions must await further theoretical and computational advances. This is clearly not the case, as many theoretical approaches are directly applicable to biological systems, and even incomplete theories and limited-timescale simulations have been capable of producing accurate models and predictions. The second is related to the first: that simulations alone hold the key to understanding macromolecular behavior. At some future date such methods, applied blindly, may indeed be capable of reproducing and reliably predicting the results of certain experimental studies. But used this way they offer no insight, and we will continue to be faced with the difficulty of abstracting the principles of the phenomenon under study in order to go still further.

Apart from detailed atomic-level simulations, considering the structures of macromolecules and their complexes in terms of distances, areas, and volumes can lead to fruitful insights. Our experience with everyday objects underlies numerous heuristic approaches to measuring such geometric properties of macromolecules and interpreting them. If these approaches are sometimes less than rigorous, common sense (or common experience) provides checks and feedback that tend to prevent the worst sorts of errors. Further, the biochemical results that we wish to interpret are themselves subject to error. As mentioned in the preceding section, experimentally observed properties generally represent averages in time and space over a great number of instances of the macromolecules in question in many somewhat different conformations. Sample homogeneity can also be a concern. Experimental error can thus be substantial, typically a factor of 2 or more in protein-protein binding affinities [39], and sample variability may trump the theoretical shortcomings of a particular heuristic.

But we must also go beyond heuristics. The systematic errors that can find their way into calculated geometric values using simple but inexact approaches can be substantial [14]. Further, formal definitions of geometric properties facilitate generalization, allowing one to obtain new insights by incorporating analyses from related fields, as in the case of Voronoï diagrams [56]. The need for rigorous algorithms for calculating geometrical properties becomes especially critical when addressing problems with a higher dimensionality, such as those encountered in relating different conformations of the macromolecule in the $3N$ dimensional conformational space. Here our day-to-day experience can be misleading, and real intuition is limited.

At a higher level of abstraction, topology permits describing shapes in a qualitative but rigorous manner. In macromolecules, one might wish to identify tunnels and voids—the former perhaps providing direct access to an active site, the latter indicating poor packing or trapping solvent or other small molecules. Geometry provides the quantitative measures that complement such topological descriptions. These in turn reflect physical interactions, so that the particular observed geometries and topologies become a fingerprint of the most favourable arrangements of atoms in biological constructions. In selecting the components of living systems, Biology has imposed a bias on the physics and chemistry of macromolecules. The approaches presented in this chapter are aimed at better analyzing and interpreting this bias.

1.1.4 Chapter Overview

Outline

This chapter covers two main topics. The first is concerned with the modeling of macromolecular complexes at the atomic scale. As mentioned above, the PDB contains relatively few structures of complexes compared to unbound proteins. A major goal of biological modeling, known as *docking*, consists of predicting the structure of a complex from the structures of the unbound partners. Docking methodologies have been developed for such purposes, and an ongoing, open community experiment (called CAPRI) has been set up in which crystallographers and NMR groups furnish unreleased structures of complexes as targets for docking prediction [36]. Docking, and flexible docking in particular, is a difficult problem, and its successful resolution will require a detailed understanding of the principles underlying protein complex formation. The goal of Sect. 1.2 is precisely to contribute to such an understanding.

The second topic is concerned with low-resolution modeling, particularly relevant to the study of protein complexes involving from on the order of 10 to the order of 100 polypeptide chains. Modeling these complexes is especially challenging due to their plasticity (their composition may change over time) and their flexibility, and using complementary data is often compulsory in designing models. These data are often noisy and ambiguous, and the work presented in Sect. 1.3 aims precisely at dealing with such uncertainties.

Interestingly, while the questions addressed in these two domains are fundamentally different, the concepts and the constructions used all pertain to the realm of Voronoï diagrams.

Public

As just discussed, the goal of this chapter is to show that the more precise the mathematical models used to investigate macromolecular systems, the sharper the biological and biophysical conclusions that can be derived. It should thus be of interest for those designing structural models, as they will find recent geometric and topological developments. It may also furnish a resource to those wishing to use the various programs accompanying the contributions described in this chapter as more than black boxes. Beyond structural biology, this text should also be of interest to a broad audience in particular in physics and engineering, where simulation methods involving collections of balls and simplicial complexes are being used. Computers scientists interested in geometric and topological modeling will find good reasons to believe that some of the fundamental concepts they have developed over the past decades are indeed of high interest to the biological community. In particular, we hope to provide some orientation in a world where mathematically well-posed questions are not commonplace, and where designing models is actually the central issue. We also hope to stimulate work, in particular in the realm of curved Voronoï diagrams and α-shapes, in which a number of algorithmic and complexity questions are open. Finally, we might hope that students interested in the life sciences in general will get a feeling for what a protein is, the nature of its complexes, as well as some ongoing developments concerning the use of geometry to better define their properties.

We have striven to limit the prerequisites necessary for reading this chapter to a minimum. In particular, the numerous geometric constructions used are illustrated, and the reader is invited to reproduce them. This can be facilitated with the help of the IPE editor, which encapsulates so-called ipelets, allowing one to run advanced geometric algorithms from the Computational Geometry Algorithms Library (see *on-line resources*).

1.2 Modeling Atomic Resolution Complexes

1.2.1 Challenges

Formally, binary complexes are the simplest examples of macromolecular interactions. Yet they can be quite challenging, especially when the conformations of the partners change upon association, which encompasses a great number of biological systems. Increasingly, atomic level structures of the unbound macromolecular components of such complexes are available in public data repositories such as the PDB. A significant number of complex structures exist as well, allowing one to characterize protein-protein interfaces with the objective of empirically inferring the rules governing complex formation (e.g., [32]). The goals of biophysical modeling then include being able to understand the structure, energetics and dynamics of a complex, all in relation to its biological function. In this section we will discuss the characterization of complex structures but not their prediction (e.g., docking), although the methods used here may certainly be applied to that end. Geometrical encoding of the structures and the interface between the components does however provide means for interpreting dynamic and energetic properties of the complex. We will see that Voronoï diagrams, Delaunay triangulation and the associated α-shape provide descriptors of molecules that can be correlated to experimental quantities.

1.2.1.1 Experimental Measurements Relevant for Macromolecular Modeling

Numerous quantitative experimental measures have been developed to describe and qualify molecular complexes. We present several of these here.

Structural data and its interpretation. Experimentally-determined protein structures are stored as PDB entries. Along with details concerning the experimental setup, the biochemical or biological source of the macromolecule, and the sequences of the polypeptide chains, the data for each atom is stored. Each atom is associated with the chain identifier, the residue number and amino acid type, and the atom name. The x, y, and z values are given in Å (0.1 nm) to three decimal places in order to maintain correct bond lengths and angles, but the actual experimental precision is much lower. The data also contains the fractional occupancy of each atom position (alternative positions may be defined from the crystallography), along with the thermal B-factor (in units of Å^2). The latter quantity describes the region containing a particular fraction of the probability of finding the atom and is related to the mean square displacement of the atom by $B = (8/3)\pi^2 < \delta^2 >$ for the isotropic (spherical) case. Anisotropic temperature factors may also be supplied in the form of the principal axes of an ellipsoid. In proteins, B-factors typically range from 5 to over 60 Å^2, so that the positional uncertainty (>0.2 Å) is much larger than the precision that the coordinate entries would lead one to believe.

Crystallization is a process which obliges molecules to come together, forming interfaces that may have no significance in Nature. One important goal when interpreting crystal structures of protein complexes is to be able to distinguish a biological interface from a banal crystal packing interface. Assignments of biological interfaces given in the PDB file itself are often wrong [37, 42]. The true interface can be determined by experiments in solution, or by capturing the essential features of protein-protein interfaces to a sufficient degree that predictions can be made.

Absolute and relative affinities. One of the most fundamental biochemical quantities in protein-protein interactions is the affinity. The affinity of an association reaction $A + B \rightarrow AB$ refers to the ratio $K = c_{AB}/c_A c_B$, in which the c's refer to the concentration of free A, free B and of their complex in an equilibrium aqueous solution. (As the affinity constant K has units of inverse concentration, care must be taken in comparing values obtained using different concentration scales.) K characterizes a given associating system; at a given temperature it is a constant and constitutes a constraint on the concentrations of the different species. High values of K indicate a strong tendency for A and B to form the complex. Experimental estimates for the affinity are typically obtained from isothermal titration calorimetry or surface plasmon resonance measurements.

The affinity is related to the free energy change ΔG° for the association (per molecule of complex). Under standard state conditions, $\Delta G^\circ = -kT \ln Kc$, in which k is Boltzmann's constant and T the temperature in degrees Kelvin, and the zero of the free-energy scale is set by the reference concentration c, which is usually specified as 1M, or 1 mole of the component in question per liter of solution. (Recall that 1 mole is about 6×10^{23} molecules.) Measured free-energy changes for protein-protein interactions are generally in the range of about -7 to -18 kcal/mole [39], from which it can be seen that the concentration at which half the protein is tied up in complexes ranges from the μM to the fM range.

Rationalizing the absolute binding free energy is a complex task. Site-directed mutagenesis consists of changing the amino-acid identity of a single residue in the wild-type protein sequence (primary structure) to another, usually alanine. The effect of this mutation on the binding affinity is quantified using the difference in binding energies $\Delta\Delta G^\circ$ (note that this corresponds to the logarithm of the ratio of the corresponding binding affinities). Systematic measurements of $\Delta\Delta G^\circ$ for mutating each residue in a single protein-protein complex allows the identification of so-called *hotspot residues* that contribute disproportionately to the free energy of association [50]. While providing invaluable information on the importance of specific residues, this technique does not, however, directly convey information on the correlations and dependencies that may exist between the residues of a binding patch [55].

Residue conservation. Bioinformatics and molecular dynamics investigations have shown that residue conservation and dynamics are distinct in the interface region compared to the surface as a whole [46, 63], and that the interface core, as opposed to the rim, appears to have a particular importance [5, 17]. Hotspot residues may also be interpreted in light of the degree of conservation seen at these positions in the protein sequence. Such conservation refers implicitly to a classification of known proteins sharing a common ancestor and similar structures, and, importantly, function. The latter is particularly important for assessing protein-protein interactions, as function is predicated on the character of the partner protein as well. For analyses of residue conservation, sequences of related proteins from different organisms must be gathered and aligned; this step involves positing that the interaction and the interface are conserved in the different organisms (further discussion on this point can be found elsewhere [43, 54, 66]). After alignment, an estimation of the degree of conservation can be made at each position. Amino acids that are critical to the stability of a complex are often found to be more or less invariant (conserved) [35]. A common way of assessing this conservation is through the Shannon entropy s calculated for each position i in the alignment, using the frequencies of appearance of the 20 natural amino-acids (aa) at that position: $s_i = -\sum_{aa} f_{i,aa} \ln f_{i,aa}$. Positions at which an amino acid is more conserved exhibit a more peaked distribution, with a consequently lower entropy. Indirect prediction of hotspot residues from measures of residue conservation can thus help rationalize the affinity of a protein-protein interaction. We shall also see that certain geometric constructions can help unravel the relationship between residue conservation and structural features of a complex.

1.2.1.2 Physical Modeling of Macromolecules

The biological, biochemical, and biophysical data just presented can be approached by different theoretical means. A classical approach is to use physical chemical modeling of the atomic interactions in the system.

Energy functions and forcefields. The set of coordinates of all particles in the system constitutes the configuration X. Statistical mechanics shows how knowledge of the potential energy $E(X)$ of the macromolecular system can be used to deduce relative probabilities of different configurations of the system via the Boltzmann factor $exp^{-E(X)/kT}$. High potential energies are associated with low probability configurations, and vice versa.

It is thus important to be able to calculate the potential energy as accurately as possible. It is currently difficult to envision ab initio quantum mechanical approaches to modelling entire proteins and their dynamics. Thankfully, empirical forcefield approaches using the classical approximation can provide remarkably accurate results, e.g. [62]. The relevant particles whose coordinates are required in order to define a configuration X are thus limited to the set of atoms forming the macromolecule and the solvent included in the system. The forcefield approach provides an empirical energy model that is parameterized by referring to molecules whose structures and physical properties are known to a high degree of accuracy. The applicability of the forcefield to, say, a protein depends on the transferability of the parameters. The potential energy is usually formulated as a sum of many terms, which is designed to take into account as many physical interactions as possible. Non-covalent interactions such as the van der Waals interaction are typically treated using a Lennard-Jones potential. In this function the interaction energy is negative at distances on the order of the sum of the vdW radii but climbs steeply as the distance of approach of the two atoms is reduced, while at longer distances the energy tends towards zero. For covalently bonded atoms, the parameters are often simple Hooke's-law style force constants associated with deviation of a bond distance or angle from a reference value; this value is itself another parameter. Dihedral angles are used to model the energy of rotation about a covalent bond, for which a term is constructed having the number of minima chosen as a function of the bond order. Functional forms can be found in texts, such as [41].

Electrostatic interactions between charged atoms or groups present somewhat of a special case. The simplest approach is also the most demanding computationally: the protein and the solvent are both represented explicitly, using a large number of molecules of water together with dissolved ions, in order to model as faithfully as possible an entire region of aqueous solution containing the protein. For accurate energy calculations this approach requires on the order of $10\times$ more atoms of solvent than of the macromolecule itself. A second approach treats the solvent as a continuum dielectric and resorts to solving the Poisson-Boltzmann equation to obtain solvation energies. It is then necessary to define the dielectric boundary between the bulk and the protein itself, and to ignore the molecular nature of the water itself. The dielectric boundary may be defined as the union of vdW spheres representing the atoms in the macromolecule. The vdW volume for the molecule as a whole is smaller than the sum of the volumes of its atoms, because atoms joined by a covalent bond lie closer together than the sum of their vdW radii. However, there are inevitably a large number of empty spaces (packing defects) in the interior of the macromolecule as well. In the corresponding electrostatic calculation, these interstitial volumes are formally added to the bulk solvent—even if they provide too little space for a physical solvent molecule to lodge [21]. Other approaches rely on the solvent accessible surface (SAS) to define the dielectric boundary, as we shall detail in Sect. 1.2.3. In short, the SAS consists of enlarging the atoms, so as to guarantee that only truly solvent-accessible cavities in the macromolecule are counted for the water contribution (Fig. 1.2a).

All-atom simulations. Once one has a suitable potential energy function, together with a set of initial positions of the atoms of the macromolecule (e.g., the PDB structure) and a set of pseudo-random starting atom velocities consistent with a given temperature, the equations of motion for the system can be solved numerically.

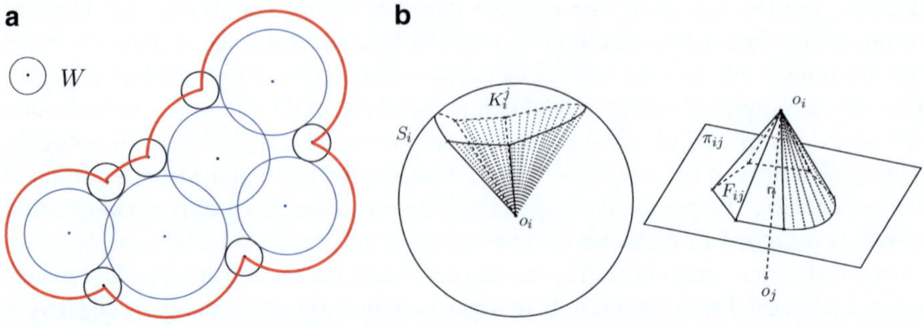

Fig. 1.2 Molecular surfaces and volumes. (**a**) The solvent accessible surface, in *red*, is defined by tracing the locci of points of a water probe W rolling on the boundary of the van der Waals model, in *blue*. (**b**) Tiling the volume of a restriction with two types of pyramids: *left*, a pyramid topped with a spherical cap; *right*, a pyramid with flat base

From this a new set of atomic positions and velocities are obtained, applicable for a short time ("a short time" here being on the order of a fs). This procedure can be repeated ad infinitum in order to simulate the thermal motions of the macromolecule.

Such a simulation of a protein in thermal equilibrium allows one to obtain information regarding the detailed dynamics of the macromolecule as well as thermodynamic information. The simulation provides a way of approximating the partition function of the system, which is directly related to the free energy. The affinity is a natural target for such studies. Many factors are known to contribute to the free energy change, including desolvation of the two surfaces that will form the interface, net changes in hydrogen bonding, electrostatic interactions, and other more detailed contributions. Molecular dynamics simulations can in principle be used to take all of these effects into account in calculating affinities, but conformational changes are still particularly difficult to handle due to the relevant timescales involved. $\Delta\Delta G^{\circ}$ values can be targeted if the mutant complexes can be reasonably assumed to have structures similar to the wild-type protein, so that many contributions to the free energy change cancel in a first approximation.

Even in calculating affinities from MD simulations, it is often necessary to call upon geometric surface calculations in order to take into account solvation energies. One such method is the MM/PBSA method [29], a commonly used "endpoint" thermodynamic approach to affinity calculations [48]. This method calculates the proteins' internal energy contribution to the free-energy change explicitly, and exploits an implicit solvent approach that relies on the estimation of the macromolecular surface along with electrostatic terms. An advantage of such approaches is that solvent entropy is taken into account in large part by the implicit solvent approaches, relieving the burden of extensive sampling of explicit solvent in the MD simulations. It might be pointed out that accuracy can be improved by adding critical waters (e.g., [8]) back into the system explicitly [67].

1.2.1.3 Geometrical Modeling

Many molecular dynamics simulations can be performed without any light being shed on questions such as why one type of conformational change predominates over another or why two flexible proteins associate in a particular way. But the results of these blind calculations can of course be used to arrive at new hypotheses. Geometrical reasoning becomes important once again as we wish to model the nature of the macromolecular complex interfaces and their organization from a different perspective, that of the 3D shapes involved. Such modeling does not require intensive conformational sampling, but complements the simulation methods that provide it. It is usually based on an experimental structure, which is used as the de facto reference point for subsequent geometrical and biophysical reasoning.

Most analyses start with a space-filling representation of the macromolecule itself. High resolution, atomic-level models can be made using the van der Waals model. While an atom nucleus, composed of protons and neutrons, is only a few femtometers across, the atom dimensions are on the order of Å. This reflects the region of space around the nucleus in which there is a high probability for finding the electrons. In the van der Waals model this region is represented by a ball. Two non-bonded atoms can only get so close to each other before the interaction of their electron clouds engenders a prohibitive repulsive energy. This can be used to define their atomic radii (vdW radii), which can be measured experimentally by deviations from ideal gas behavior or from actual atom separations seen in small-molecule crystals [52].

Lower resolution representations are also used in many studies, including embedding the molecule in a grid [47] or defining its shape by a sum of spherical harmonics [57]. Methods based on the Voronoï diagrams offer alternatives that provide a different attribution of regions of space to the atoms or residues of the protein. They can be useful in describing the protein's 3D properties, and can be extended to the characterisation of macromolecular interfaces themselves.

We change speed now as we explore Voronoï diagrams and related constructions in detail.

1.2.2 Affine Voronoï Diagrams and α-Shapes

Voronoï diagrams and spatial partitions. Consider a set of sites (points, spheres, polygons, etc) in 3D, and a *generalized distance* to these sites. In simple terms, the Voronoï diagram is the partition of the space into Voronoï regions, defined as follows: the Voronoï region of a site consists of the points in space having this site as nearest neighbor for the distance considered. The most classical Voronoï diagram is certainly that of points under the Euclidean distance. But atoms have unequal size, as the van der Waals radii of the atoms found in bio-molecules range from 1 Å for hydrogen to 2 Å for sulfur. (We note in passing that since hydrogen atoms are not often reported in crystal structures, modeling may be carried out using the so-called group radii [19], in which the atomic radii are slightly expanded to account for the missing hydrogen atoms.) To handle different radii, the following generalization is resorted to.

Consider a collection $\mathcal{S}$ of n balls $\{B_i\}_{i=1,\ldots,n}$, each representing an atom. The ball $B_i(a_i, r_i)$ is centered at a_i and has radius r_i, and its bounding sphere is denoted S_i. The overall volume occupied by the n balls, also called the *space-filling diagram* or simply molecule in our context, is defined by $\mathcal{F} = \cup_i B_i$. To define the associated Voronoï diagram, recall that the *power distance* from a point x with respect to the sphere S_i is defined by $\pi(x, S_i) = ||\ x - a_i\ ||^2\ - r_i^2$. Denoting E^3 the usual three dimensional Euclidean space, the Voronoï diagram of $\mathcal{S}$ equipped with the power distance consists of the Voronoï regions:

$$\text{Vor}(S_i) = \{x \in E^3 \text{ such that } \pi(x, S_i) \leq \pi(x, S_j), \forall S_j \neq S_i\}. \qquad (1.1)$$

Note that the Voronoï cells partition the space into convex regions, each assigned to one of the input balls, namely an atom in our case. The Voronoï diagram is an example of *cell complex*, as it is made of cells (Voronoï regions, Voronoï faces, Voronoï edges, and Voronoï vertices), which satisfy the following two conditions (1) every face of a cell is also a face, and (2) the intersection of two cells is either empty or a common face of both.

Delaunay diagram and privileged neighbors. The Delaunay diagram Del($\mathcal{S}$) is the dual of the Voronoï diagram, in the following sense. Whenever a collection of $k+1$ Voronoï cells have a non-empty intersection, that is,

$$\cap_{i \in I=\{i_0,\ldots,i_k\}} \text{Vor}(S_i) \neq \emptyset, \qquad (1.2)$$

one reports the convex hull of the centers of the $k+1$ balls defining these regions into Del($\mathcal{S}$). To fully understand this construction, recall that a geometric k-simplex is the convex hull of $k+1$ affinely independent points. For example, a 0-simplex is a point, a 1-simplex is a line-segment, a 2-simplex is a triangle, a 3-simplex is a tetrahedron, etc. Generically, that is if no four points are co-circular in 2D and no five points co-spherical in 3D, the convex hull of the points involved in Eq. (1.2) is a k-simplex. Phrased differently, a k-simplex corresponds to the non-void intersection of exactly $k+1$ Voronoï regions, a property known as the Delaunay-Voronoï duality. The terminology used to describe this duality in 3D is presented in Table 1.1, and a 2D illustration is presented on Fig. 1.3a.

The Voronoï or equivalently the Delaunay diagram of n balls in 3D has quadratic $O(n^2)$ complexity in the worst-case, and can be computed in expected $O(n \log n + k)$ time, with k the size of the output, that is the number of simplices of the Delaunay triangulation [9]. This complexity, which depends on the size of the output, is called *output-sensitive*.

Practically, elaborate algorithms have been designed to compute the Delaunay triangulation, both from a combinatorial and a numerical standpoints—the latter to make the computation robust to degeneracies and numerical rounding errors.

Table 1.1 Duality between the k-simplices of the 3D Delaunay triangulation and the Voronoï faces of dimension $3 - k$

$k = 0$	Delaunay vertex	Voronoï region
$k = 1$	Delaunay edge	Voronoï facet
$k = 2$	Delaunay triangle	Voronoï edge
$k = 3$	Delaunay tetrahedron	Voronoï vertex

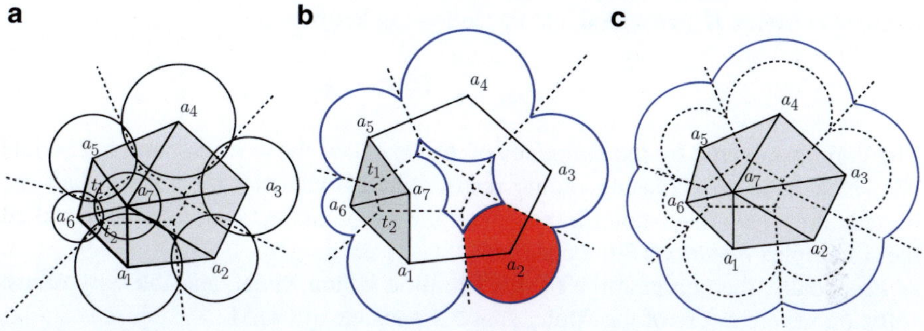

Fig. 1.3 A fictitious 2D molecule with seven atoms. (**a**) The Voronoï diagram in *dashed-lines* and the dual Delaunay triangulation in *solid lines*. Note that the Delaunay triangulation contains simplices of dimension zero (*points*), one (*line-segments*), and two (*grey triangles*). (**b**) The space-filling diagram $\mathcal{F}$ is the region delimited by the two *blue curves*, and the α-complex K_α contains seven vertices, nine line-segments and two triangles. The restriction of the atom centered at a_2 is presented in *red*. (**c**) The same molecule whose atoms have been grown. Note that the α-complex now contains all the triangles of the Delaunay triangulation, and that the void in the middle vanished

For example, the 3D Delaunay (regular) triangulation of the Computational Geometry Algorithms Library (CGAL), see www.cgal.org, handles about 10^5 points per second on a desktop computer.

α-complex: partition of the domain into restrictions. Since selected Voronoï regions are unbounded and our focus is on atoms, it is actually beneficial to consider the *restriction* of an atom to its Voronoï region, that is $R_i = B_i \cap \text{Vor}(B_i)$ (Fig. 1.3b). An elementary property stipulates that the volume $\mathcal{F}$ of the molecule and its boundary $\partial\mathcal{F}$ decompose into the contributions of restrictions, namely:

$$\mathcal{F} = \cup_i(\text{Vor}(S_i) \cap B_i); \quad \partial\mathcal{F} = \cup_i(\text{Vor}(S_i) \cap S_i). \tag{1.3}$$

In dealing with restrictions, it is convenient to consider balls whose radius is a function of a scaling parameter, so as to facilitate multiscale studies. For a real value α, define the grown ball $B_i[\alpha]$ as the ball whose squared radius has been enlarged by α, that is:

$$B_i[\alpha] = B_i(a_i, \sqrt{r_i^2 + \alpha}). \tag{1.4}$$

These grown balls define the domain $\mathcal{F}_\alpha = \cup_i B_i[\alpha]$ called the *space-filling diagram*. It is easily checked that the Voronoï diagram of the grown balls $\mathcal{S}_\alpha$ matches that of $\mathcal{S}$. The restriction of a grown ball is defined as $R_i[\alpha] = B_i[\alpha] \cap \mathrm{Vor}(B_i)$, and these restrictions also partition $\mathcal{F}_\alpha$, as specified by Eq. (1.3) mutatis mutandis. Restrictions can be used to define the analogue of Eq. (1.2), resulting in a simplicial complex called the α-complex, which is a subset of the Delaunay triangulation $\mathrm{Del}(\mathcal{S})$ [23]. More precisely, given a set of restrictions identified by their indices $I = \{i_0, \ldots, i_k\}$, one (generically) finds the corresponding k-simplex in the α-complex K_α provided that the following holds:

$$\cap_{i \in I = \{i_0, \ldots, i_k\}} R_i[\alpha] \neq \emptyset. \tag{1.5}$$

The domain covered by the simplices of K_α is called the α-*shape*, and is denoted W_α. Note that in increasing α, the restrictions associated with a given ball are nested, and so are the α-complexes. The finite set of distinct α-complexes is called the α-complex *filtration* (Fig. 1.3a, b).

Practically, the computation of this filtration is non-trivial, and the only robust software we are aware of the Alpha_shape_3 package of CGAL.

α-complex and topological features: cavities and tunnels. The quantities defined so far are atom-centric, in the sense where they provide information on a given atom and its neighbors. Remarkably, selected features of the α-complex also encode global features of the molecule. This is illustrated on Fig. 1.3b, where the cavity in the middle of the seven atoms is mirrored by the cavity delimited by the edges and triangles of the α-complex—for $\alpha = 0$ in this case since balls have not been grown. To make a precise statement, the α-shape W_α and the space-filling diagram $\mathcal{F}_\alpha$ have the same homotopy type [24].

In molecular modeling, two global topological features of utmost interest are cavities and tunnels. Mathematically, such features are defined in terms of generators of so-called homology groups, and efficient algorithms exist to compute them for collections of balls. Practically, cavities found in the interior of a macromolecule are important since they may contain small or solvent molecules. As for tunnels, also called channels, they typically provide passages from the bulk solvent to an active site.

α-complex and multi-scale analysis. As mentioned above, real data are often plagued with uncertainties, and the question arises to decide whether a particular feature is noise or not. This can be done with α-shapes as follows. Upon growing atoms as specified by Eq. (1.4), topological features (connected components, tunnels and cavities) appear and disappear. For example, in moving from the situation of Fig. 1.3b to that of Fig. 1.3c, the inner void disappears. In particular, one can define the stability of a feature as its life-time interval, in α units, a piece of information easily computed from the α-complex filtration. (Notice, though, that the growth model consisting of adding α to the squared radius does not have a straightforward physical motivation.)

1.2.3 Molecular Surfaces and Volumes

Assume that the balls in S represent a van der Waals model. The van der Waals surface is clearly defined as the boundary $\partial\mathcal{F}$ of the union of these balls. From a modeling perspective, this surface raises two difficulties. The first is related to the fact that most macromolecules are found in an aqueous environment, so that a vdW model generally delimits a number of tiny cavities which are not accessible to solvent molecules—recall our discussion of electrostatics in Sect. 1.2.1.2. The second stems from the fact that non covalent interactions account for the structure of macromolecular complexes. Given that the distance between non-covalently bonded atoms is strictly larger than the sum of their vdW radii, a vdW model does not inform us about such contacts in a complex—atoms from two chains in contact do not intersect.

Both problems are solved resorting in the Solvent Accessible Model, which consists of expanding the atomic radii, thus mimicking a continuous layer of solvent molecules. This fills meaningless cavities and recovers contacts between interacting atoms in a complex. More precisely, let W be a water probe, i.e. a ball representing a water molecule, and denote r_w its radius. (Note that this is a coarse representation focused on the oxygen atom of the water molecule; neither the hydrogen atoms nor the polarity of the molecule are represented.) To define a SAS model from a vdW model, one rolls the probe W over the vdW surface, tracing the locii of points visited by the center of W. Equivalently, the SAS surface is defined as the boundary of the union of the expanded balls $\{B_i(a_i, r_i + r_w)\}$, see Fig. 1.2a. An atom contributing to this surface is called *accessible*, and *buried* if not. A typical value for r_w is 1.4 Å.

As just discussed, both the vdW and the SAS surfaces are defined as the boundary of a collection of balls. Moreover, as already seen with Eq. (1.3), the description of such a surface as well as its enclosed volume only require computing the restrictions of balls. The information to compute the geometry of restriction is actually contained in the α-complex for $\alpha = 0$ [2].

For the boundary $\partial\mathcal{F}$ of the union, one actually builds from the 0-complex a representation of $\partial\mathcal{F}$ which is a cell complex. Its 2-cells are spherical polygons, also called caps; if two such cells intersect, they share a circle arc supported by the intersection circle of their defining spheres. Its 1-cells are circle arcs; if two such arcs intersect, they share one vertex defined by the intersection of (generically) three spheres. To represent the volume $\mathcal{F}$ using its partitioning into restrictions, following Eq. (1.3), one actually resorts to a tiling of each restriction R_i using two types of pyramids depicted on Fig. 1.2b. In particular, adding up the (signed) volumes of such pyramids allows one to compute the volume of R_i and thus of $\mathcal{F}$ with a controlled accuracy. The proofs can be found in [14], and the corresponding program, Vorlume, is available at http://cgal.inria.fr/abs/Vorlume/.

1.2.4 Modeling Interfaces

Consider a binary complex involving three molecular species, say A and B for the two partners, and W for the water molecules. As mentioned in Sect. 1.2.1.1, a typical PDB entry lists the coordinates of the atoms making up these species. To model a binary complex, one wishes to identify the interface atoms, including water molecules that may be present between the two partners, and to describe the geometry and the topology of the interface in a manner best accounting for biological and biophysical properties. An interface model is therefore of phenomenological nature, and the rest of this section presents our Voronoï based model [14], whose implementation is available from http://cgal.inria.fr/abs/Intervor.

Interface atoms. Identifying interface atoms can be done resorting to a distance threshold, typically in the range 4–7 Å. But such a strategy tends to overcount atoms in the inner regions of the interface, as seen from Fig. 1.5a. A common alternative consists of selecting as interface atoms those losing solvent accessibility in the complex, as illustrated in Fig. 1.4b. However, interface atoms will be missed if they have zero solvent accessibility [16].

To avoid these drawbacks, we define a threshold-independent interface model using the Voronoï diagram of the atoms of the three species A, B, W of the complex in the SAS model. The construction is based on pairs of adjacent restrictions, which, by the definition of the α-complex, correspond to edges in the α-complex of the SAS model for $\alpha = 0$.

Let an *interface water molecule* be a water molecule whose restriction has neighboring restrictions of type A and B. As opposed to *bulk* water molecules, such molecules are exactly sandwiched between the partners. As illustrated on Fig. 1.5b, our Voronoï interface model involves the pairs of neighboring restrictions of type $[A; B]$ or $[A; W]$ or $[B; W]$, with W interface water molecules. Note that in addition to the two atoms in contact, each pair also defines the Voronoï facet, also called *tile*, separating the Voronoï regions of these two atoms. Tiles of a given type define the eponymous interface, namely the AB (AW and BW) interface for tiles of type AB (AW and BW). Tiles AB define direct contacts between the partners, while tiles AW and BW tiles define contacts between A and B mediated by interface water molecules. Moreover, the union of tiles AW and BW defines the $AW - BW$ interface, and the union of the $AW - BW$ and AB interfaces defines the ABW interface. The latter separates the partners and gives a global overview of the interaction area, regardless of the role played by water molecules. These notions are illustrated on Fig. 1.6, the role of water molecules being illustrated by Fig. 1.6c. A key property of this interface model is to precisely encode the geometry and the topology of the interface.

Topology of the interface. The connectivity between the tiles allows one to define the number of connected components or patches of the interface. Of particular interest are the number of connected components of the AB interface, and if more than one, the role of water molecules (possibly) sealing these connected components

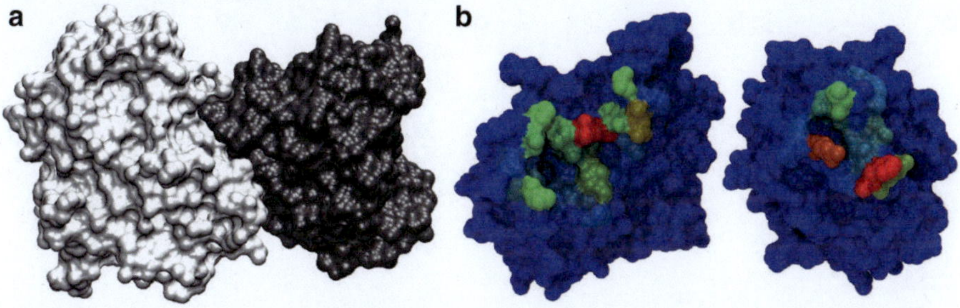

Fig. 1.4 An example binary protein complex and its interface (PDB id: 1a2k). (**a**) The complex is defined by two components: chains A and B, colored *white*, and chain C, colored *grey*. (**b**) Burial of Solvent Accessible Surface in the complex (SAS, see definition in Sect. 1.2.3). The two components have been separated, and each is rotated about its vertical axis so as to display the binding patches. Amino acids are colored from *blue* to *red* in order of increasing burial in the interface, evaluated as their SAS in the separated component minus the SAS in the complex. *Blue* indicates zero buried area, corresponding to surface left free in the complex

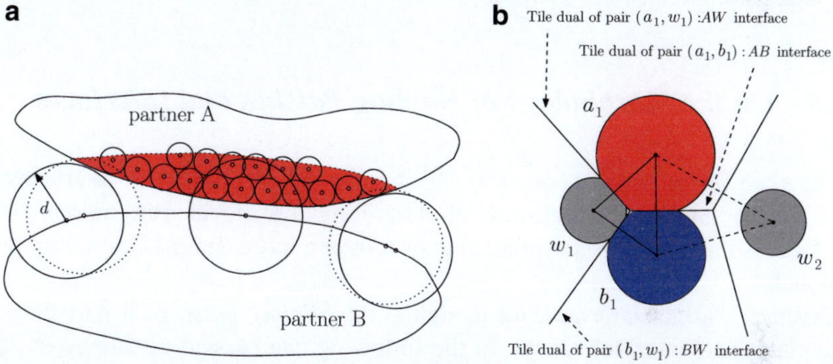

Fig. 1.5 Modeling interfaces of macro-molecular complexes. (**a**) The distance based definition, which consists of selecting for a given atom all atoms of the partner within a distance threshold d, imposes a bias towards convex regions. (**b**) The α-shape based definition consists of selecting pairs of incident restrictions

together. Another important topological descriptor is the number of boundaries of the AB interface, since all but the outer boundary correspond to water molecules trapped in between the partners.

Geometry of the interface. From a geometric standpoint, a simple yet important parameter of the interface is the surface area of the AB and $AW - BW$ interfaces, defined as the sum of the areas of their constituting tiles. The surface area is a good descriptor of the specificity of interactions [17]. Another important parameter is the curvature of the ABW interface. Since the ABW interface is a cell complex consisting of Voronoï tiles, its *extrinsic*, or mean, curvature, is directly encoded in the dihedral angles defined by incident Voronoï facets—two incident Voronoï facets define a hinge. From a biological standpoint, curvier interfaces indeed generally

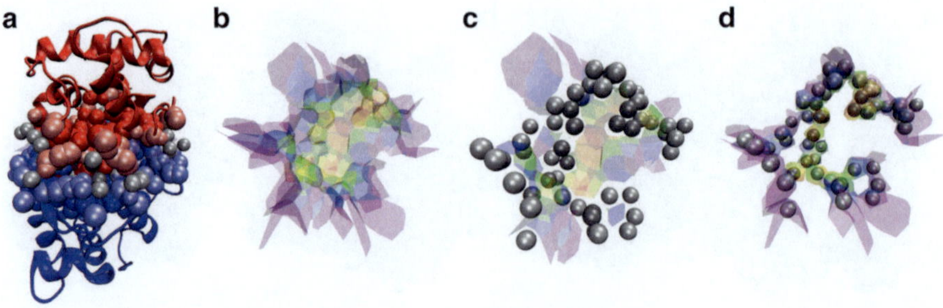

Fig. 1.6 The Voronoï interface of complex PDB 2dqa, the color code of Voronoï facets being the Shelling Order defined in Sect. 1.2.5. (**a**) The two polypeptide chains and the interface atoms (**b**) Top view of the ABW interface (**c**) The AB interface and the solvent molecules involved in the $AW - BW$ interface. (**d**) The $AW - BW$ interface and the solvent molecules

correspond to specific interactions—the more curved a surface the more difficult it is to find a complementary surface.

1.2.5 On the Morphology of Binding Patches and Interfaces

We have so far provided geometric and topological descriptors of interfaces, but these hardly answer the question of describing the global morphology of the interface: is the interface isotropic, or does it have a more complex geometry, like a dumbbell?

Given a Voronoï interface, let us define the *binding patch* of a partner as the collection of its interface atoms. In the following, we present a strategy to define the morphology of interfaces and binding patches. In a nutshell, the idea consists of partitioning the object of interest (Voronoï interface or binding patch) into concentric shells.

Shelling order. The ABW Voronoï interface consists of Voronoï tiles. Some of them are clearly outer tiles as they bound the whole interface, while others are located in the center, a distinction known as the *rim-core* model in molecular modeling [18]. However, this boolean model does not convey the fact that describing the position of a tile within the interface is a gradual process. This observation motivates the definition of the *Shelling Order* (SO), which we present in abstract terms as it actually applies to cell complexes in general, and to the Voronoï interface as well as binding patches in particular.

Consider a cell complex $\mathcal{C}$ consisting of d-dimensional cells incident across $(d-1)$-dimensional cells. For the ABW Voronoï interface, the cells are the Voronoï tiles, and two incident tiles share a Voronoï edge. For the SAS surface, possibly restricted to a binding patch, the cells are spherical caps of the boundary of the union, and two such cells share a circle arc. Note that cells and their incidence define a graph G whose nodes are the cells, and edges correspond to incidences between cells.

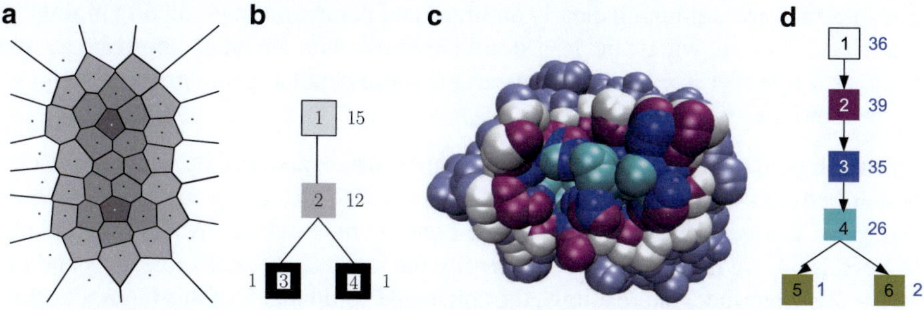

Fig. 1.7 Shelling Voronoï interfaces and binding patches. (**a**) A fictitious Voronoï interface consisting of Voronoï tiles arranged into three concentric shells. Compare with Fig. 1.6b for a real example. (**b**) The shelling tree of (**a**), the integers indicating the number of tiles associated to a tree node. (**c**) The atoms of the binding patch of the antibody partner in the complex with PDB id 1vfb.pdb. The *white atoms* represent the outer rim i.e. the atoms with a null SO. (**d**) The shelling tree of (**c**), the integers indicating the number of atoms associated to a tree node

Now term a cell a *boundary cell*, provided at least one of its $(d-1)$ faces does not have any incident cell in $\mathcal{C}$, and assign a value of one to such cell. The SO of a cell c is one plus the smallest number of edges of G required to reach c from a boundary cell. See Fig. 1.7a, b for the Voronoï interface, and binding patches, respectively.

Shelling graphs and global morphological features. The SO defines an integer valued function on cells. To integrate this information at the level of the complex $\mathcal{C}$, define a *shell* as a maximal connected component of cells having an identical SO. Also, term two shells of *incident* if they contain two incident shells whose SO differ of one unit. We arrange shells into a *shelling graph*, whose roots correspond to shells at depth one, and whose parent-child relationship encodes the incidence between shells [45]. In particular, a branching point correspond to a split of a shell into sub-shells while increasing the SO. See Fig. 1.7 for an illustration.

1.2.6 Success Stories

The Voronoï diagram has been applied to macromolecules since the early 1970s [56], when it was first used to define volumes and packing densities in proteins. Since then it has found many applications in structural biology [53], but we will focus on applications related to macromolecular complexes in a rapid, non-exhaustive survey.

Identifying biological protein-protein interfaces. A recurring problem in biological structure determination by crystallographic techniques is the proper identification of the biological interface, as opposed to the crystal contacts that are simple artefacts of crystallization conditions. The problem is widespread [37,42], and must obviously be resolved before meaningful biological interpretation of the structural data is undertaken. Voronoï tesselations have been used in classifying crystal interfaces in terms of their biological likelihood, in an approach called DiMoVo [7].

The interface area, although clearly an important parameter in predicting biological interfaces, was shown to be less dominant here than in other approaches, and assumes a role that is complementary to the more detailed geometrical descriptors of the interface.

Solvation properties. Molecular recognition is often mediated by water molecules bound between the partners; such molecules contribute to lowering the potential energy of the system through hydrogen bonding and van der Waals interactions. In Sect. 1.2.4, we have seen how to identify the interfacial water molecules and the $AW-BW$ interface. Interestingly, the connected components of this interface allow one to single out different hydration patterns, from isolated water molecules to large networks of interfacial water [16]. The connected components of the AB interface correspond to previously identified binding patches [17], which are themselves connected to the decomposition of the whole interface into modules of amino-acids [55].

However, the affinity measures the difference in free energy between the complex itself and the unbound proteins. Indeed, association is accompanied by desolvation of the regions of the unbound protein surfaces that will become the protein-protein interface. Affinity predictions have been investigated in two directions. First, the power diagram has been used to characterize the protein-water interface in general. A technical comment is in order here concerning the difficulties faced by Voronoï models to handle large or unbounded cells, which arise in this case if the solvent structure is undefined—this situation is common, for the model of a protein structure does not in general specify the solvent structure unambiguously. While the sole Voronoï diagram cannot cope with such uncertainty, information contained in the α-shape allows restriction of these cells [16]. Of course, crystal structures typically do feature many first-shell water molecules. Gerstein and Chothia [28] thus used the power diagram to calculate both protein and solvent atomic densities at protein-water interfaces in crystals, which showed a volume increase of protein atoms near the interface together with a corresponding volume reduction in the solvating water. Shape information was also incorporated, notably in the dependence of the densities on the concave or convex regions in the protein surface. Second, models of binding patches have been used to investigate the correlations between structural parameters and dissociation free energies. In [45], the weighted average of the shelling order of the atoms of the binding patch has been shown to outperform all other contenders to predict the binding affinities of the binding affinity benchmark [39].

On the morphology of an interface: core and rim models. An important topic when analyzing interfaces consists of unraveling the relationship between the morphology of an interface and its biological and biophysical properties.

Simple geometric concerns coupled with residue conservation analyses of a database of biological protein-protein complexes of known structure led Chakrabarti and Janin [17] to suggest that biological protein-protein interfaces are organized into a *rim*, consisting of residues for which all atoms are to some degree accessible to solvent, and the *core*, consisting of residues in which any atom is fully buried. Residue composition of the core region was observed to be significantly different from that of the free protein surface. Subsequent studies have suggested core residues are more likely to include hotspot residues [32].

While the core-rim distinction is inherently binary, the shelling-order introduced in Sect. 1.2.5 measures the depth of an interfacial atom with an integer. The SO has been used to refine the understanding of the several physico-chemical properties, based on statistics gathered on 18 homodimers and 36 heterodimers protein complexes. We have seen that the composition of the core and the rim of an interface are different [17], and one would tend to believe that charged or polar amino-acids tend to be located near the rim, where electrostatic interactions with the solvent are favored. However, the correlation between SO and the electrostatic properties of amino acids is not statistically significant in general [11]. It has also been shown in [31], based on a statistical meta-analysis, that conserved residues tend to locate in the interface core. (Conservation of a residue refers to its repeated presence at a particular position within a non redundant set of protein sequences. The reader is referred to Sect. 1.2.1 for the Shannon entropy based evaluation of conservation.) Replacing the binary core-rim partitioning by the SO confirms this finding at the dataset level, but also shows that the conclusion does not hold in general on a per-complex basis [11]. Finally, another important property of interfacial residues is their *dryness*, i.e., their isolation from mobile solvent. Using all-atom molecular dynamics simulations on the aforementioned 54 complexes, it was shown in [46] that dryness is correlated to residue conservation. But as shown in [11], the dryness of a residue is in fact determined by its SO. Thus dryness can be evaluated from a mere Voronoï interface calculation, as opposed to a costly molecular dynamics simulation.

This is a case in which appropriate geometric analysis naturally spotlights the important biological determinants of protein-protein interactions.

Predicting the structure of protein complexes. The power diagram has also been used in deriving new scoring functions to be used in ranking candidate solutions for the structure of a protein-protein complex from docking studies. In one such study, amino-acid residue centroids and synthetic solvent molecule positions were used to create the Voronoï descriptions, and parameters such as the cell volumes, interface areas, and centroid-to-centroid distances of known interfaces were integrated into the final scoring function using machine learning techniques [6].

The more general α-complex has also been employed, obviating the need for solvent-atom positions, notably in defining volume derivatives of the macromolecule [26]. Such results should also prove important in improving implicit-solvent models in molecular simulations, in which modifications in the coordinates of the protein must be taken into account.

1.3 Modeling Large Assemblies

Having dealt with binary protein complexes, we now consider the problem of modeling large assemblies, that is, complexes involving on the order of hundreds of polypeptide chains. In doing so, we shall focus on the Nuclear Pore Complex

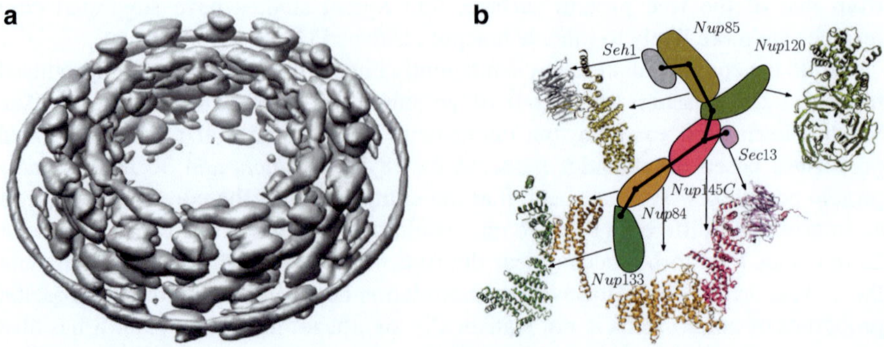

Fig. 1.8 The nuclear pore complex at different scales. (**a**) Level set surfaces representing all the proteins of the NPC. The outer ring anchors the NPC in the nuclear membrane, while the inner channel serves the nucleo-cytoplasmic transport. The diameter of the NPC is circa 100 nm. (**b**) An atomic resolution model of the so-called Y-complex, a sub-complex of nuclear pore complex. (Adapted from [38])

(NPC), which is illustrated at two different scales on Fig. 1.8. First, we explain why the reconstruction of large assemblies such as the NPC yields ambiguous results; second, we present a geometric modeling paradigm that accommodates ambiguities on the shapes and positions of proteins within an assembly; finally, we present selected tools allowing one to quantitatively bridge the gap between global-ambiguous models, and local—atomic-resolution ones.

1.3.1 Challenges

1.3.1.1 From Atoms to Assemblies: Jumping Across Three Orders of Magnitude

The largest protein complex known to date in eukaryotic cells is the nuclear pore complex (NPC), which raises prototypical difficulties for modeling large assemblies.

The NPC consists of about 456 protein instances of 30 protein types. Where appropriate, we speak of protein types and protein instances (or types and instances for short), instead of proteins. It is a cylindrical 100 nm-wide channel between the nucleus and the cytoplasm, with a lumen of circa 40 nm. It is involved in the passive diffusion of small molecules, and the active transport of large ones, with RNA moving from the nucleus into the cytoplasm, and selected proteins synthesized in the cytoplasm moving back into the nucleus. The NPC has eightfold axial symmetry, and an approximate twofold rotational symmetry between the nucleoplasmic and the cytosolic halves, yielding 16 so-called *half-spokes*. Its architecture has been abstracted into four concentric cylinders [34], which are

1. The channel cylinder, containing proteins having unstructured regions (filaments) that regulate active transport;
2. The adapter cylinder, which occupies an intermediate position between channel and scaffold;
3. The coat cylinder, which defines the scaffold of the NPC;
4. The pore membrane cylinder, anchoring the NPC in the nuclear membrane.

Making a precise statement on the stoichiometry of the protein types in the NPC is difficult per se, as the proteins present indeed change along the life cycle of the cell. During cell division, with concomitant nuclear division, selected NPC proteins diffuse back into the cytoplasm and join the new nuclei at a later stage.

This plasticity, together with the large number of proteins involved and their inherent flexibility, makes the reconstruction of such a molecular assembly a highly challenging endeavor.

As of now, two types of structural results are available for the NPC, as illustrated on Fig. 1.8. On the one hand, atomic resolution structures of monomers and dimers have been resolved, from which plausible sub-complexes of the NPC have been modeled. One example is the Y-complex, an heptameric assembly found in the coat cylinder, making up the scaffold of the NPC [61]. Another complex reconstructed at atomic resolution is the so-called T-complex, whose unstructured filaments are believed to participate in active transport across the NPC [60]. On the other hand, putative global models of the NPC in its entirety have also been reconstructed by integrating data providing complementary information [3, 4]. A variety of data can be used in these models, and the following three deserve a special mention. First, the overall shape of the assembly is typically inferred from cryo-electron microscopy. Second, the 3D location of particular proteins can be captured thanks to immuno electron-microscopy, a process which consists of tracking under an electron microscopy proteins coated with antibodies that have been conjugated with gold particles. Third, particular interactions between pairs of proteins can be inferred from proteomics experiments, in particular Tandem Affinity Purification (TAP) data.

Such data are usually noisy and/or ambiguous, as we shall discuss in an example in the next section.

1.3.1.2 Low Resolution and Ambiguous Data: The Example of TAP data

As discussed in Sect. 1.1.1, structural information for assemblies can be obtained from cryoEM. Such data provide information at the scale of the assembly, yet because of their noisy nature complementary information is needed in order to exploit them. Of particular interest are proteomics data, which provide information on the interaction between protein types, which we illustrate with Tandem Affinity Purification (TAP) data.

TAP experiments give access to all types of proteins found in all complexes that contain a prescribed protein type, say R. More precisely, the method consists of the following steps. First, a fusion protein is created by modifying the gene for

R: coding sequences for two sticky tags are added, separated by a sequence coding for a protease cleavage site. Upon introducing this engineered gene into a host cell, the modified protein gets expressed and takes its place in its usual complexes—assuming that there is no hindrance induced by the tags themselves. On lysing the cell, the protein complexes containing protein R are retrieved thanks to two affinity purification steps. Each purification step consists of capturing the complexes on an affinity purification column thanks to one of the sticky tags. Between the first and the second purification steps, the complexes hooked on the first column are released thanks to addition of a protease which cuts the linker containing the first sticky tag at the level of the cleavage site. This reveals the remaining sticky tag for the second purification step. Upon completing these purification steps and dismantling the complexes during electrophoresis, one obtains a gel with one band per protein type. Mass spectrometry is then used to identify the protein types present.

This list of protein types obtained, also called a pullout, calls for two comments. First, one does not know whether the list of interacting types corresponds a single complex or to several complexes. For example, a list (R, S, T) obtained by tagging R may correspond to a single complex containing the three species, or to two binary complexes respectively involving (R, S) and (R, T). Second, no information on the stoichiometry of protein instances within a complex is available. Despite these inherent combinatorial ambiguities, TAP data are of prime interest for the reconstruction of large assemblies: knowing that protein instances participate in a complex imposes distance restraints between them.

1.3.1.3 Reconstruction by Data Integration: Procedure and limitations

To deal with the ambiguities just discussed, the authors of [4] proposed a reconstruction strategy based on three ingredients:

- A model M for the protein instances of the NPC consisting of balls. Note that a model involving n balls with prescribed radii defines a $3n$ dimensional space corresponding to the xyz coordinates of all centers.
- A scoring function measuring the coherence between a particular model and the experimental data.
- An optimization procedure aiming at finding the models which best comply with the data.

Scoring function. Consider a particular type of experimental data, such as those described in Sect. 1.3.1.1. In short, a restraint is a real-valued function measuring the discrepancy between the model M and these data: a quadratic penalty is applied the farther the model is from the data; if the model complies with the data, the restraint is zero. Let us consider the following three examples:

- A cryoEM envelope can be used to make sure that the model M does not protrude from this envelope. That is, if the one-sided Hausdorff distance between the model and the envelope is beyond some threshold, a penalty is applied.

- TAP data can be used to enforce proximity, i.e., a distance relationship between proteins participating in a given complex. As a protein is modeled by balls, a distance restraint entails that two balls from the two proteins must be within a distance threshold.
- Positional information gathered using immuno-EM can be used to enforce the location of specific protein instances within a prescribed region of the model.

That is to say, using one restraint R_{exp_i} for each type of experimental data, the reconstruction process aims at finding the model M minimizing the following penalty function:

$$F(M) = \sum_{\text{All experiments } exp_i} R_{exp_i}(M). \tag{1.6}$$

For the NPC, which consists of $n = 456$ instances of $p = 30$ protein types, a maximum of 1,848 balls (of fixed radius) have been used, whence an optimization problem in a space of dimension $3 \times 1,848 = 5,544$. This problem being non convex, local minima were sought using an optimization strategy based on simulated annealing and coarse-grain molecular dynamics, from which $N = 1,000$ plausible configurations were singled out [3].

Output of the reconstruction. To interpret these N structures selected, a *probability density map* was created per protein type, by collecting all instances of that protein type across the N models and blending the corresponding balls. The probability density map is a 3D matrix, each voxel being endowed with the probability of it being contained within an instance of that type. (Note that such a map should not be confused with a cryoEM map which encodes a density of matter.) Merging back all the probability densities yields a probabilistic model of the whole NPC, which is illustrated by the contour plot of Fig. 1.8a.

In a sense, the uncertainties in the various input data account for ambiguities in the shape and position of the proteins encoded in the density map(s). In what follows, we present a panoply of tools allowing one to make quantitative statements from such ambiguous models.

1.3.2 Toleranced Models and Curved Voronoï Diagrams

We wish to accommodate ambiguous protein models within a probability density map. To see which difficulties are faced, consider the fictitious map of Fig. 1.9, which corresponds to a fictitious complex involving three molecules of three balls each. The color coding clearly indicates that some regions of the map are much more likely than others. Using balls to model the proteins contained within such a map would face ambiguities regarding the locations and the radii of the balls. To ease the process, we introduce *toleranced models* and explain their connexion to curved Voronoï diagrams [12]. Recalling that the NPC consists of about 456 protein instances of $p = 30$ protein types, we shall in particular use toleranced models in

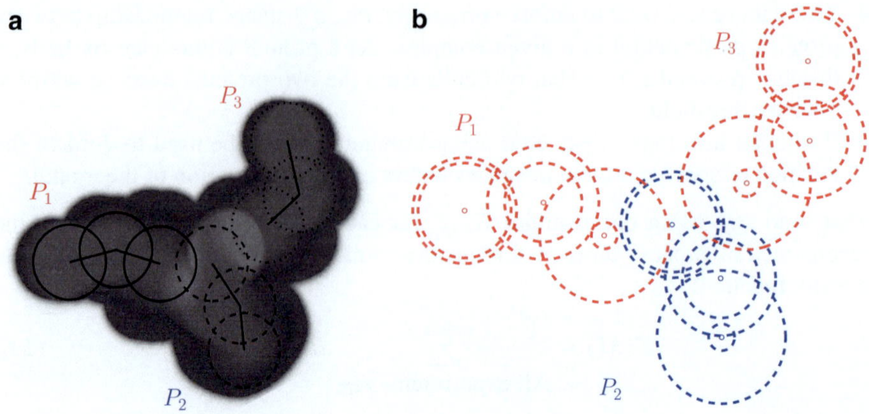

Fig. 1.9 From probability density maps to toleranced models. (**a**) Three conformations of three flexible molecules defining a fictitious assembly, and a probability density map whose color indicates the probability of a given point to be covered by a random conformation—from low (*black pixels*) to high (*white pixels*) probabilities. (**b**) A toleranced model, where each toleranced molecule consists of a set of pairs of concentric balls, inner and outer. Note that the three molecules have been assigned to two groups corresponding to *red* and *blue* molecules, thus defining a bicolor toleranced model

the so-called bicolor setting, which consists of segregating the p protein types into two families: the red and the blue groups. Typically, the red group will refer to the protein types involved in a TAP experiment or to those involved in a sub-complex.

Toleranced models. Let a *toleranced ball* $\overline{B_i}(c_i; r_i^-, r_i^+)$ be a pair of concentric balls centered at c_i, the *inner* and *outer* balls of radii $r_i^+ > r_i^-$, respectively. Inner and outer balls are respectively meant to encode high confidence regions and uncertain regions in a map. In order to deal with balls of intermediate size, we introduce a parameter $\lambda > 0$ governing a *growth process* consisting of linearly interpolating and extrapolating the radii. That is, the *grown ball* $\overline{B_i}[\lambda]$ stands for the ball centered at c_i and of radius:

$$r_i(\lambda) = r_i^- + \lambda(r_i^+ - r_i^-). \tag{1.7}$$

Note that for $\lambda = 0$ (resp. $\lambda = 1$), the grown ball matches the inner (resp. outer) ball. We define a *toleranced protein* as a collection of toleranced balls, and a *toleranced assembly* as a collection of toleranced proteins. For a given value of λ, a protein of intermediate size is denoted $\overline{P_j}[\lambda]$, and $\mathcal{F}_\lambda$ denotes the domain of the space-filling diagram, that is $\mathcal{F}_\lambda = \cup_i \overline{B_i}[\lambda] = \cup_j \overline{P_j}[\lambda]$. For a fixed λ, the topology of the domain $\mathcal{F}_\lambda$ is of utmost interest: a connected component of this domain is called a *complex*, and the domain is called a *mixture* if it involves several complexes.

Similarly to the α-shapes of Sect. 1.2.2, a toleranced model defines a 1-parameter family of shapes, except that the linear interpolation of the radius specified by

Eq. (1.7) replaces the radius growth which consists of adding α to the squared radius, as in Eq. (1.4).

Curved Voronoï diagrams. The growth process just defined is coupled to curved Voronoï diagrams and α-shapes. To see why, let p be a point belonging to the grown ball of Eq. (1.7). Denoting $\delta_i = r_i^+ - r_i^-$, observe that

$$r_i(\lambda) = \| c_i p \| \Leftrightarrow \lambda = \frac{1}{\delta_i}(\| c_i p \| - r_i^-). \tag{1.8}$$

That is, given a point p, we can define the function $\lambda(\overline{B_i}, p)$ corresponding to the value of λ such that p is on the corresponding grown ball. Denoting $\mu_i = 1/\delta_i$ and $\alpha_i = r_i^- / \delta_i$, this latter equation can be rewritten as:

$$\lambda(\overline{B_i}, p) = \mu_i \| c_i p \| - \alpha_i. \tag{1.9}$$

In this equation, the Euclidean distance is multiplicatively weighted by μ_i, and additively weighted by α_i. The associated Voronoï diagram is termed *compoundly-weighted* [51], CW for short, and consists of the partition of the 3D space into the Voronoï regions defined by:

$$\text{Vor}(\overline{B_i}) = \{p \in E^3 \text{ such that } \lambda(\overline{B_i}, p) \le \lambda(\overline{B_j}, p), \forall \overline{B_j} \ne \overline{B_i}\}. \tag{1.10}$$

Intuitively, a point p belongs to $\text{Vor}(\overline{B_i})$ if the growing ball $\overline{B_i}[\lambda]$ reaches point p before any ball $\overline{B_j}[\lambda] \ne \overline{B_i}[\lambda]$. A region $\text{Vor}(\overline{B_i})$ is bounded by curved bisectors, which are degree four algebraic surfaces. See Fig. 1.10 for a 2D example. Note that a Voronoï cell may not be (simply) connected. While a naïve algorithm has recently been developed to compute such diagrams [12], as opposed to affine Voronoï diagrams, both the complexity of CW diagrams and the design of efficient algorithms are open problems.

λ-shapes and the λ-complex. To compute complexes and mixtures in the bicolor setting, we generalize the α-shapes of Sect. 1.2.2 to the CW Voronoï diagram. For a given ball $\overline{B_i}[\lambda]$, consider its *restriction* to its Voronoï region, that is the intersection $\overline{B_i}[\lambda] \cap \text{Vor}(\overline{B_i})$. These restrictions naturally partition the domain $\mathcal{F}_\lambda$, and their connected components correspond to the aforementioned complexes.

In using the λ-complex, one needs to decide up to which value $\lambda_{\max}$ the growth process is performed. This value is defined using the following volumetric criterion. Consider a complex C, namely a connected component of $\mathcal{F}_\lambda$, and denote its volume $Vol_\lambda(C)$, that is the sum of the volumes of its restrictions in the CW Voronoï diagram. (As computing these volumes, which are bounded by degree four algebraic surfaces, is an open problem, a practical alternative consists of adding up the volumes of the restrictions in the power diagram of the grown balls, as explained in Sect. 1.2.3.) Since this complex corresponds to a list of toleranced proteins, let Vol_{ref} be the sum of the reference volumes of these proteins [33]. (The reference volume of a protein is estimated from its sequence of amino-acids. These reference

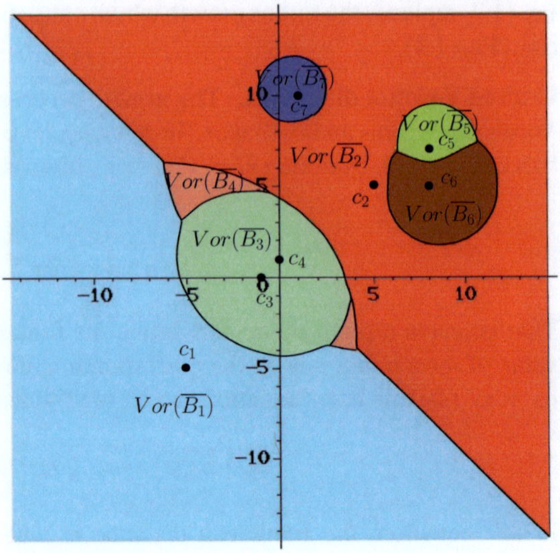

Fig. 1.10 The compoundly weighted Voronoï diagram of 7 toleranced balls in 2D: $\overline{B_1} = (-5, -5; 3, 7)$, $\overline{B_2} = (5, 5; 3, 7)$, $\overline{B_3} = (-1, 0; 4, 5)$, $\overline{B_4} = (0, 1; 2, 5)$, $\overline{B_5} = (8, 7; 2, 3)$, $\overline{B_6} = (8, 5; 3, 4)$, $\overline{B_7} = (1, 10; 1, 2)$. $Vor(\overline{B_4})$ is not connected, and so is the intersection $Vor(\overline{B_1}, \overline{B_2})$ of $Vor(\overline{B_1})$ and $Vor(\overline{B_1})$. $Vor(\overline{B_2})$ is not simply connected. δ_1 and δ_2 are maximal among all balls, and $\overline{B_1}, \overline{B_2}$ have unbounded Voronoï regions

volumes account for the spatial occupancy of amino-acids in a globular protein.) The volume ratio of the complex C at value λ is defined by:

$$r_\lambda(C) = Vol_\lambda(C) / Vol_{ref}(C). \tag{1.11}$$

Because the volume $Vol_\lambda(C)$ is monotonically increasing as a function of λ, the value $\lambda_{\max}$ of λ is set so as to upper-bound the ratio $r_\lambda(C)$. Practically, the bound is set to match the uncertainties of the probability density maps: if the volume of all voxels with a non null probability is v times larger than the volume of the proteins involved, $\lambda_{\max}$ is chosen so that $r_\lambda(C) \sim v$.

1.3.3 Stability Analysis of Multi-scale Toleranced Models

Upon growing λ, the restrictions of growing balls merge, and so do the complexes involving these restrictions (Fig. 1.11). This information is recorded in two complementary data structures.

At the complex level, we use the pairwise intersections between the restrictions involved in a complex C to define its *skeleton* graph G_C. The nodes of this graph are the toleranced proteins of C, and an edge links P_i and P_j provided that there exists two intersecting restrictions, one from P_i and one from P_j. See Fig. 1.12 for an illustration in the bicolor setting.

At the assembly level, the evolution of complexes is recorded in a directed acyclic graph called *Hasse diagram*, a special graph whose nodes are the complexes, with an edge joining (generically) two nodes when the complexes merge along the growth process. The origin (endpoint) of an edge therefore represents the birth (resp. death)

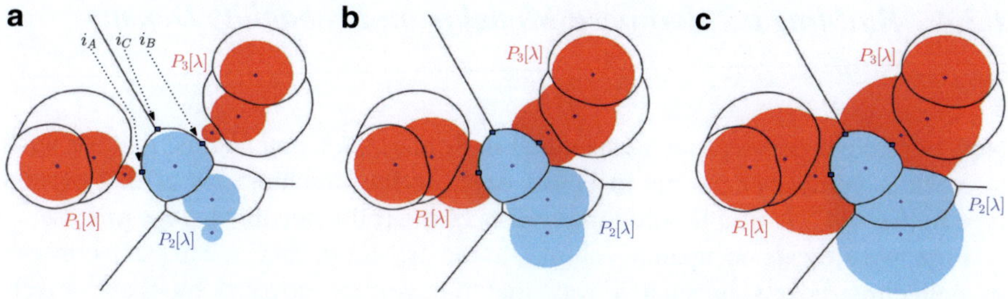

Fig. 1.11 Tracking the interactions of the three toleranced proteins of Fig. 1.9. The sub-figures (**a**, **b**, **c**) respectively show grown balls $\overline{B_i}[\lambda]$ for $\lambda = 0, 0.5, 1$. The region of the plane consisting of points first reached by a growing toleranced ball is the Voronoï region of this ball, represented by *solid lines*. Colored solid regions feature the *restrictions*—each an intersection of a growing ball and its Voronoï region. Along the growth process, the restrictions intersect in three points i_A, i_B, i_C, represented as *blue squares*

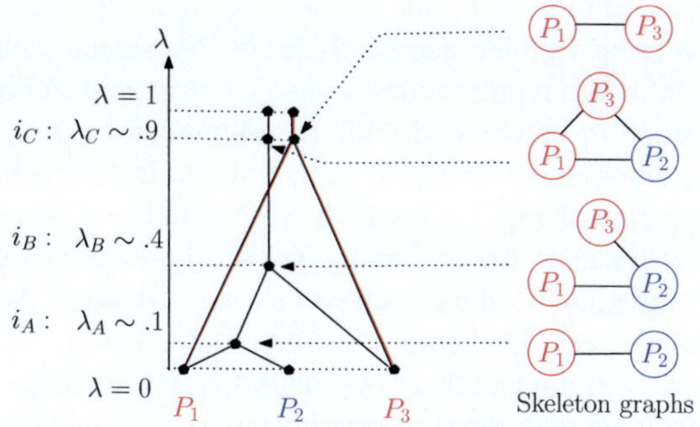

Fig. 1.12 Hasse diagrams encoding the interactions of the three toleranced proteins of Fig. 1.9. *Black tree*: all instances; *red tree*: red instances only

of a complex C. For example, at $\lambda = \lambda_b(C)$, the complex gets formed by a merge of two or more complexes; at $\lambda = \lambda_d(C)$, the complex dies by merging with at least another complex. Thus, the *lifetime* $s(C) = \lambda_d(C) - \lambda_b(C)$ provides a measure of the topological stability of the complex C. Also, the *ancestors* and *successors* of C are the complexes contained in and containing, respectively, the complex C. See Fig. 1.12 for an illustration.

In the bicolor setting, let T be the list of red protein types. A complex C of the Hasse diagram is made of instances whose types are in T. If each type of T is present exactly once in C, the complex C is termed an *isolated copy*. The number and the lifetime of isolated copies give a measure of the separability of the different copies of a complex involving all the types of T. Note that the intersection of the lifetime intervals of the different isolated copies may be empty.

We note in passing that we track the evolution of connected components, but not that of higher order homology generators. In fact, the investigation of the stability of topological features, for collection of balls and more generally for nested topological spaces, falls in the realm of *topological persistence* [25].

1.3.4 Building a Toleranced Model from Probability Density Maps

Now consider a density map as specified in Sect. 1.3.1.3, and assume that we wish to create a number of protein instances equal to the stoichiometry of that protein type in the NPC. A simple solution consists of the following three-stage process.

First, we allocate occupancy volumes to the protein instances. This step consists of collecting voxels in such a way that the volume covered by these voxels matches the estimated volume of all instances, namely Vol_{ref} multiplied by the stoichiometry. These voxels are collected by a greedy region-growing strategy that favors large values of the probability. That is, starting from local minima, a priority queue of candidate voxels based is maintained. These candidate voxels are those incident to the voxels already recruited, and they are sorted by decreasing probability. The voxel with top density is added, and the process halts when the aforementioned volume criterion is met. Second, we compute a canonical representation involving 18 toleranced balls for each instance. This number allows the construction of the four regular morphologies represented on Fig. 1.13. Consider an occupancy volume to be covered with 18 toleranced balls of identical radius. Using a principal components analysis, each volume is assigned one of the four canonical arrangements of Fig. 1.13, which correspond to a shape that is roughly isotropic, flat, semi-linear or linear. Finally, we set the inner and outer radii. For a given protein type, the inner radius is set so that the volume of the union of the 18 inner balls matches the estimated volume of the protein Vol_{ref}. The specification of the outer radius relies on the fact that the probability density maps of large proteins tend to be more accurate than those of small proteins, a feature likely related to the higher mobility of the latter. Therefore, r_i^+ is set such that the discrepancy $r_i^+ - r_i^-$ is proportional to α / r_i^-:

$$r_i^+ = \frac{\alpha}{r_i^-} + r_i^-. \tag{1.12}$$

This formula actually entails that the Hasse diagram representing the evolution of skeleton graphs depends only on the inner radii $\{r_i^-\}$, but not on the parameter α. We arbitrarily set $\alpha = 10$ and compute the whole λ-complex of the toleranced model.

As discussed in Sect. 1.3.2, the growth process is controlled by the volume ratio of Eq. (1.11); that is, it is stopped at $\lambda = \lambda_{\max}$ such that $r_{\lambda_{\max}} \sim 5$.

1.3.5 Success Stories

We now develop some insights provided by toleranced models of the NPC [13].

Qualifying the contact between protein types. Given two protein types p_i and p_j, and a stoichiometry $k \geq 1$, we wish to qualify the contacts between the instances

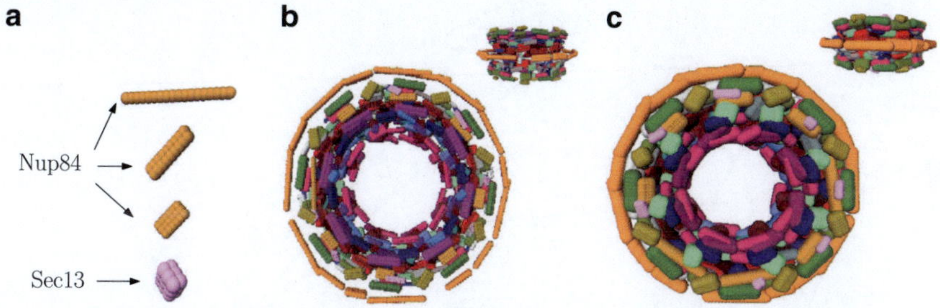

Fig. 1.13 Toleranced model of the entire NPC. (**a**) The four canonical configurations, 18 balls each, illustrated with protein types Nup84 and Sec13. (**b**) The toleranced model at $\lambda = 0$ corresponds to *inner balls*. (**c**) The toleranced model at $\lambda = 1$ corresponds to *outer balls*

of these two types. Having painted all the proteins types in red, let us consider the Hasse diagram for the range of λ-values $[0, \lambda_{\max}]$, as discussed in Sect. 1.3.4. As soon as k pairwise contacts between distinct pairs of instances of these types are observed, say at $\lambda = \lambda(p_i, p_j)$, the contact probability $p_{ij}^{(k)}$ is set as $p_{ij}^{(k)} = 1 - \lambda(p_i, p_j)/\lambda_{\max}$; if the two types make strictly less than k contacts, then $p_{ij}^{(k)} = 0$. For a given probability b, the set of *k-significant contacts* $S_b^{(k)}$ is the set of contacts such that $p_{ij}^{(k)} \geq b$ and $p_{ij}^{(k+1)} < b$.

An illustration of k-significant contacts is provided in Fig. 1.14. To appreciate the "value added" by this figure, recall that the NPC is composed of 16 half-spokes, and that the stoichiometry of all protein types is either 8, 16 or 32. In particular, for a large number of protein pairs, 16 copies of the corresponding complexes are expected. Remarkably, the size four cliques of this graph show five quadruples of proteins, two of which correspond to the intensively studied Y-complex and T-complex. The remaining ones are under scrutiny, and together with the remaining k-significant contacts, have captured the attention of biologists specializing in the architecture of the NPC.

Global assessment w.r.t. a collection of types: stoichiometry, symmetry, stability. Assume that the red proteins are instances of types prescribed in a set T, e.g. a TAP *pullout* as discussed in Sect. 1.3.1.2. The following parameters can be assessed.

- *Stoichiometry.* Analyzing the complexes of the Hasse diagram is of interest for several reasons: first, one sees whether the set T corresponds to a single complex or to a mixture of complexes; second, one can spot the copies associated with the set T; third, if T corresponds to a TAP experiment, one can check whether each complex contains the tagged protein.
- *Symmetry.* For an assembly that exhibits symmetries, one can compare the number of complexes with the expected number. For example, in the NPC, the multiplicity of selected complexes is expected to be 16.

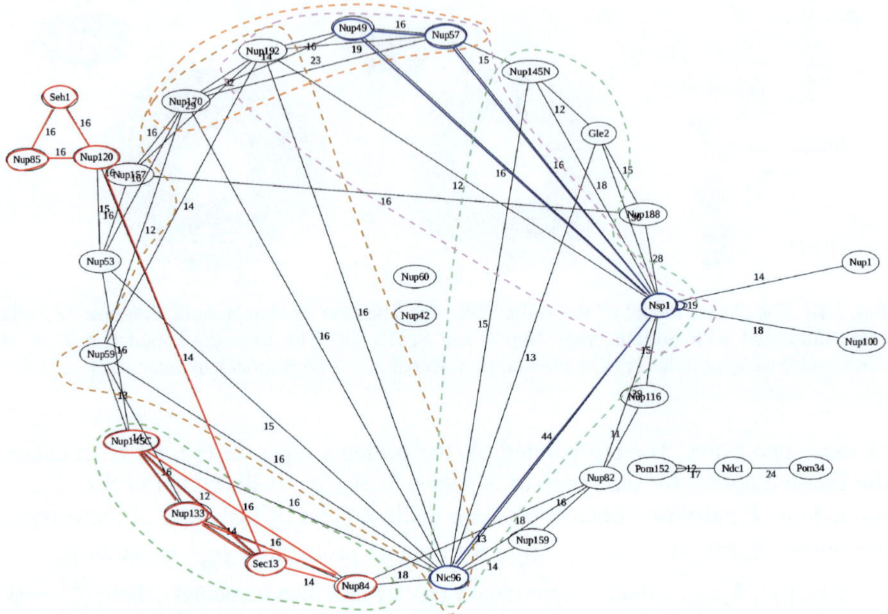

Fig. 1.14 Graph of k-significant contacts in $S_{0.65}^{(k>10)}$ for $\lambda_{\max} = 1$. The *red* and *blue* sub-graphs respectively correspond to the Y-complex and T-complex. The nodes contained in each of the five *dashed* regions define a complete sub-graph, i.e., a clique of size 4

- *Topological stability.* In Sect. 1.3.3, the stability of a complex has been defined as the difference between its birth and death dates. This information is particularly relevant when a given complex collides with another to form a larger complex. For an assembly involving a prescribed number of complexes, one expects the variation of the number of complexes as a function of λ to exhibit a plateau. Also, for an assembly with symmetries, the homogeneity of the model can be inferred from the stability of complexes featuring the same types, but located in different places.
- *Geometric accuracy.* A complex may involve the correct protein instances, but may have a loose geometry. Comparing its volume to that occupied by its constituting instances is the goal of the volume ratio of Eq. (1.11).

These analysis are illustrated in Fig. 1.15, which is concerned with a tuple T corresponding to the seven types of the Y-complex. That is, the protein instances painted in red correspond to the seven types involved in the Y-complex. Interestingly, eight isolated copies of the Y-complex are observed in the Hasse diagram, out of 16 expected. This observation shows that contacts between protein instances belonging to several copies of the Y-complex can prevail over contacts within the isolated copies. Equally importantly, the variation of the number of connected components shows that, upon termination, the growth process leaves two red

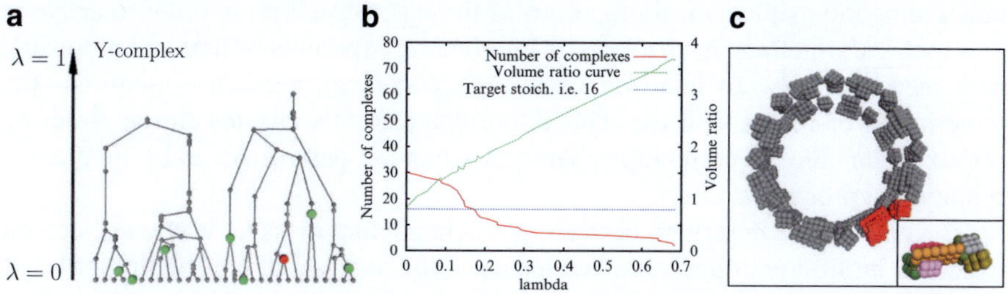

Fig. 1.15 Global assessment for the Y-complex. (**a**) The Hasse diagram representing the evolution of the connected components. Fat nodes correspond to isolated copies. (**b**) Evolution of the number of complexes and volume ratio r_λ as a function of λ. (**c**) The complex corresponding to the *red fat* node of the Hasse diagram presented in (**a**)

connected components. These components correspond to the so-called rings of the NPC, whose structure is still under controversy. In fact, analysing the contacts accounting for the closure of the two rings provides useful information, and is currently being used to discuss hypotheses about the structure of these two rings.

Local assessment w.r.t. a 3D model. Assume now that we wish to compare a complex C against a model T, which may come from a crystal structure or which may have been designed in-silico—such as the model for the Y-complex of Fig. 1.8b. Assume that the model T comes with the pairwise contacts between its constituting proteins. As discussed in Sect. 1.3.3, we maintain the skeleton graph of C, which precisely encodes the contacts between the toleranced proteins of C. Thus, comparing C against its model T boils down to comparing two graphs. The two standard operations to do so consist of computing the Maximal Common Induced Sub-graph and the Maximal Common Edge Sub-graph of the two graphs [15].

These operations are used to compare the contacts encoded in complexes of the toleranced model against those present in putative models. In particular, they have allowed the design of a new 3D template for the T-complex [13].

1.4 Outlook

While investigating methodological developments for docking [20], Michael Connolly claimed "*Geometry is not everything but it is the most fundamental thing*" Indeed, geometrical reasoning applies everywhere in Biology: in formulating the equations used to model the basic physical forces between atoms, in representing the shapes of the macromolecules themselves, in describing their interacting surfaces, and modeling of the structure of complexes, both small and large. Many of these approaches rely on simple models of the macromolecule, be it a van der Waals model, a Voronoï diagram, or a low-resolution assembly of balls representing pseudo-atoms. But geometric reasoning further allows one to build strong statements concerning the mathematical properties of these constructions in order to arrive at new metrics, which can in turn serve in interpreting experimental data and proposing new measurements. To build on Connolly's statement, geometry conditions the magnitudes of forces, and the models presented in this chapter aim at fostering our understanding of correlations between structural parameters and biological—biophysical properties.

The approaches described here are just a beginning in many ways, and can be extended in different directions. For instance, the analysis of the shelling order of an interface forms a framework for examining new relations between the interface structure and the affinity of a protein-protein interaction, or the $\Delta\Delta G$ of binding obtained from alanine scanning, that is more rational than the simple core-rim model. It is also easy to envisage development of such metrics in the context of new scoring functions for predictive protein-protein docking, or for discriminating biological contacts from crystal contacts. Complexes involving proteins and small molecules such as pharmaceutical agents, or their interaction with nucleic acids such as DNA were not explored in this chapter, but such systems also provide opportunities, albeit with geometrical constraints of a different nature.

Still, a concept that we did not approach directly using the techniques deployed here is the nature of biological macromolecular dynamics. As described in the introduction, such dynamics underlie virtually all biological interactions, and complementary geometric and topological concepts should prove instrumental for investigating them. As we mentioned in the Introduction, a molecule with n atoms enjoys $d = 3n - 6$ conformational degrees of freedom, and a conformational transition is a path through the d-dimensional hyper-surface representing the energy landscape of the system. Learning the salient features of such landscapes and inferring thermodynamical and kinetic quantities relies on statistical, geometric and topological analysis of high-dimensional point clouds, and the evolution of dynamical systems defined on such landscapes. We anticipate that these and future developments will shed new light on the deterministic behaviour of macromolecular complexes, despite the high dimensionality of the systems, and help us better discern Nature's strategies.

1.5 Online Resources

1.5.1 Databases

The Protein Databank (PDB)
http://www.rcsb.org
http://www.ebi.ac.uk/pdbe/
http://www.pdbj.org/
The PDB is the reference database for structure of proteins and complexes. The PDB actually refers to the database of structures itself, and three different interfaces

are offered, with different tools and connexions with complementary informtion in structural biology, biology, and bioinformatics.

The IMGT/3D-structure database
http://www.imgt.org/3Dstructure-DB
This database provides annotated versions of complexes from the immune system, retrieved on a weekly basis from the PDB.

The Unified Data Resource for 3-Dimensional Electron Microscopys
http://emdatabank.org
The EMDataBank is the sister of the PDB, for cryo-electron microscopy models.

Protein–protein docking Benchmark
http://zlab.umassmed.edu/benchmark
The docking benchmark contains annotations regarding the three-dimensional structures (X-ray and NMR) of protein-protein complexes together with the corresponding structures of the unbound components.

Protein-Protein Interaction Affinity Database
http://bmm.cancerresearchuk.org/~bmmadmin/Affinity/
This database provides crystal structures of protein complexes together with those of the unbound partners, as well as experimentally measured binding affinities. See also [39].

1.5.2 Software

1.5.2.1 Molecular Dynamics Simulation

Several programs are commonly used to simulate macromolecular dynamics at atomic resolution in either implicit or explicit solvent. Among the most commonly used are

Amber
http://ambermd.org
Amber consists of a molecular force field, together with a suite of programs allowing users to perform molecular dynamics simulations.

Charmm
http://www.charmm.org
Chemistry at HARvard Macromolecular Mechanics is one of the oldest molecular simulation packages.

NAMD
http://www.ks.uiuc.edu/Research/namd
Not (just) Another Molecular Dynamics program is a molecular dynamics code with strong emphasis on parallel calculations, that in addition can be linked to the Visual Molecular Dynamics visualization environment (see below).

GROMACS

http://www.gromacs.org/

GROMACS (GROningen MAchine for Chemical Simulations) is a molecular dynamics simulation package originally developed in an open source mode and released under the GPL licence.

1.5.2.2 Macromolecular Docking

Predictive protein-protein docking (using the structures of the unbound proteins as starting point) can be investigated using either full atomic-resolution or coarse-grained, lower-resolution models. A sampling of current resources includes

Zdock

http://zdock.bu.edu/cgi/help.cgi

Performs a convolution of three-dimensional grid representations of the two component proteins via translational fast Fourier transforms (FFTs) in order to maximize the interaction score.

Hex

http://hex.loria.fr/

Represents each component (protein or DNA) using expansions of real orthogonal spherical polar basis functions, allowing the use of FFT correlations to search both translation and rotational space for optimal docking geometries.

Haddock

http://www.nmr.chem.uu.nl/haddock/

Minimizes a penalty function that additionally incorporates restraints based on experimental data concerning interacting residues in the actual complex, as derived for example from NMR, mutagenesis, or biochemical data.

P-tools/Attract

http://ptoolsdocking.sourceforge.net/

Energy minimization of the interaction energy using either atomic-level or coarse-grained representations of the component proteins. Incorporates some protein flexibility via the addition of collective coordinate descriptions or multicopy representations of alternative loop conformations, for example.

1.5.2.3 Molecular Visualization Environments

Visual Molecular Dynamics (VMD)

http://www.ks.uiuc.edu/Research/vmd/

VMD is a molecular visualization environment for displaying, animating and analyzing large biomolecular systems using 3-D graphics and built-in TCL scripting.

PyMOL
http://www.pymol.org/
A python based molecular visualization environment, offering functionalities similar to those of VMD.

1.5.2.4 Methodology

The Computational Geometry Algorithms Library (CGAL)
http://www.cgal.org
Developing geometric applications in general and computational structural biology applications in particular is a difficult endeavor for two reasons. First, the numerical rounding errors inherent to any floating point representation must be controlled, in order for the program to follow to correct decision tree. Second, the code design itself must be flexible enough to accommodate various numerical strategies but also combinatorial structures. These two constraints underpin the C++ library CGAL, which has become the reference library for efficient and robust geometric software.

The Ipe extensible drawing editor
http://ipe7.sourceforge.net/
IPE is a vector graphics editor, which carries two main advantages over competitors such as xfig or inkscape for figures involving mathematical formulae and/or ellaborate geometric constructions: first, it directly allow typing LaTeX formulae; second, it is coupled to CGAL for constructions such as Delaunay and Voronoï diagram.

1.5.2.5 Atomic Resolution Models

A number of programs have been made available to analysis protein—protein interactions [65]. In the sequel, we focus on the software implementing the recent developments presented in this chapter.

Intervor
http://cgal.inria.fr/abs/Intervor
Intervor is the application implementing the Voronoï interface model presented in Sect. 1.2.4.

Vorlume
http://cgal.inria.fr/abs/Vorlume
Vorlume is a program (the only one to the best of our knowledge), computing a certified estimate of the surface area and volume of a union of balls. See Sect. 1.2.3.

vorpatch
http://cgal.inria.fr/abs/vorpatch
vorpatch is the application computing a topological encoding of a binding patch based on the shelling tree presented in Sect. 1.2.5.

compatch

http://cgal.inria.fr/abs/compatch

compatch is the program comparing two binding patches, either from a geometric or topological standpoint, based on the shelling tree computed by vorpatch. Refer to Sect. 1.2.5.

voratom

http://cgal.inria.fr/abs/voratom

voratom is a software suite meant to create and explore toleranced models. The applications provided allow performing the segmentation of a 3D density map, creating a toleranced model from occupancy volumes, exploring the Hasse diagram of a toleranced model, and checking the complexes associated with nodes of the Hasse diagram against 3D templates.

SITUS

http://situs.biomachina.org/

SITUS is a package for modeling atomic resolution structures by combining low-resolution density maps (electron microscopy, tomography, or small angle X-ray scattering) on the one hand, and atomic models on the other hand.

Acknowledgements Joël Janin is acknowledged for insightful discussions.

List of Acronyms

PDB	Protein Data Bank
cryoEM	cryo-electron microscopy
TAP	Tandem Affinity Purification

References

1. S.A. Adcock and J.A. McCammon. Molecular dynamics: survey of methods for simulating the activity of proteins. *Chem Rev*, 106(5):1589–1615, May 2006.
2. N. Akkiraju and E. Edelsbrunner. Triangulating the surface of a molecule. *Discrete Applied Mathematics*, 71(1):5–22, 1996.
3. F. Alber, S. Dokudovskaya, L.M. Veenhoff, W. Zhang, J. Kipper, D. Devos, A. Suprapto, O. Karni-Schmidt, R. Williams, B.T. Chait, M.P. Rout, and A. Sali. Determining the architectures of macromolecular assemblies. *Nature*, 450(7170):683–694, Nov 2007.
4. F. Alber, F. Forster, D. Korkin, M. Topf, and A. Sali. Integrating diverse data for structure determination of macromolecular assemblies. *Ann. Rev. Biochem.*, 77:11.1–11.35, 2008.
5. R.P. Bahadur and M. Zacharias. The interface of protein-protein complexes: analysis of contacts and prediction of interactions. *Cell Mol Life Sci*, 65(7–8):1059–1072, Apr 2008.
6. J. Bernauer, J. Azé, J. Janin, and A. Poupon. A new protein-protein docking scoring function based on interface residue properties. *Bioinformatics*, 23(5):555–562, Mar 2007.
7. J. Bernauer, R.P. Bahadur, F. Rodier, J. Janin, and A. Poupon. DiMoVo: a Voronoi tessellation-based method for discriminating crystallographic and biological protein-protein interactions. *Bioinformatics*, 24(5):652–658, Mar 2008.

8. T.N. Bhat, G.A. Bentley, G. Boulot, M.I. Greene, D. Tello, W. Dall'Acqua, H. Souchon, F.P. Schwarz, R.A. Mariuzza, and R.J. Poljak. Bound water molecules and conformational stabilization help mediate an antigen-antibody association. *Proc Natl Acad Sci U S A*, 91(3):1089–1093, Feb 1994.
9. J-D. Boissonnat and M. Yvinec. *Algorithmic Geometry*. Cambridge University Press, UK, 1998. Translated by Hervé Brönnimann.
10. A.M. Bonvin. Flexible protein-protein docking. *Curr Opinion in Struct Biol*, 16:1–7, 2006.
11. B. Bouvier, R. Grünberg, M. Nilges, and F. Cazals. Shelling the voronoi interface of protein-protein complexes reveals patterns of residue conservation, dynamics, and composition. *Proteins*, 76(3):677–692, Aug 2009.
12. F. Cazals and T. Dreyfus. Multi-scale geometric modeling of ambiguous shapes with toleranced balls and compoundly weighted α-shapes. In B. Levy and O. Sorkine, editors, *Symposium on Geometry Processing*, pages 1713–1722, Lyon, 2010. Also as Inria Tech report 7306.
13. T. Dreyfus, V. Doye, and F. Cazals. Assessing the reconstruction of macromolecular assemblies with toleranced models. Proteins: structure, function, and bioinformatics, 80(9), 2012.
14. F. Cazals, H. Kanhere, and S. Loriot. Computing the volume of union of balls: a certified algorithm. *ACM Transactions on Mathematical Software*, 38(1), 2011.
15. F. Cazals and C. Karande. An algorithm for reporting maximal c-cliques. *Theoretical Computer Science*, 349(3):484–490, 2005.
16. F. Cazals, F. Proust, R. Bahadur, and J. Janin. Revisiting the Voronoi description of protein-protein interfaces. *Protein Science*, 15(9):2082–2092, 2006.
17. P. Chakrabarti and J. Janin. Dissecting protein-protein recognition sites. *Proteins*, 47(3):334–343, May 2002.
18. P. Chakrabarti and J. Janin. Dissecting protein-protein recognition sites. *Proteins*, 47(3):334–43, 2002.
19. C. Chotia and J. Janin. Principles of protein-protein recognition. *Nature*, 256:705–708, 1975.
20. M.L. Connolly. Shape complementarity at the hemoglobin a1b1 subunit interface. *Biopolymers*, 25:1229–1247, 1986.
21. F. Dong and H-X. Zhou. Electrostatic contribution to the binding stability of protein-protein complexes. *Proteins*, 65(1):87–102, Oct 2006.
22. S. Dutta and H.M. Berman. Large macromolecular complexes in the protein data bank: a status report. *Structure*, 13:381, 2005.
23. H. Edelsbrunner. Weighted alpha shapes. Technical Report UIUCDCS-R-92-1760, Dept. Comput. Sci., Univ. Illinois, Urbana, IL, 1992.
24. H. Edelsbrunner. The union of balls and its dual shape. *Discrete Comput. Geom.*, 13:415–440, 1995.
25. H. Edelsbrunner and J. Harer. *Computational topology: an introduction*. AMS, 2010.
26. H. Edelsbrunner and P. Koehl. The weighted-volume derivative of a space-filling diagram. *Proc Natl Acad Sci U S A*, 100(5):2203–2208, Mar 2003.
27. J. Frank. *Three-dimensional electron microscopy of macromolecular assemblies: visualization of biological molecules in their native state*. Oxford University Press, USA, 2006.
28. M. Gerstein and C. Chothia. Packing at the protein-water interface. *Proc Natl Acad Sci U S A*, 93(19):10167–10172, Sep 1996.
29. H. Gohlke and D.A. Case. Converging free energy estimates: MM-PB(GB)SA studies on the protein-protein complex Ras-Raf. *J Comput Chem*, 25(2):238–250, Jan 2004.
30. M. Grünwald, C. Dellago, and P.L. Geissler. Precision shooting: Sampling long transition pathways. *J Chem Phys*, 129(19):194101, Nov 2008.
31. M. Guharoy and P. Chakrabarti. Conservation and relative importance of residues across protein-protein interfaces. *PNAS*, 102(43):15447–15452, Oct 2005.
32. M. Guharoy and P. Chakrabarti. Conserved residue clusters at protein-protein interfaces and their use in binding site identification. *BMC Bioinformatics*, 11:286, 2010.
33. Y. Harpaz, M. Gerstein, and C. Chothia. Volume changes on protein folding. *Structure*, 2:641–649, 1994.

34. K-C. Hsia, P. Stavropoulos, G. Blobel, and A. Hoelz. Architecture of a coat for the nuclear pore membrane. *Cell*, 131(7):1313–1326, 2007.
35. Z. Hu, B. Ma, H. Wolfson, and R. Nussinov. Conservation of polar residues as hot spots at protein interfaces. *Proteins*, 39(4):331–342, Jun 2000.
36. J. Janin. Protein-protein docking tested in blind predictions: the CAPRI experiment. *Mol Biosyst*, 6(12):2351–62, Dec 2010.
37. J. Janin. X-ray study of protein-protein complexes and analysis of interfaces. In M. Zacharias, editor, *Protein-Protein Complexes: Analysis, Modeling and Drug Design*. Imperial College Press, London, 2010.
38. M. Kampmann and G. Blobel. Three-Dimensional Structure and Flexibility of a Membrane-Coating Module of the Nuclear Pore Complex. *Nat. Struct. Mol. Biol.*, 16(7):782–788, 2009.
39. P.L. Kastritis, I.H. Moal, H. Hwang, Z. Weng, P.A. Bates, A.M.J.J. Bonvin, and J. Janin. A structure-based benchmark for protein-protein binding affinity. *Protein Sci*, Jan 2011.
40. C.L. Lawson, M.L. Baker, C. Best, C. Bi, M. Dougherty, P. Feng, G. van Ginkel, B. Devkota, I.r Lagerstedt, S.J. Ludtke, R.H. Newman, T.J. Oldfield, I.R., G. Sahni, R. Sala, S. Velankar, J. Warren, J.D. Westbrook, K. Henrick, G. J Kleywegt, H.M. Berman, and W. Chiu. Em-databank.org: unified data resource for cryoem. *Nucleic Acids Res*, 39(Database issue):D456–D464, Jan 2011.
41. Andrew R. Leach. *Molecular modelling: principles and applications*. Pearson Prentice Hall, 2009.
42. Emmanuel D Levy. Piqsi: protein quaternary structure investigation. *Structure*, 15(11):1364–1367, Nov 2007.
43. Olivier Lichtarge and Mathew E Sowa. Evolutionary predictions of binding surfaces and interactions. *Curr Opin Struct Biol*, 12(1):21–27, Feb 2002.
44. J. Liu and B. Rost. Sequence-based prediction of protein domains. *Nucleic Acids Res*, 32(12):3522–3530, 2004.
45. N. Malod-Dognin, A. Bansal, and F. Cazals. Characterizing the morphology of protein binding patches. Proteins: structure, function, and bioinformatics, in press, 2012.
46. I. Mihalek, I. Res, and O. Lichtarge. On itinerant water molecules and detectability of protein-protein interfaces through comparative analysis of homologues. *J Mol Biol*, 369(2):584–595, Jun 2007.
47. J. Mintseris, B. Pierce, K. Wiehe, R. Anderson, R. Chen, and Z. Weng. Integrating statistical pair potentials into protein complex prediction. *Proteins*, 69(3):511–520, Nov 2007.
48. M.A. Miteva, C.H. Robert, J.D. Maréchal, and D. Perahia. Receptor flexibility in ligand docking and virtual screening. In M. Miteva, editor, *In-silico lead discovery*. Bentham Science Publishers, 99–117, 2011.
49. S.K. Mohan and C. Yu. The il1alpha-s100a13 hetero-tetrameric complex structure: A component in the non-classical pathway for interleukin 1alpha secretion. *J Biol Chem*, Jan 2011.
50. K.L. Morrison and G.A. Weiss. Combinatorial alanine-scanning. *Curr Opin Chem Biol*, 5(3):302–307, Jun 2001.
51. A. Okabe, B. Boots, K. Sugihara, and S.N. Chiu. *Spatial Tessellations: Concepts and Applications of Voronoi Diagrams (2nd Ed.)*. Wiley, 2000.
52. L. Pauling. *General Chemistry*, volume 5. Dover, 1988.
53. A. Poupon. Voronoi and voronoi-related tessellations in studies of protein structure and interaction. *Curr Opin Struct Biol*, 14(2):233–241, Apr 2004.
54. Antonio Rausell, David Juan, Florencio Pazos, and Alfonso Valencia. Protein interactions and ligand binding: from protein subfamilies to functional specificity. *Proc Natl Acad Sci U S A*, 107(5):1995–2000, Feb 2010.
55. D. Reichmann, O. Rahat, S. Albeck, R. Meged, O. Dym, and G. Schreiber. The modular architecture of protein-protein binding interfaces. *PNAS*, 102(1):57–62, 2005.
56. F.M. Richards. The interpretation of protein structures: total volume, group volume distributions and packing density. *J Mol Biol*, 82(1):1–14, Jan 1974.
57. D.W. Ritchie and V. Venkatraman. Ultra-fast fft protein docking on graphics processors. *Bioinformatics*, 26(19):2398–2405, Oct 2010.

58. P.W. Rose, B. Beran, C. Bi, W.F. Bluhm, D. Dimitropoulos, D.S. Goodsell, A. Prlic, M. Quesada, G.B. Quinn, J.D. Westbrook, J. Young, B. Yukich, C. Zardecki, H.M. Berman, and P.E. Bourne. The rcsb protein data bank: redesigned web site and web services. *Nucleic Acids Res*, 39(Database issue):D392–D401, Jan 2011.
59. L. Salmon, G. Bouvignies, P.R.L. Markwick, and M. Blackledge. Nmr provides a quantitative description of protein conformational flexibility on physiologically important timescales. *Biochemistry*, Mar 2011.
60. N. Schrader, P. Stelter, D. Flemming, R. Kunze, E. Hurt, and I.R. Vetter. Structural basis of the nic96 subcomplex organization in the nuclear pore channel. *Molecular cell*, 29(1):46–55, 2008.
61. H.S. Seo, Y. Ma, E.W. Debler, D. Wacker, S. Kutik, G. Blobel, and A. Hoelz. Structural and functional analysis of nup120 suggests ring formation of the nup84 complex. *PNAS*, pages 14281–14286, 2009.
62. D.E. Shaw, P. Maragakis, K. Lindorff-Larsen, S. Piana, R.O. Dror, M.P. Eastwood, J.A. Bank, J.M. Jumper, J.K. Salmon, Y. Shan, and W. Wriggers. Atomic-level characterization of the structural dynamics of proteins. *Science*, 330(6002):341–346, Oct 2010.
63. G.R. Smith, M.J.E. Sternberg, and P.A. Bates. The relationship between the flexibility of proteins and their conformational states on forming protein-protein complexes with an application to protein-protein docking. *J Mol Biol*, 347(5):1077–1101, Apr 2005.
64. T.A. Steitz. A structural understanding of the dynamic ribosome machine. *Nat Rev Mol Cell Biol*, 9(3):242–53, Mar 2008.
65. N. Tuncbag, G. Kar, O. Keskin, A. Gursoy, and R. Nussinov. A survey of available tools and web servers for analysis of protein-protein interactions and interfaces. *Briefings in bioinformatics*, 10(3):217, 2009.
66. A. D. Wilkins, R. Lua, S. Erdin, R. M. Ward, and O. Lichtarge. Sequence and structure continuity of evolutionary importance improves protein functional site discovery and annotation. *Protein Sci*, 19(7):1296–1311, Jul 2010.
67. S. Wong, R.E. Amaro, and J.A. McCammon. Mm-pbsa captures key role of intercalating water molecules at a protein-protein interface. *J Chem Theory Comput*, 5(2):422–429, Feb 2009.

Chapter 2
Modeling and Analysis of Gene Regulatory Networks

Gilles Bernot, Jean-Paul Comet, Adrien Richard, Madalena Chaves, Jean-Luc Gouzé, and Frédéric Dayan

2.1 Introduction

In many ways, a living cell can be compared to a complex factory animated by molecular nanomachines, mainly proteins complexes. Hence it is easy to conceive that the expression of proteins, which are cellular effectors, cannot be constant. On the contrary, it is highly dependent on the general context; environmental conditions (pH, temperature, oxygenation, nutrient availability), developmental stage of an organism (fetal spectrum of proteins differ from adult proteins in mammals), response to a stress (UV irradiation, presence of a chemical toxic, osmotic pressure alteration) and even diseases (cancer, attack of a pathogen) are examples of contextual changes in the level of protein expression.

In order to understand this cellular state plasticity, a simplified view of this machinery, following general transfers of information according to the central dogma of molecular biology, is the sequence of events: (1) stimulation via a signaling pathway (e.g. presence of an environmental stimulation, followed by internal transduction of the signal), (2) effective stimulation of a transcription factor, (3) activation of the transcription of a particular gene, (4) production of messenger RNA (mRNA) (see Fig. 2.1), (5) translation of mRNA, i.e. production of a functional

G. Bernot (✉) · J.-P. Comet · A. Richard
I3S – UMR 6070 CNRS/UNSA, Algorithmes-Euclide-B, 2000 Route des Lucioles, B.P. 121, 06903 Sophia Antipolis, France
e-mail: bernot@unice.fr; comet@unice.fr; richard@unice.fr

M. Chaves · J.-L. Gouzé
Inria Sophia Antipolis Méditerranée, Biocore project-team, 2004 Route des Lucioles, 06902 Sophia Antipolis, France
e-mail: madalena.chaves@inria.fr; jean-luc.gouze@inria.fr

F. Dayan
SOBIOS SA, 2229 Route des Crêtes, 06560 Valbonne Sophia Antipolis, France
e-mail: frederic.dayan@sobios.com

F. Cazals and P. Kornprobst (eds.), *Modeling in Computational Biology and Biomedicine*, DOI 10.1007/978-3-642-31208-3_2,

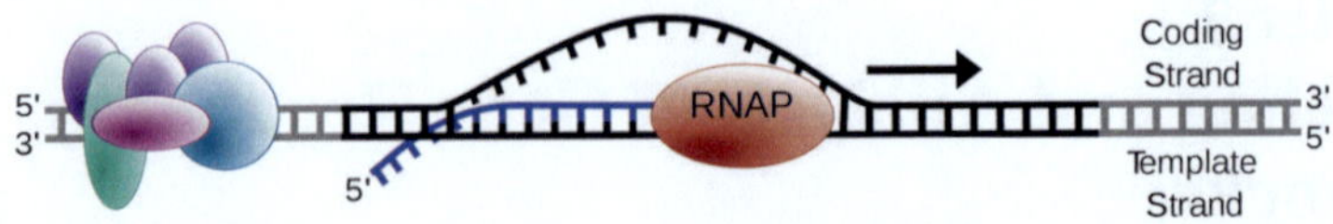

Fig. 2.1 Gene transcription (steps (3) and (4) in the Introduction): the enzyme RNA polymerase (RNAP) binds to DNA (*black* double strand) and produces a strand of RNA messenger (*blue* strand). This blue strand is a complementary copy of a sequence of DNA code (Image taken from [2])

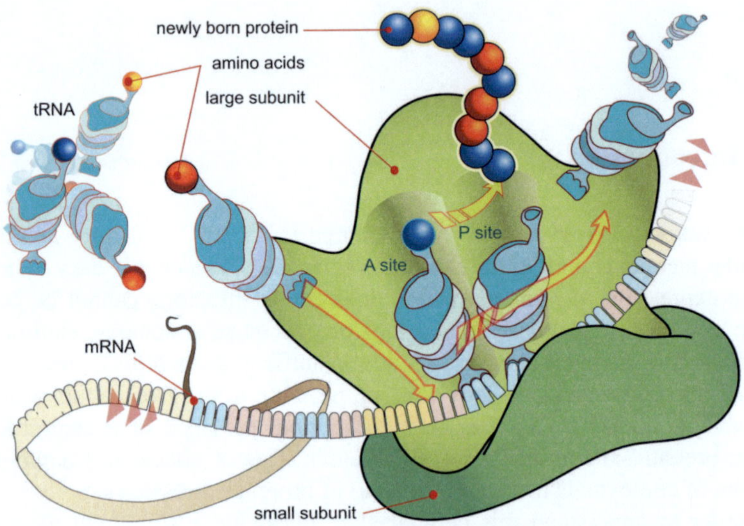

Fig. 2.2 Translation and protein synthesis (step (5) in the Introduction): a ribosome (two *green* units) is a large complex involving specific RNA (ribosomic RNA) complexed with proteins, synthesizing a polypeptide chain from a messenger RNA. Such a chain may form a protein on its own, or may contribute to a multimeric protein, see also Chap. 1 (Image taken from [1])

protein (see Fig. 2.2). Transcription factors are particular proteins that can recognize DNA motifs on the genome and consequently stimulate the transcription of a precise gene. A recognition motif is a short DNA sequence that is specific to a certain transcription factor. The transcription factor itself can be considered as a sensor of the cellular context.

2.1.1 Biological Systems and Experimental Techniques

A first example is provided by the Hypoxia Inducible Factor-1 (HIF-1), which is stimulated when oxygen pressure decreases: chemically, low intracellular oxygen concentration impairs the hydroxylation of the HIF-1α subunit, which leads to stabilization of this transcription factor (hypoxia signaling pathway). As a consequence, it recognizes Hypoxia Response Elements which are DNA motifs associated to a series of genes involved in adaptation to low pO_2. Among these genes, *erythropoietin* (or *epo*) is a well known inducer of red blood cells production; thus the low oxygen signal leads to secretion of the EPO protein and ultimately to an increase in erythropoiesis. This whole process aims at compensating poor oxygenation. In contrast, under normal pO_2 conditions, this stimulation of the *epo* gene is absent.

Another classical example, for bacterial cells, is the *lac* operon. An operon is a typical structure of bacterial genomes. It can be viewed as a "pack" of genes that are regulated by a unique promoter sequence. For instance, the *lac* operon is composed of the *lacZ*, *lacY* and *lacA* genes. In an environment with no glucose but with lactose available, the *lac* operon genes are transcribed, leading to lactose consumption as a source of energy. In that case, the level of transcription factors does not change directly, but the efficiency of transcription is regulated by a lactose repressor protein (impairing transcription in the absence of lactose) and by a Catabolite Activator Protein (which favors transcription in the absence of glucose).

These examples illustrate the complexity of gene regulation networks (GRN) for eukaryotic as much as prokaryotic cells. From an experimental point of view, biologists can access different intermediaries of these networks: genomic data (presence/absence of a DNA motif, complete sequence determination, mutations), mRNA quantification (large scale semi-quantitative screening with DNA arrays, or more focused and more precise analysis with Quantitative Polymerase Chain Reaction), quantitative gene activity measurements (luciferase reporter genes), quantitative protein detection (use of specific antibodies, fluorescent fusion proteins), or even molecular interactions estimation (semi-quantitatively with Förster/Fluorescence Resonance Energy Transfer, double hybrid, co-precipitation). Dynamics can also be followed thanks to tools like time-lapse microscopy or Fluorescence Recovery After Photo-bleaching microscopy on living cells (for an overview of some of these techniques see [15, 28]).

2.1.2 Mathematical Modeling

Therefore, large amounts of data, of more or less qualitative nature, are now available; one of the main challenges of molecular biology is to develop methodologies for using these data to address biological questions. Because of the complexity of the networks, it is necessary to design models, describing the dynamical functioning of the GRN. Indeed, the expression of genes, the concentrations of mRNA and proteins evolve with respect to time, and possibly converge toward some steady state, some periodic behavior or some other complex dynamical attractor. The emergence of these patterns from the dynamical interactions between the elements of the network, and the comparison with experimental data, will provide new keys to the comprehension of molecular biology, and enable scientists to solve important problems.

Yet, the methods for design and analysis of GRN are still quite recent. The Boolean and discrete approaches go back to works of Kauffmann (1969) and Thomas (1973), the continuous differential approach was applied by Goodwin (1963) to GRN, and the piecewise affine models were initiated by Glass and Kauffman [22]; for references see the review by de Jong [16]. There are still many open problems, mainly due to the large number of elements in a network. It is now possible to describe the behavior of a network with dozens of genes, but what about networks with several thousands of genes? These are still not attainable, even with the power of present computers and algorithmic methods.

As we have seen, the choice of a modeling approach is dependent on the type and amount of experimental data available, and on the nature of the biological questions asked by the modeler. In this chapter, two fundamental strategies will be detailed: continuous models and logical models. The first one gives quantitative predictions but needs quantitative biological data in order to fit parameters. The second is mainly based on a correct description of the logical links between biological entities (and is for instance particularly adapted to DNA array data that describe if a given gene is on/off). Nevertheless, both approaches can be used either to simulate biological phenomena or to predict properties that are intrinsically linked to the structure of the model, such as oscillatory or switch behaviors.

2.1.3 Chapter Overview

Public

This short introduction to GRN modeling is directed at Master level students whose background is either in the biological or the mathematical sciences.

Outline

A short overview of the main mathematical tools and concepts is provided, both on continuous (ordinary differential equations or hybrid systems) (see Sect. 2.2) and discrete (see Sect. 2.3) formalisms. For each type of formalism, simple examples of how to model genetic networks are worked out in more detail. Some successful applications of these methodologies to complex networks are also described.

2.2 Continuous and Hybrid Models of Genetic Regulatory Networks

The concentrations of molecular species (such as proteins or messenger RNAs) change in response to cellular signals. In this section, the concentrations are assumed to vary in a continuous manner, and their dynamical behavior will be described by systems of ordinary differential equations or the more abstract piecewise affine

(PWA) systems. These models often assume an homogeneous distribution of the molecules over a selected volume of space and describe, for instance, the dynamics of the concentration of some protein in a population of cells.

2.2.1 Challenges

The choice of appropriate variables is one of the first steps in the construction of a model for a biological network. The network is made of nodes (proteins, or RNA) and the edges usually describe the fact that some biochemical species acts positively or negatively on the variation with respect to time of some other biochemical species. Each variable (node) will play a different role in the behavior of the system, and have different degrees of relevance. Some variables can be measured experimentally, and are thus easier to compare to the model. Other variables may be easier to control from the exterior. Large systems of differential equations will require the introduction of a large number of parameters which will be unknown and should be estimated. In general, from a theoretical point of view, large dimensional systems are difficult to analyze and can only be studied through numerical simulations. Therefore, a fundamental step is the development of model reduction methods to simplify large networks and obtain more tractable systems of lower dimension, which are more easily studied in detail.

Two classical examples are the "positive" and "negative" feedback loops, formed by variables (proteins, for instance) that influence one another in a closed circuit, or loop. A circuit with two proteins that mutually repress or activate each other is a positive loop; if one of the interactions is a repression and the other an activation, then the circuit is a negative loop. Each of these two motifs appears frequently in GRN, and has a well known dynamical behavior; they can be combined with other motifs to represent the dynamics of complex regulatory networks. The negative loop is a system that generates oscillatory behavior, while the positive loop generates one or two stable steady states, and will be analyzed in detail in the next sections.

2.2.2 Mathematical Tools

This section quickly summarizes some basic mathematical results that will be useful in the analysis of systems of ordinary differential equations. For further details see, for instance [18].

2.2.2.1 Analysis of Two-Dimensional Systems

Consider a system with two variables, $x = (x_1, x_2)^t$, where each x_i represents the concentration of some molecular species, and x evolves in the region of space where all coordinates are either zero or positive (to be called *the positive orthant*). Define $\mathbb{R}_+ = [0, +\infty)$ and the two-dimensional positive orthant as $\mathbb{R}^2_+ = \mathbb{R}_+ \times \mathbb{R}_+$. The evolution of variable x_i along time is governed by a known function $f_i : \mathbb{R}^2_+ \to \mathbb{R}_+$, which depends on both variables. Given initial values $x_0 = (x_{10}, x_{20})$, solutions $x_i(t; x_{10}, x_{20})$ for $i = 1, 2$ may be found by solving the *initial value problem*:

$$\dot{x}_1 = f_1(x_1, x_2), \quad x_1(0) = x_{10},$$
$$\dot{x}_2 = f_2(x_1, x_2), \quad x_2(0) = x_{20}.$$

A sufficient condition to guarantee that this problem has a *unique solution* is that the functions f_1 and f_2 are continuous and have bounded, continuous derivatives, with respect to both variables. The positive orthant is invariant for this system if: whenever $x_{i0}(0) \geq 0$, then $x_i(t; x_{10}, x_{20}) \geq 0$ for all $t \geq 0$ ($i = 1, 2$). The following condition guarantees *invariance of the positive orthant*:

$$x_i = 0 \Rightarrow f_i(x_1, x_2) \geq 0, \ i = 1, 2, \tag{2.1}$$

which means that, at the boundary of the positive orthant, the vector field is either zero or points towards the interior of the orthant, thus preventing the variables to decrease to negative numbers. From now on, it will be assumed that functions f_i satisfy the required conditions, and that solutions of the initial value problem exist, are unique, and non-negative.

For most systems the f_i are nonlinear functions, and it is not possible to obtain closed form solutions of the initial value problem. However, qualitative analysis of the phase space can give a very good idea of the general behavior of the solutions. The signs of the vector field $(f_1(z), f_2(z))$ at each point $z \in \mathbb{R}^2_+$ indicate the direction of the solution at that point: for example, if $f_1(z) < 0$ and $f_2(z) > 0$, then the variable x_1 will decrease and x_2 will increase whenever a solution passes through the point z. The *nullclines* are curves that delimit regions of the plane where the sign of the vector fields is constant:

$$\text{Nullcline } i\text{:} \quad \Gamma_i = \{x \in \mathbb{R}^2_+ : f_i(x) = 0\}.$$

For an example see Fig. 2.5. The points of intersection of the nullclines are called the equilibria or *steady states* of the system:

$$x^* = (x_1^*, x_2^*) \in \mathbb{R}^2_+ : \quad f_1(x_1^*, x_2^*) = 0 \text{ and } f_2(x_1^*, x_2^*) = 0.$$

A steady state is a configuration of the system where both variables remain constant, and may be stable or unstable. To characterize this stability property suppose a small perturbation is applied to the initial condition, when $x(0) = x^*$. If the solution always returns back to x^* after a while, then the steady state x^* is stable; if the solution moves away from x^* without returning to the point, then the steady state x^* is unstable. The *basin of attraction of* x^* is the set of points $x_0 \in \mathbb{R}^2_+$ such that the solution $x(t; x_0)$ converges to x^* as time approaches infinity. The *stability of a steady state* x^* can be determined by computing the *Jacobian matrix and its eigenvalues,* $\lambda_\pm$ at that point:

$$J(x) = \begin{pmatrix} \frac{\partial f_1}{\partial x_1} & \frac{\partial f_1}{\partial x_2} \\ \frac{\partial f_2}{\partial x_1} & \frac{\partial f_2}{\partial x_2} \end{pmatrix}.$$

The *steady state* x^* *is locally stable* if all eigenvalues of $J(x^*)$ have a strictly negative real part: $\mathrm{Re}(\lambda_\pm) < 0$. For two-dimensional systems, the stability can also be established by looking at the trace and the determinant of the Jacobian matrix:

$$\mathrm{tr}(J(x)) = \frac{\partial f_1}{\partial x_1} + \frac{\partial f_2}{\partial x_2}, \qquad \det(J(x)) = \frac{\partial f_1}{\partial x_1}\frac{\partial f_2}{\partial x_2} - \frac{\partial f_2}{\partial x_1}\frac{\partial f_1}{\partial x_2}.$$

The steady state x^* is locally stable if: $\mathrm{tr}(J(x^*)) < 0$ and $\det(J(x^*)) > 0$. Geometrically speaking, the equilibria in dimension two can be classified into saddle (one positive and one negative real eigenvalue), stable sink (two real negative eigenvalues), unstable sink (two real positive eigenvalues), stable focus (two complex conjugate eigenvalues with negative real part), unstable focus (two complex conjugate eigenvalues with positive real part), plus the non-generic cases.

2.2.2.2 Analysis of n-Dimensional Systems

This analysis can be extended to general systems of ordinary differential equations. Consider now a system with n variables $x = (x_1, \ldots, x_n)^t \in \mathbb{R}^n_+$, $f = (f_1, \ldots, f_n)^t$ with $f : \mathbb{R}^n_+ \to \mathbb{R}^n$ and

$$\dot{x} = f(x), \quad x(0) = x_0. \tag{2.2}$$

For large n, it becomes difficult to perform the stability analysis for a general set of parameters, and so the steady states, the Jacobian matrix and its egenvalues will typically be computed numerically, for given sets of parameters. As for the two-dimensional systems, existence and uniqueness of solutions of Eq. (2.2) are guaranteed by sufficient conditions on f: each f_i is continuously differentiable. The invariance of the positive orthant may be checked by condition in Eq. (2.1) for $i = 1, \ldots, n$.

The *nullclines* corresponding to each variable can be similarly computed: $\Gamma_i = \{x \in \mathbb{R}^n_+ : f_i(x) = 0\}$. The *steady states* are given by all points x^* such that $f_i(x^*) = 0$, for $i = 1, \ldots, n$. The *Jacobian matrix* is again obtained by computing the partial derivatives of f_i:

$$J(x) = \begin{pmatrix} \frac{\partial f_1}{\partial x_1} & \cdots & \frac{\partial f_1}{\partial x_n} \\ \vdots & & \vdots \\ \frac{\partial f_n}{\partial x_1} & \cdots & \frac{\partial f_n}{\partial x_n} \end{pmatrix}.$$

The eigenvalues of the Jacobian matrix at equilibria are computable, at least numerically. *Local asymptotic stability* of x^* arises if all eigenvalues have a strictly negative real part. *Global stability* may be established through a Lyapunov function (but they are not easy to find). A Lyapunov function for system $\dot{x} = f(x)$ is a continuously differentiable function $V : \mathbb{R}^n_+ \to \mathbb{R}_+$ satisfying $V(x) \geq 0$ for all $x \in \mathbb{R}^n_+$ with $V(x) = 0$ if and only if $x = x^*$, and $\frac{\partial V}{\partial x}\dot{x} \leq 0$.

2.2.2.3 Different Timescales: Tikhonov's Theorem

Systems in the form of Eq. (2.2) whose variables evolve at different timescales can often be simplified. The main idea is to separate the system into "fast" and "slow" variables, and assume that the "fast" variables reach a (quasi) "steady state". This method allows reducing system in Eq. (2.2) to a new system with less variables, but with essentially the same dynamical behavior. This method can be applied only under appropriate conditions (briefly stated below) which are known as Tikhonov's Theorem (see, for instance, [27]). Let $x \in \mathbb{R}^p_+$, $y \in \mathbb{R}^q_+$, and $\varepsilon \ll 1$ be a small real number. Consider a system of the form

$$\begin{cases} \dot{x} = f(x, y, \varepsilon), \\ \varepsilon\dot{y} = g(x, y, \varepsilon), \\ (x(0), y(0)) = (x_0, y_0), \end{cases} \tag{2.3}$$

with f and g sufficiently smooth, under the following hypotheses:

- **H1** (slow manifold): there exists a unique solution, $y = \tilde{g}(x)$, sufficiently smooth, of $g(x, y, 0) = 0$; the matrix $\partial g/\partial y(x, \tilde{g}(x), 0)$ has all eigenvalues with strictly negative real part;
- **H2** (reduced system): the scalar system $\dot{x} = f(x, \tilde{g}(x), 0)$, $x(0) = x_0$ has a solution $x^0(t)$ on an interval $[0, T]$ $(0 < T < \infty)$;
- **H3**: y_0 is in the basin of attraction of the steady state $\tilde{g}(x_0)$ of the fast system $\dot{\xi} = g(x, \xi, 0)$.

If hypotheses H1-H3 are satisfied, the system in Eq. (2.3) admits a solution $(x^\varepsilon(t), y^\varepsilon(t))$ on $[0, T]$; in addition, $\lim_{\varepsilon \to 0^+} x^\varepsilon(t) = x^0(t)$ and $\lim_{\varepsilon \to 0^+} y^\varepsilon(t) = y^0(t) = \tilde{g}(x^0(t))$, uniformly on time on any closed interval contained in $(0, T]$.

The variables y are "faster", since $\dot{y}$ evolves very rapidly when compared to $\dot{x}$. Hypothesis H1 means that y evolves rapidly to a *quasi steady state* value, $y = \tilde{g}(x)$, depending only on x. This quasi steady state evolves on the slow time scale.

2.2.2.4 General Piecewise Affine Systems

The model has the general form

$$\dot{x}_i = f_i(x) - \gamma_i x_i, \qquad 1 \leq i \leq n, \tag{2.4}$$

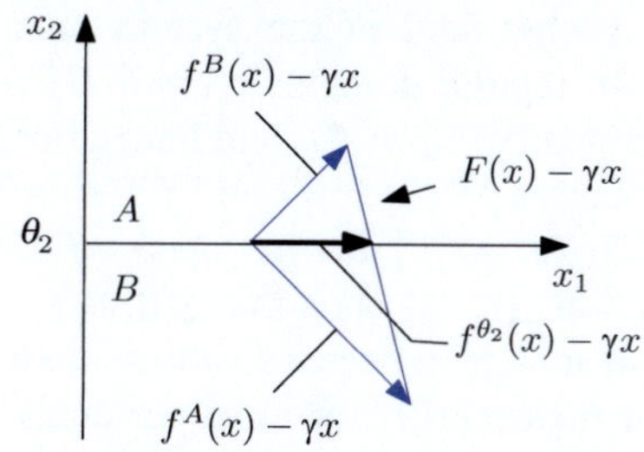

Fig. 2.3 Sliding mode solution. These type of solutions may occur when the vector fields in regions A and B point in opposite directions (f^A, f^B). At the boundary of A and B (the segment $x_2 = \theta_2$), the vector field may be defined as a convex combination of the two vector fields: $F(x) = \alpha f^A(x) + (1 - \alpha) f^B(x)$. The values of α range between $[0, 1]$, forming the convex hull cone. A sliding mode solution, with vector field $f^{\theta_2}(x)$, can be found by setting $x_2 = \theta_2$ and $F_2(x_1, \theta_2) - \gamma\theta_2 = 0$, and computing the appropriate value for α

where $x = (x_1, \ldots, x_n)^t$ is a non-negative vector of variables. The non-negative quantities $f_i(x)$ and $\gamma_i x_i$ represent production and loss (or transformation) rates for each variable x_i. The functions $f_i : \mathbb{R}^n_+ \to \mathbb{R}_+$ will be constant in rectangular regions of the state space whose boundaries will be called *thresholds*. The $(n-1)$-dimensional hyperplanes defined by these thresholds partition the state space into hyper-rectangular regions which are called *domains* or boxes (see an example in Sect. 2.2.3.3). For any domain D, the function $f(x) = (f_1(x), \ldots, f_n(x))$ is constant for all $x \in D$, and it follows that the PWA system can be written as an affine vector field $\dot{x} = f^D - \gamma x, \ \ x \in D$ where f^D is constant in D.

The point $\phi(D) = \gamma^{-1} f^D \in \Omega$ is called the *focal point* for the flow in D, and globally attracts the trajectories until they reach the boundaries of the domain. The focal points define the possible transitions associated with the domain D; *the transition graph* describes these transitions and gives the qualitative behavior of the system. This graph can be efficiently computed, and its properties analyzed (see the example in Sect. 2.2.3.3).

On the thresholds, the solutions have to be appropriately defined, typically through a construction due to Filippov. This construction considers all the solutions, as if the step function could take all the values of the interval $[0, 1]$ on the threshold. To be more explicit, let $n = 2$ and consider two regular domains, A and B, separated by one threshold ($x_2 = \theta_2$), as in Fig. 2.3. Let $\overline{\text{co}}$ denote the closed convex hull of a set of vector fields. We define the differential inclusion

$$\dot{x} \in H(x), \tag{2.5}$$

with

$$H(x) = \overline{\text{co}} \left\{ f^D(x) - \gamma x : \ D^s \in \partial D \right\}, \text{if } x \in D^s, \text{ a switching domain}$$

where $f^D - \gamma x$ is the vector field of the system on a regular domain D. In Fig. 2.3, there are only two regular domains (A and B) whose boundary contains the switching domain defined by $x_2 = \theta_2$, and hence the convex hull $H(x)$ is the cone formed by the two vectors $f^A(x) - \gamma x$ and $f^B(x) - \gamma x$.

A solution of Eq. (2.5) on $[0, T]$ *in the sense of Filippov* is an absolutely continuous (w.r.t. t) function $\xi(t; x_0)$ such that $\xi(0; x_0) = x_0$ and $d\xi/dt \in H(\xi)$ for almost all $t \in [0, T]$. It may give rise to a sliding mode along the plane defined by the threshold. See for instance [9] for a deeper analysis of solutions of PWA systems.

2.2.3 *Methodological Developments*

In general, there are n molecular species in a system, $x = (x_1, \ldots, x_n)^t \in \mathbb{R}^n_+$, and the changes in the concentration of species i result from a balance between production or synthesis processes ($g_i(x) \geq 0$) and degradation or transformation ($d_i(x) \geq 0$) into other species. Each species is thus characterized by an equation of the form:

$$\frac{dx_i}{dt} = g_i(x) - d_i(x). \tag{2.6}$$

The main problem is the choice of appropriate mathematical expressions of $g_i(x)$ and $d_i(x)$. These will depend on the modeling formalism and should reflect the observed dynamical behavior. For instance, for PWA systems, $g_i(x) = f_i(x)$ and $d_i(x) = \gamma_i x_i$, as seen in Eq. (2.4).

2.2.3.1 Modeling Transcription and Translation

In the classical sequence of events, transcription is initiated upon binding of a transcription factor to the gene. Besides transcription factors, other proteins may bind to specific sites of the gene and contribute to enhance (respectively, reduce) the transcription rate. Such proteins are called *activators* (respectively, *repressors*). In general, the binding of m molecules of protein A to the promoter site of a gene (D) to form a new complex (C), is represented as

$$D + mA \underset{k_2}{\overset{k_1}{\rightleftarrows}} C, \tag{2.7}$$

and can be modeled through the *law of mass-action*, where each reaction rate is proportional to a product of concentrations according to the stoichiometry:

$$\begin{cases} \dot{C} &= k_1 D A^m - k_2 C, \\ \dot{D} &= -\dot{C}. \end{cases} \tag{2.8}$$

If no other reactions take place, there is conservation of mass $D + C = D_T$ (the total amount of promoter sites, free or bound, remains constant), which can be used to reduce the number of variables. One can view A as an external variable. Using the observation that binding processes are typically faster than transcription, the system in Eq. (2.8) can be simplified through a process similar to Tikhonov's method, by setting the equations for C and D at "quasi steady state" ($\dot{C} \approx 0$), to obtain:

$$\begin{cases} C &= D_T \frac{A^m}{\theta_A^m + A^m}, \\ D &= D_T - C = D_T \frac{\theta_A^m}{\theta_A^m + A^m}, \end{cases} \tag{2.9}$$

with $\theta_A = (k_2/k_1)^{1/m}$. The amount of mRNA molecules (denoted M) produced may depend both on the concentration of DNA sites bound to an activator or repressor, and on the amount of free DNA sites. For simplicity, it will be assumed that the effect of activators and repressors can be modeled independently. Since binding of A to D is fast, the most natural form for the production of mRNA is a linear dependence on D and C. In addition, mRNA decays at a constant rate γ_M, which yields the equation:

$$\dot{M} = \alpha_0 D + \alpha_1 C - \gamma_M M. \tag{2.10}$$

In the case of an activator, the contribution of C to mRNA production is much larger than that of D, which can be represented by specifying $\alpha_1 \gg \alpha_0$. Defining $\kappa_0 = \alpha_0 D_T$ and $\kappa_1 = (\alpha_1 - \alpha_0) D_T$, substitution of the quasi-steady state expressions for C and D in Eq. (2.9) into Eq. (2.10) yields:

$$\dot{M} = \kappa_0 + \kappa_1 \frac{A^m}{\theta_A^m + A^m} - \gamma_M M. \tag{2.11}$$

In the case of a repressor, the contribution of C to mRNA production is much smaller than that of D ($\alpha_1 \ll \alpha_0$). Now define $\kappa_0 = \alpha_1 D_T$ and $\kappa_1 = (\alpha_0 - \alpha_1) D_T$, to obtain:

$$\dot{M} = \kappa_0 + \kappa_1 \frac{\theta_A^m}{\theta_A^m + A^m} - \gamma_M M \tag{2.12}$$

In either case, the parameter κ_0 ($\ll \kappa_1$) denotes a residual or basal activity. For further modeling aspects see also [3].

The next step is translation, or protein synthesis from mRNA. This is itself a sequence of several intermediate steps, and can be modeled as a linear function of the mRNA, together with a natural decay term:

$$\dot{P} = \kappa_2 M - \gamma_P P. \tag{2.13}$$

The translation and transcription system in Eqs. (2.10)–(2.13) can be further simplified by using the fact that mRNA degrades faster than protein, or $\gamma_M \gg \gamma_P$.

Consider the case of an activator as in Eq. (2.11) and change the time variable to $\tau = \gamma_P t$, to obtain:

$$\begin{cases} \frac{dM}{d\tau} & = \frac{\kappa_0}{\gamma_P} + \frac{\kappa_1}{\gamma_P} \frac{A^m}{\theta_A^m + A^m} - \frac{\gamma_M}{\gamma_P} M, \\ \frac{dP}{d\tau} & = \frac{\kappa_2}{\gamma_P} M - P. \end{cases} \tag{2.14}$$

For a fixed value of A, Tikhonov's theorem can now be applied with $y = M, x = P$, $\varepsilon = \gamma_P/\gamma_M$ and with $f(x, y, \varepsilon) = \frac{\kappa_2}{\gamma_P} y - x$, $g(x, y, \varepsilon) = \frac{\kappa_0}{\gamma_M} + \frac{\kappa_1}{\gamma_M} \frac{A^m}{\theta_A^m + A^m} - y$. Substituting the quasi steady state expression for mRNA into the protein Eq. (2.14), and rewriting the system in the original time variable, obtains the reduced system:

$$\dot{P} = \tilde{\kappa}_0 + \tilde{\kappa}_1 \frac{A^m}{\theta_A^m + A^m} - \gamma_P P, \tag{2.15}$$

where $\tilde{\kappa}_0 = \kappa_2 \kappa_0 / \gamma_M$ and $\tilde{\kappa}_1 = \kappa_2 \kappa_1 / \gamma_M$. This yields a dynamical equation for the protein concentration, directly dependent on the amount of activator (A). From now on, all the intermediate steps (the binding of A to the promoter and synthesis of mRNA) can be left out of the model.

The expression $h^+(x, \theta, m) = x^m/(\theta^m + x^m)$ (or *Hill function*) is known to fit well to synthesis and activity rates. Similarly, the inhibition function can be represented as: $h^-(x, \theta, m) = 1 - h^+(x, \theta, m) = \theta^m/(\theta^m + x^m)$. For gene regulatory networks, the exponent m is considered to be "large" ($m \geq 2$), according to experimental data [40]. Note that the qualitative form of $h^+(x, \theta, m)$ remains essentially unchanged for $m \geq 2$, with the same maximal and half-maximal values ($\max(h^-) = 1$ and $h^\pm(\theta, \theta, m) = 1/2$), the only difference being the steepness of the function around the value θ. For large m, the parameter θ has therefore a special meaning: it is a threshold value below which there is practically no activity and above which activity is (almost) maximal. In the limit as m tends to infinity, the Hill function becomes a step function, as described in Sect. 2.2.3.3.

2.2.3.2 Continuous Differential Systems for Genetic Network Models

To illustrate the modeling and analysis of complex GRN, consider a regulatory motif that appears frequently in genetic networks: two genes that mutually inhibit themselves or, more precisely, the protein A encoded by gene a represses transcription of gene b, and vice-versa (Fig. 2.4). The concentration of each protein can be described by $\dot{x}_j = \kappa_j M_j - \gamma_j x_j$, and each mRNA by an expression as in Eq. (2.12):

$$\dot{M}_j = \kappa_{j0} + \kappa_{j1} h^-(x_i, \theta_i, m_i) - \gamma_{Mj} M_j, \quad \text{for } j, i \in \{1, 2\} \text{ and } j \neq i. \tag{2.16}$$

Using the quasi-steady state assumption for the protein and mRNA equations, the system can be reduced to the dynamics of the protein concentrations, $\dot{x}_i = f_i(x_1, x_2)$ with (renaming constants):

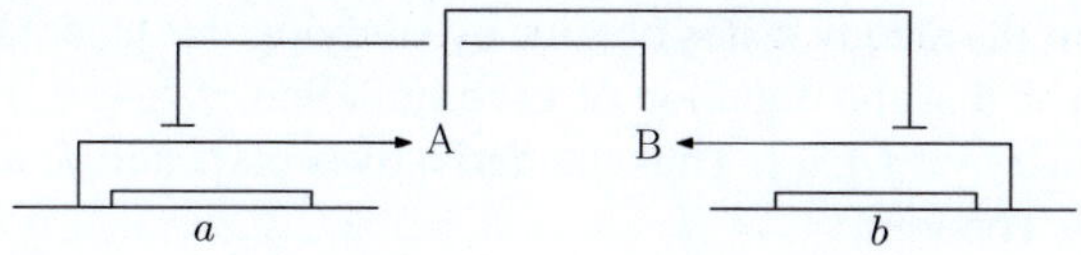

Fig. 2.4 Mutual inhibition between two genes. The white rectangles represent genes a and b, which are transcribed and then translated into the respective proteins, A and B (this is represented by the *arrows*). Each of these proteins inhibits the transcription of the gene of the other protein (A⊣B, B⊣A)

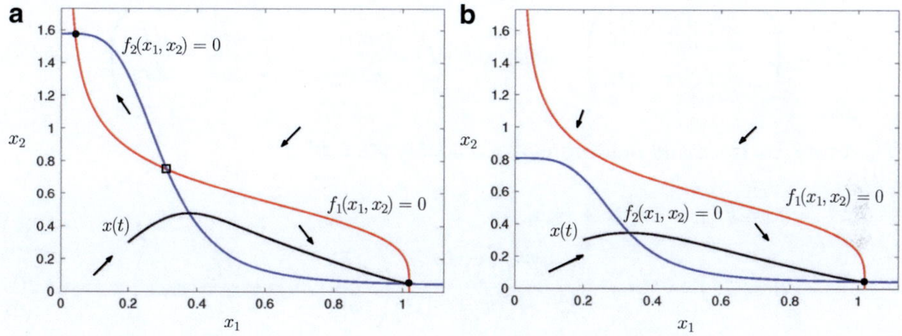

Fig. 2.5 Phase plane for the system of Eq. (2.17), for two different sets of parameters: (**a**) bistability, or (**b**) a single steady state. Stable steady states are marked by a *black circle*, and the unstable steady state by an *open rectangle*. In each region delimited by the nullclines ($f_1(x) = 0$ in *red*, $f_2(x) = 0$ in *blue*), the sign of the vector field of each coordinate is unchanged. One solution is shown in *black*. Parameter values for case (**a**): $\kappa_{10} = 0.02$, $\kappa_{11} = 1$, $\theta_1 = 0.3$, $\gamma_1 = 1$, $\kappa_{20} = 0.05$, $\kappa_{21} = 2$, $\theta_2 = 0.6$, $\gamma_1 = 1.3$, $m_1 = m_2 = 4$; for case (**b**) the only difference is $\kappa_{21} = 1$

$$\begin{cases} \dot{x}_1 &= \kappa_{10} + \kappa_{11} \frac{\theta_2^{m_2}}{\theta_2^{m_2} + x_2^{m_2}} - \gamma_1 x_1, \\ \dot{x}_2 &= \kappa_{20} + \kappa_{21} \frac{\theta_1^{m_1}}{\theta_1^{m_1} + x_1^{m_1}} - \gamma_2 x_2. \end{cases} \tag{2.17}$$

(Note that, in this section, the function f_i denotes the full right-hand side of the $\dot{x}_i$ equation, as in Sect. 2.2.2.1.) The dynamics of this system depend on the values of the parameters. The nullclines and steady states for system in Eq. (2.17) are represented in Fig. 2.5. Two distinct generic cases can be identified:

(a) m_i large, $\theta_i < (\kappa_{i0} + \kappa_{i1})/\gamma_i$, for all $i = 1, 2$. There are two stable and one unstable steady states. The stable steady states can be intuitively understood: one of the proteins is present at high concentration hence inhibiting transcription of the second gene, and implying that the second protein is only present in low concentration. This co-existence of two stable steady states is called *bistability*;
(b) m_i large, $\theta_i > (\kappa_{i0} + \kappa_{i1})/\gamma_i$, for some $i = 1, 2$. There is a unique steady state, which is stable.

The stability of the steady states obtains by studying the Jacobian matrix. In this example, it is clear that the equation of each nullcline, $f_i(x_1, x_2) = 0$, implicitly defines a function $x_2 = \tilde{f}_i(x_1)$. Thus the derivatives of f_i and $\tilde{f}_i$ are related by the Implicit Function Theorem:

$$0 = \frac{\partial f_i}{\partial x_1} + \frac{\partial f_i}{\partial x_2}\frac{dx_2}{dx_1} = \frac{\partial f_i}{\partial x_1} + \frac{\partial f_i}{\partial x_2}\frac{d\tilde{f}_i}{dx_1}. \tag{2.18}$$

The Jacobian matrix and its determinant can thus be written:

$$J(x) = \begin{pmatrix} -\frac{\partial f_1}{\partial x_2}\frac{d\tilde{f}_1}{dx_1} & \frac{\partial f_1}{\partial x_2} \\ -\frac{\partial f_2}{\partial x_2}\frac{d\tilde{f}_2}{dx_1} & \frac{\partial f_2}{\partial x_2} \end{pmatrix} \quad \text{and} \quad \det(J) = \frac{\partial f_1}{\partial x_2}\frac{\partial f_2}{\partial x_2}\left(\frac{d\tilde{f}_2}{dx_1} - \frac{d\tilde{f}_1}{dx_1}\right). \tag{2.19}$$

Therefore, its trace and determinant at a steady state x^* are

$$\begin{aligned} \mathrm{tr}(J^*) &= -(\gamma_1 + \gamma_2), \\ \det(J^*) &= \gamma_2 \kappa_{11} \frac{m_2 \theta_2^{m_2} (x_2^*)^{m_2 - 1}}{(\theta_2^{m_2} + (x_2^*)^{m_2})^2} \left(\frac{d\tilde{f}_2}{dx_1}(x^*) - \frac{d\tilde{f}_1}{dx_1}(x^*)\right), \end{aligned} \tag{2.20}$$

where $d\tilde{f}_i/dx_1(x^*)$ denote the slope of the curves $\tilde{f}_i$ at x^*. It is clear that the trace is always negative. For the steady states near one of the axis (one of the proteins at low concentration), it holds that $0 > d\tilde{f}_2/dx_1(x^*) > d\tilde{f}_1/dx_1(x^*)$, and therefore the determinant is positive – these are stable steady states. The middle steady state is unstable, since the opposite inequality holds and the determinant is negative.

This example is also known as *the bistable switch*, as only an external stimulus can force the system to evolve, or switch, from one steady state to the other (see discussion on Sect. 2.2.4).

2.2.3.3 Piecewise Affine Systems for Genetic Network Models

As seen above, the analysis of the dynamics of a dynamical system described by differential equations can be quite complicated in dimension greater than two. We are looking for a more algorithmic approach, easily implementable on a computer. We will consider a *qualitative* description of the bistable switch, corresponding to the case $m_i \to \infty$ where sigmoidal functions h^- become step functions. This is an approximation of the "real" system, done for an easier comprehension.

The formalism is as described in Sect. 2.2.2.4. The functions f_i now represent the dependence of the rate of synthesis of a protein encoded by gene i on the concentrations x_j of the other proteins in the cell. The term $\gamma_i x_i$ represents the degradation rate of protein x_i. The functions $f_i : \mathbb{R}^n_+ \to \mathbb{R}_+$ can be written as

$$f_i(x) = \sum_{l \in I} \kappa_{il} b_{il}(x), \tag{2.21}$$

where $\kappa_{il} > 0$ is a rate parameter, $b_{il} : \mathbb{R}_+^n \to \{0,1\}$ is a boolean-valued regulation function, and I is an index set. The regulation functions b_{il} capture the conditions under which the protein encoded by gene i is synthesized at a rate κ_{il}. These conditions are written down as combinations (sums of products) of step functions $s^+, s^- : \mathbb{R}_+ \times \mathbb{R}_+ \to \{0,1\}$, where $s^+(x_j, \theta_j^i) = 1$ if $x_j > \theta_j^i$, and $s^+(x_j, \theta_j^i) = 0$ if $x_j < \theta_j^i$, and $s^-(x_j, \theta_j^i) = 1 - s^+(x_j, \theta_j^i)$. The parameters θ_j^i are threshold concentrations.

This class of PWA systems was first introduced by Glass and Kauffman [22], and is widely used for modeling genetic regulatory networks [9, 17, 22, 34]. Step functions are not defined at threshold points, but solutions of the system "across" or "along" a threshold can still be defined in the sense of Filippov, as the solutions of differential inclusions, as shown in Sect. 2.2.2.4 and Fig. 2.3.

In the PWA formalism, the bistable system in Eq. (2.17) is defined inside the (invariant) set $\Omega = [0, \kappa_1/\gamma_1] \times [0, \kappa_2/\gamma_2]$. Assuming for the sake of simplicity that $\kappa_{10} = \kappa_{20} = 0$, one gets the equations:

$$\begin{cases} \dot{x}_1 &= \kappa_1 s^-(x_2, \theta_2) - \gamma_1 x_1, \\ \dot{x}_2 &= \kappa_2 s^-(x_1, \theta_1) - \gamma_2 x_2. \end{cases} \tag{2.22}$$

The space of state variables Ω is now divided into four boxes, or *regular domains*, where the vector field is uniquely defined:

$$\begin{aligned} B_{00} &= \{x \in \mathbb{R}_+^2 : 0 < x_1 < \theta_1,\ 0 < x_2 < \theta_2\} \\ B_{01} &= \{x \in \mathbb{R}_+^2 : 0 < x_1 < \theta_1,\ \theta_2 < x_2 < \kappa_2/\gamma_2\} \\ B_{10} &= \{x \in \mathbb{R}_+^2 : \theta_1 < x_1 < \kappa_1/\gamma_1,\ 0 < x_2 < \theta_2\} \\ B_{11} &= \{x \in \mathbb{R}_+^2 : \theta_1 < x_1 < \kappa_1/\gamma_1,\ \theta_2 < x_2 < \kappa_2/\gamma_2\}. \end{aligned}$$

In addition, there are also *switching domains*, where the system is defined only as a differential inclusion, corresponding to the segments where each of the variables is at a threshold ($x_i = \theta_i$ and $x_j \in [0, \kappa_j/\gamma_j]$). In each of the four regular domains, the differential system is affine, and simple to study. In B_{00} for instance $\dot{x}_1 = \kappa_1 - \gamma_1 x_1$, $\dot{x}_2 = \kappa_2 - \gamma_2 x_2$, and the solution can easily be written and converges exponentially towards a steady state $(\kappa_1/\gamma_1, \kappa_2/\gamma_2)$. If we suppose that $\theta_i < \frac{\kappa_i}{\gamma_i}$, then this steady state is *outside* B_{00}, and the solution will switch to another system when it crosses one of the thresholds. This succession of possible transitions will result in a *transition graph*, describing the possible sequences of boxes.

For the bistable switch, there are two classical stable steady states, P_1 and P_2, and an unstable Filippov equilibrium point, P_3, analogous to a saddle point (see Fig. 2.6):

$$P_1 = \left(\frac{\kappa_1}{\gamma_1}, 0\right), \quad P_2 = \left(0, \frac{\kappa_2}{\gamma_2}\right), \quad P_3 = (\theta_1, \theta_2).$$

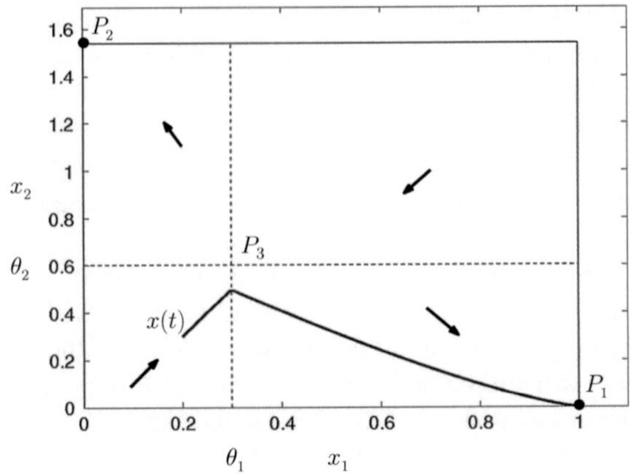

Fig. 2.6 Phase plane for the piecewise linear system of Eq. (2.22), with all parameters as in Fig. 2.5a (except $\kappa_{10} = \kappa_{20} = 0$ and $m_i = \infty$). The nullclines cannot be defined as in the continuous model (2.17) but, instead, the threshold values θ_1, θ_2 divide the plane into four rectangular regions, where the vector field is constant. There are still two stable steady states (P_1,P_2), but the unstable steady state is now defined as an unstable Filippov equilibrium (P_3). One solution is shown in black, which may be compared to that shown in Fig. 2.5a

P_1 and P_2 belong to the boundary of their respective domains (B_{10} and B_{01}), so that any trajectory entering one of these domains remains there. In contrast, trajectories starting in B_{00} or B_{11} will switch to another domain. This leads to the following transition graph for the bistable switch:

$$\begin{array}{ccc} \mathbf{01} & \longleftarrow & 11 \\ \uparrow & & \downarrow \\ 00 & \longrightarrow & \mathbf{10} \end{array}$$

where P_1 is represented by $\mathbf{10}$, P_2 by $\mathbf{01}$, and P_3 is not represented in this diagram, as it is located in the middle, at the boundary of the four regular domains. This discrete abstraction (in the sense of hybrid systems) is a qualitative description of the behavior of the dynamical system. It can be used to check some qualitative properties of the system. Software exist that are able to compute the graph and check some of its properties, with *model checking* techniques.

2.2.3.4 Towards Control of Genetic Networks

An important problem is to be able to lead the system to a prescribed behavior. In control theory, the input represents the actions that a user (here a biologist) is able to exert on the system. From an experimental point of view, one common manipulation is to change the synthesis rate of messenger RNA by addition of a plasmid (a small unit of DNA that replicates within a cell independently of the chromosomal DNA). Therefore, it is reasonable to suppose that the input will act on the synthesis rates. The system with inputs can thus be written:

$$\dot{x}_1 = u\kappa_1 s^-(x_2, \theta_2) - \gamma_1 x_1, \quad \dot{x}_2 = u\kappa_2 s^-(x_1, \theta_1) - \gamma_2 x_2. \tag{2.23}$$

Moreover, different hypotheses can be made concerning the control u: it can take continuous values, or only qualitative values, e.g. two values $u_{\text{min}}, u_{\text{max}}$ and the value $u = 1$ corresponding to no control. The effective application of the control also depends on the measurements we are able to do, which can be continuous or qualitative. The ultimate goal can be to render the point P_1 (or P_2) stable in the whole space (see, for instance, [10]). This poses new and original problems to both mathematicians (to design the input laws) and to biologists (to implement them in the cell). This approach is used in synthetic biology, where artificially built biological circuits realize some tasks [21].

2.2.4 *Success Stories*

2.2.4.1 The Bistable Switch

The example studied in Sect. 2.2.3.2 is a frequently observed motif in biological networks (for instance, the system governing transcription of the *lac* operon contains a similar positive loop). These positive loops are typically observed composed with other motifs to form larger networks, but a very successful experiment by Gardner et al. in [21] showed that such a system can be synthetically implemented in a cell. Gardner et al. constructed plasmids containing two genes coding for proteins that repress each other, and inserted these plasmids in a strain of the bacterium *Escherichia Coli*. Transcription of each gene could be further controlled by an inducer, so that the whole system could be re-set. A mathematical model was used to determine appropriate intervals for some of the parameters, such as the maximal transcription rates, and concentrations of inducers. The experiments measure the expression of one of the genes, which we will call A, and show that the synthetic circuit indeed behaves as a bistable switch: following induction with inducer 1, gene A will be highly expressed, and stably maintain this expression for several hours after the inducer has been removed. Re-setting the system by means of inducer 2 will cause gene A to lower its expression to near zero and remain thus (until a new inducer is applied). Therefore, the synthetically constructed system has the capacity to stably exist in two distinct modes, corresponding to high or low expression of gene A, while the inducers are used to force the system to switch between these two modes. This is a clear observation of bistability in a genetic network.

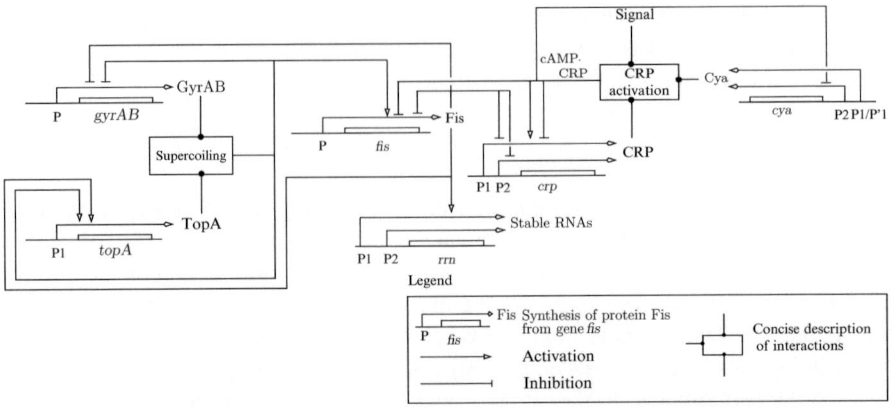

Fig. 2.7 Genetic network, including proteins and regulations that come into play during a nutritional stress response in *E.coli*: CRP activation module (Cya, CRP, Fis), DNA topology module (GyrAB, TopA, Fis), stable RNA output module (Rrn) (Adapted from [34])

2.2.4.2 The Carbon Starvation Response in *Escherichia coli*

One of the successful applications of the PWA formalism is the study of the genetic network that regulates nutritional stress response in *Escherichia Coli*. The model was developed by Ropers et al. [34] to describe the dynamics of a family of genes that regulate the carbon starvation response in *E.coli* (Fig. 2.7): *crp* (x_c), *cya* (x_y), *fis* (x_f), *gyrAB* (x_g), *topA* (x_t), and *rrn* (x_r). Nutritional stress is represented by an input $u \in \{0, 1\}$: $u = 0$ if carbon is present (no stress), and $u = 1$ in the absence of carbon. The PWA equations are shown in Table 2.1, and their mathematical study can be found in [24].

For the case $u = 1$, the asymptotic dynamics of the system in Table 2.1 satisfies:

$$
\begin{aligned}
x_c(t) \to \frac{\kappa_c^1 + \kappa_c^2 + \kappa_c^3}{\gamma_c} > \theta_c^3 > \theta_c^2; \quad & x_y(t) = \theta_y^3 \text{ (in finite time)}; \\
x_f(t) \to 0; \quad x_g(t) = \theta_g^2 \text{ (in finite time)}; , \quad & x_t(t) \to 0.
\end{aligned} \tag{2.24}
$$

Therefore, solutions converge to an equilibrium point in the sense of Filippov. In practice, there are sliding modes along the planes $x_g = \theta_g^2$ and $x_y = \theta_y^3$.

For the case $u = 0$, the asymptotic dynamics of the system in Table 2.1 can be reduced to the equations on x_g and x_f with:

1. $x_c(t) \to \frac{\kappa_c^1}{\gamma_c}$ and $x_y(t) \to \frac{\kappa_y^1+\kappa_y^2}{\gamma_y}$, after some finite time;
2. Sliding mode along the plane $x_t = \theta_t^1$ with the solution eventually jumping down to the region $x_t < \theta_t^1$, and staying there;

Table 2.1 Example piecewise affine model and parameter inequalities

$$\dot{x}_c = \kappa_c^1 + \kappa_c^2 s^-(x_f,\theta_f^2)s^+(x_c,\theta_c^1)s^+(x_y,\theta_y^1)s^+(u,\theta_u) + \kappa_c^3 s^-(x_f,\theta_f^1) - \gamma_c x_c$$
$$\dot{x}_y = \kappa_y^1 + \kappa_y^2[1 - s^+(x_c,\theta_c^3)s^+(x_y,\theta_y^3)s^+(u,\theta_u)] - \gamma_y x_y$$
$$\dot{x}_f = \kappa_f^1[1 - s^+(x_c,\theta_c^1)s^+(x_y,\theta_y^1)s^+(u,\theta_u)]s^-(x_f,\theta_f^5)$$
$$+\kappa_f^2 s^+(x_g,\theta_g^1)s^-(x_t,\theta_t^2)s^-(x_f,\theta_f^5) \times [1 - s^+(x_c,\theta_c^1)s^+(x_y,\theta_y^1)s^+(u,\theta_u)] - \gamma_f x_f$$
$$\dot{x}_g = \kappa_g^1[1 - s^+(x_g,\theta_g^2)s^-(x_t,\theta_t^1)]s^-(x_f,\theta_f^4) - \gamma_g x_g$$
$$\dot{x}_t = \kappa_t^1 s^+(x_g,\theta_g^2)s^-(x_t,\theta_t^1)s^+(x_f,\theta_f^4) - \gamma_t x_t$$
$$\dot{x}_r = \kappa_r^1 s^+(x_f,\theta_f^3) + \kappa_r^2 - \gamma_r x_r$$

$$0 < \theta_c^1 < \frac{\kappa_c^1}{\gamma_c} < \frac{\kappa_c^1+\kappa_c^2}{\gamma_c} < \theta_c^2 < \theta_c^3 < \frac{\kappa_c^1+\kappa_c^3}{\gamma_c}$$
$$0 < \theta_y^1 < \frac{\kappa_y^1}{\gamma_y} < \theta_y^2 < \theta_y^3 < \frac{\kappa_y^1+\kappa_y^2}{\gamma_y}$$
$$0 < \theta_f^1 < \frac{\kappa_f^1}{\gamma_f} < \theta_f^2 < \theta_f^3 < \theta_f^4 < \theta_f^5 < \frac{\kappa_f^1+\kappa_f^2}{\gamma_f}$$
$$0 < \theta_g^1 < \theta_g^2 < \frac{\kappa_g}{\gamma_g}$$
$$0 < \theta_t^1 < \theta_t^2 < \frac{\kappa_t}{\gamma_t}$$

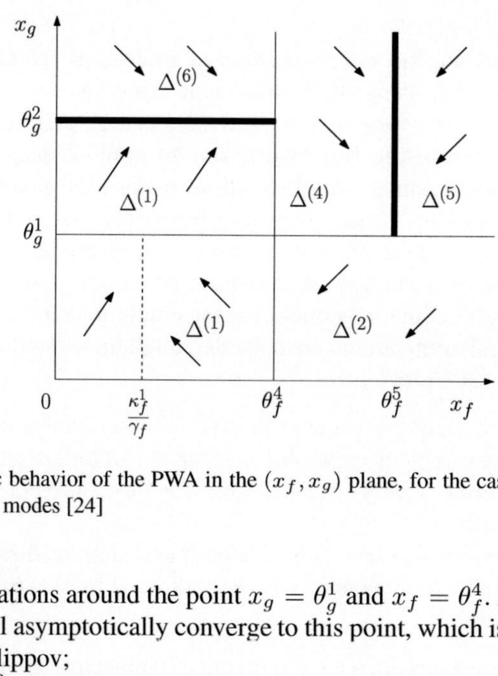

Fig. 2.8 Asymptotic behavior of the PWA in the (x_f, x_g) plane, for the case $u = 0$. *Thick black lines* indicate sliding modes [24]

3. Damped oscillations around the point $x_g = \theta_g^1$ and $x_f = \theta_f^4$. It is shown that all trajectories will asymptotically converge to this point, which is an equilibrium in the sense of Filippov;
4. $x_r(t) \to \frac{\kappa_r^1+\kappa_r^2}{\gamma_r}$ following the solution x_f.

There are also sliding modes along the segments: $x_g = \theta_g^2$ with $x_f < \theta_f^4$ and $x_g > \theta_g^1$ with $x_f = \theta_f^5$ (Fig. 2.8).

The PWA formalism allowed a more rigorous analysis of the complex network of carbon starvation response in *Escherichia Coli*. Major participants were identified as well as their roles in the presence or absence of nutritional stress. This PWA network could be further analyzed using a constraint-based method [14] to evaluate and improve the structure of the network under the parameter constraints on Table 2.1.

This example illustrates the sequence of events delineated in the introduction: the external or environmental signal is the presence ($u = 0$) or absence ($u = 1$) of a nutrient source, which may trigger the activation of the transcription of some genes, with production of the respective mRNA and proteins. Depending on the external signal, the response of the system is different, leading to low (respectively, high) expression of gene *fis* if nutrient is absent (respectively, present). Similar conclusions hold for the remaining genes, and many of the predictions have been experimentally observed.

2.3 Discrete Models of GRN

2.3.1 Challenges

In the previous section, the concentrations of molecular species are handled in continuous frameworks, using differential equations and it is shown that *some regions* (domains) of the space of state variables can be identified so that a more abstract continuous modeling framework can be applied, namely the piecewise affine systems. Going further into abstraction, many biological questions can be answered by only looking at the successive boxes that the cells under study can follow, forgetting the precise state in each box. Such models are called *discrete* models, as the state of a variable at a given time can be described by an integer value: the number of the interval containing the continuous state.

There are several motivations to consider qualitative models that forget the precise continuous state into a box:

- Numerous biological questions are themselves of qualitative nature;
- The *in vivo* measurement capabilities offer a resolution that does not allow to validate or refute a very precise value for the continuous parameters of a differential equation;
- Discrete descriptions can be easily modeled and simulated by computers, even when hundreds of variables are involved, allowing to experiment large genetic networks *in silico*.

These biological considerations have motivated the discrete approach proposed by René Thomas (presented in the next section).

Perhaps more importantly, discrete models can be studied using powerful techniques from computer science:

- *Combinatorial approaches*, often based on graph theory, are able to establish general laws about the link between the form of the interaction graph and the dynamic behavior of the system (e.g. there are behaviors that are unreachable for some interaction graphs);

- *Formal logics*, mainly temporal logics, can be used to automatically perform complex reasonnings about a given discrete model, so that discrete parameter values can often be *deduced* from behaviors observed *in vivo*.

All in all, discrete models are particularly well suited to perform qualitative reasonning in a computer aided manner and they help biologists to elucidate open questions about the functioning of many gene networks. It finally appears that computer *reasonning* capabilities are at least as useful as *simulation* capabilities. Discrete modeling is consequently able to provide the biologists with quick helpful information about open problems (possible behaviors, refutation of hypotheses, missing variables or missing interactions...); they are able to quickly prune some inconsistent intervals for the parameters in continuous models; they are also able to *suggest experimental plans* optimized to check a biological hypothesis.

2.3.2 *Methodological Developments*

2.3.2.1 René Thomas' Logical Method

R. Thomas' logical method consists in modeling the qualitative behavior of a gene network under the form of a finite state transition graph. This state transition graph is built from the interaction graph of the network together with logical parameters that describe the combined effects of switch-like interactions.

More precisely, the starting point of Thomas logical method is an *interaction* (or *regulatory*) *graph* G. The vertices, denoted from 1 to n, correspond to genes, and each arc $i \rightarrow j$ is associated with sign s_{ij} (Fig. 2.9). If s_{ij} is positive (resp. negative), it means that the protein encoded by i activates (resp. inhibits) the synthesis of the protein encoded by j. For every vertex i, we denote by G_i the set of *regulators* of i, that is, the set of vertices j such that $j \rightarrow i$ is an arc of G, and we denote by T_i the set of vertices regulated by i.

The **first step** of the logical method consists in associating with every vertex i a natural number b_i, called the *bound* of i, such that: $b_i \leq \text{card}(T_i)$, and $b_i > 0$ if T_i is not empty. Then, $X_i = \{0, 1, \ldots, b_i\}$ corresponds to the possible *(concentration) levels* for the protein encoded by i, and $X = \prod_i X_i$ corresponds to the set of possible *(discrete) states* for the system.

The **second step** consists in associating with each interaction $i \rightarrow j$ an integer $t_{ij} \in X_i$, $t_{ij} > 0$, called the *logical threshold* of the interaction $i \rightarrow j$. It is required that, for every i, and for every integer $l \in X_i$, $l > 0$, there exists at least one interaction $i \rightarrow j$ such that $t_{ij} = l$ (condition C1). Then, at state $x = (x_1, \ldots, x_n) \in X$, we say that a regulator j of i is a *resource* of i if: $x_j \geq t_{ji}$ and $s_{ji} = +$ (effective activator), or $x_j < t_{ji}$ and $s_{ji} = -$ (ineffective inhibitor). In other words, j is a resource of i when its concentration level x_j "favors" the synthesis of the protein encoded by i. The set of resources of i at state x is denoted by $\omega_i(x)$. See Fig. 2.10 for an illustration.

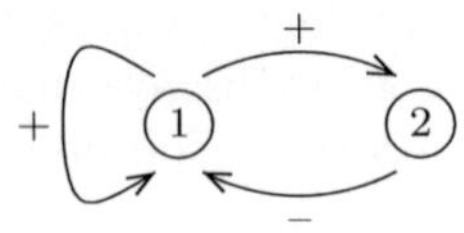

Fig. 2.9 An interaction graph

a

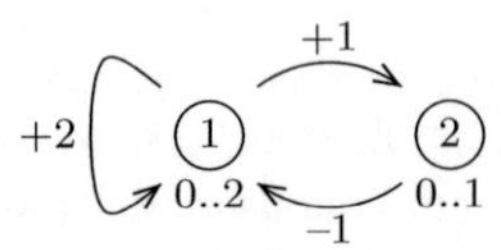

b

x	$\omega_1(x)$	$\omega_2(x)$
$(0,0)$	$\{2\}$	$\emptyset$
$(0,1)$	$\emptyset$	$\emptyset$
$(1,0)$	$\{2\}$	$\{1\}$
$(1,1)$	$\emptyset$	$\{1\}$
$(2,0)$	$\{1,2\}$	$\{1\}$
$(2,1)$	$\{1\}$	$\{1\}$

Fig. 2.10 Logical thresholds and resources. (**a**) The interaction graph of Fig. 2.9 together with the bounds $b_1 = 2$ and $b_2 = 1$, and the logical thresholds $t_{11} = 2$, $t_{12} = 1$ and $t_{21} = 1$. (**b**) The table gives the set or resources $\omega_i(x)$ of $i = 1, 2$ according to the state x of the system

The ***third step*** *consists in associating with every vertex i and every set of regulators $\Omega \subseteq G_i$ a logical parameter $K_{i,\Omega} \in X_i$, in such a way that: for all i, and for all subsets Ω and Ω' of G_i, if $\Omega \subseteq \Omega'$ then $K_{i,\Omega} \leq K_{i,\Omega'}$* (condition C2). Intuitively, $K_{i,\Omega}$ is the level toward which i evolves (focal level) when Ω is the set of resources of i. In other words, at state x, the level of i is: increasing if $x_i < K_{i,\omega_i(x)}$; stable if $x_i = K_{i,\omega_i(x)}$; and decreasing if $x_i > K_{i,\omega_i(x)}$. The signs of the interactions of G are taken into account through the condition C2, which states that the focal level of i increases ($K_{i,\Omega} \leq K_{i,\Omega'}$) when its resources increase ($\Omega \subseteq \Omega'$), that is, when there are more activators and less inhibitors (so that the resources of i favor effectively the synthesis of the protein encoded by gene i).

More precisely, once logical parameters have been given, the behavior of the system is described by a directed graph, called *asynchronous state graph*, and defined by: the set of vertices is X; for every state x and every vertex i such that $x_i \neq K_{i,\omega_i(x)}$, there is an arc (or *transition*) from x to the state x' defined by:

$$x'_i = \begin{cases} x_i + 1 \text{ if } x_i < K_{i,\omega_i(x)} \\ x_i - 1 \text{ if } x_i > K_{i,\omega_i(x)} \end{cases} \quad \text{and} \quad x'_j = x_j \quad \forall j \neq i.$$

See Fig. 2.11 for an illustration.

If every variable is stable at state x (that is if $x_i = K_{i,\omega_i(x)}$ for every i), then x has no outgoing transition in the asynchronous state graph, and it corresponds to a *stable state* of the system. More generally, the *attractors* of the system are the smallest non-empty subsets of states $A \subseteq X$ that we cannot leave, that is, such that for every transition $x \to y$ of the state graph, if $x \in A$ then $y \in A$. So $\{x\}$ is an attractor if and only if x is a stable state. Attractors that are not stable states (attractors of size at least two) are called *cyclic attractors*, because once the system is inside such an attractor, it cannot reach a stable state, and thus, it necessarily describes sustained

a

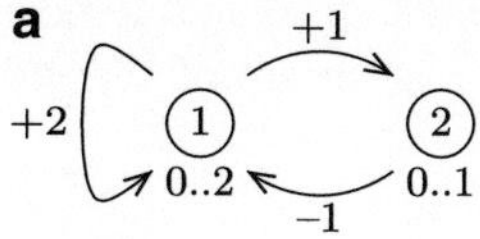

$K_{1,\emptyset} = 0$ $\quad K_{2,\emptyset} = 0$
$K_{1,\{1\}} = 2$ $\quad K_{2,\{1\}} = 1$
$K_{1,\{2\}} = 2$
$K_{1,\{1,2\}} = 2$

b

x	$K_{1,\omega_1(x)}$	$K_{2,\omega_2(x)}$
$(0,0)$	2	0
$(0,1)$	0	0
$(1,0)$	2	1
$(1,1)$	0	1
$(2,0)$	2	1
$(2,1)$	2	1

c

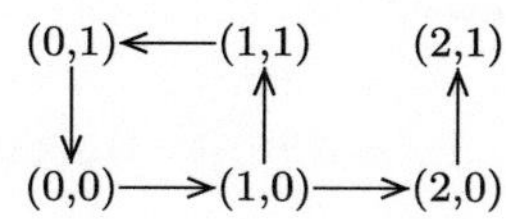

Fig. 2.11 (**a**) The interaction graph of Fig. 2.9 together with bounds, logical thresholds and logical parameters. (**b**) The table gives the focal point of $i = 1, 2$ according to the state of the system. This table results from the one of Fig. 2.10 and the parameter values given in (**a**). (**c**) The asynchronous state graph resulting from the data given in (**a**). This asynchronous state graph can be easily built from the table given in (**b**)

oscillations. It is easy to show that from any initial state, there always exists a path leading to an attractor (and so, there always exists at least one attractor). It is in this weak sense that one can consider that attractors perform an attraction. The state graph of Fig. 2.11 contains a unique attractor, the stable state $(2, 1)$, and indeed, from every initial state, there exists a path leading to this unique attractor.

2.3.2.2 Relationships with the Continuous Approaches

The dynamics of a network whose interaction graph is G may be described, using the piecewise affine model (see Sects. 2.2.3.3 and 2.2.2.4), by the following system:

$$\dot{x}_i = \kappa_i + \sum_{j \in G_i} \kappa_{ji} \cdot s^{(s_{ji})}(x_j, \theta_{ji}) - \gamma_i \cdot x_i \qquad (i = 1, \ldots, n), \tag{2.25}$$

where: κ_i and γ_i are the "basal" synthesis rate and the degradation rate of i; G_i is the set of regulators j of i; κ_{ji}, θ_{ji} and $s_{ji} \in \{+, -\}$ are the synthesis rate, the quantitative threshold and the sign associated with the interaction $j \to i$; s^+ and s^- are the step functions defined in Sect. 2.2.2.4. We will now describe how to obtain, from the quantitative parameters $\kappa_i, \kappa_{ij}, \gamma_i$ and θ_{ij}, the qualitative parameters b_i, t_{ij} and $K_{i,\Omega}$ describing an asynchronous state graph abstracting the system (2.25). First, for all i, let $\Theta_i = \{\theta_{ij} | i \in G_j\}$ be the set of "out-going" quantitative thresholds of i, and set

$$b_i = card(\Theta_i) \qquad \textbf{(first step)}. \tag{2.26}$$

Then, consider the resulting set of discrete states $X = \prod_{i=1}^{n}\{0, 1 \ldots, b_i\}$, and the discretization mapping

$$d : \mathbb{R}^n \to X, \quad d(x) = (d_1(x_1), \ldots, d_n(x_n)), \quad d_i(x_i) = \text{card}(\{\theta \in \Theta_i | x_i \geq \theta_i\}).$$

Using this discretization map, let us define the logical thresholds and logical parameters as follows: for every i, every j such that $i \to j$ is an arc of G, and every $\Omega \subseteq G_i$,

$$t_{ij} = d_i(\theta_{ij}) \quad \textbf{(second step)}, \qquad K_{i,\Omega} = d_i(\kappa_i + \sum_{j \in \Omega} \kappa_{ji}/\gamma_i) \quad \textbf{(third step)}. \tag{2.27}$$

In this way, conditions C1 and C2 are satisfied, and Snoussi [35] proved that the resulting asynchronous state graph has the following property[1]:

Theorem 1 ([35]). *There exists a transition from the discrete state $x \in X$ to the discrete state $x' \in X$ if and only if there exists a solution ξ of the differential system (2.25), and $t' \in \mathbb{R}$, such that $d(\xi(t)) = x$ for all $0 \leq t < t'$ and $d(\xi(t')) = x'$ (i.e. there exists a solution ξ starting in the regular domain $d^{-1}(x)$ that reaches the regular domain $d^{-1}(x')$). Consequently: all solutions ξ such that, for all t, there exists at most one i such that $\xi_i(t) \in \Theta_i$, (and thus almost all solutions), are represented by a path in the asynchronous state graph.*

So each differential system of the form (2.25) is abstracted by an asynchronous state graph that can be built from G using the logical method. And conversely, thanks to the conditions C1 and C2, every asynchronous state graph S built from G with the logical method (from the parameters b_i, t_{ij} and $K_{i,\Omega}$) is the abstraction of an infinite set of differential systems of the form (2.25). (More precisely, S is an abstraction of (2.25) for every κ_{ij}, γ_i and θ_{ij} verifying the equalities (2.26) and (2.27).) Thus, the logical method may be seen as a *constructive* method for abstracting the infinite number of differential systems of the form (2.25) into a *finite* number of asynchronous state graphs.

2.3.2.3 Positive and Negative Circuits

In practice, while G may often be established from experimental data, (see for instance the database RegulonDB [20] about gene interactions in *E. coli*,) the bounds b, the logical thresholds t and the logical parameters K remain most often unknown. An interesting question is then: *which dynamical properties of a gene network can be inferred from its interaction graph, in the absence of information on the value of these parameters?* This question can be partially solved by studying positive and negative circuits of G.

A *positive* (resp. *negative*) *circuit* of G is a directed cycle containing an even (resp. odd) number of negative arcs. The interaction graph of Fig. 2.9 contains a positive circuit of length one ($1 \to 1$) and a negative circuit of length two ($1 \to 2 \to 1$).

[1] Provided that $(\kappa_i + \sum_{j \in \Omega} \kappa_{ji}/\gamma_i) \notin \Theta_i$ for all $\Omega \subseteq G_i$.

René Thomas highlighted the predominant (dynamical) role of positive and negative circuits by stating the following two rules [39]: (1) *A necessary condition for the presence of several stable states is the presence of a positive circuit in G.* (2) *A necessary condition for the presence of sustained oscillations is the presence of a negative circuit in G.*

These rules are "transversal" to the considered modeling framework in the sense that they have been proved for differential models [12, 23, 26, 29, 36–38], Boolean models [4,5,30] and discrete models [32,33]. The obvious interest of these two rules is that they relate the rather simple information contained in the interaction graph G of a network to its much more complex dynamical behaviors. In addition, multistationarity and sustained oscillations are, from a biological point of view, important dynamical properties, respectively associated with differentiation processes and homeostasis phenomena.

Inside Thomas' logical method, Thomas' rules take the following form:

Theorem 2 ([32, 33]).

1. *If G has no positive circuit, then for all bounds b, logical thresholds t, and logical parameters K, the resulting asynchronous state graph has at most one attractor.*
2. *If G has no negative circuit, then for all bounds b, logical thresholds t, and logical parameters K, the resulting asynchronous state graph has no cyclic attractor.*

In fact, if G has no positive (resp. negative) circuit, then every associated asynchronous state graph contains a unique attractor (resp. at least one stable state). These are simple consequences of the above theorem and the basic observation, already mentioned, that a state graph has always at least one attractor.

Most often, real interaction graphs contains positive and negative circuits, so that the previous theorem cannot be applied to obtain information on the dynamics of the system. However, the following theorem, which extends the first point of the previous theorem, can always be used (in the worst case, take $I = \{1, \ldots, n\}$).

Theorem 3 ([4,31]). *If I is a set of vertices such that every positive circuit of G has at least one vertex in I, then the asynchronous state graph resulting from the bounds b, logical thresholds t and logical parameters K contains at most $\prod_{i \in I}(b_i + 1)$ attractors.*

This theorem shows that the number of attractors is small when positive circuits are highly connected. The number of positive circuits is not the relevant parameter: if there is one million of positive circuits, but if all these circuits contain a vertex i with $b_i = 1$, then there are at most two attractors. Note also that the upper bound is tight in some cases. For instance, if G consists in a single vertex (vertex 1) with a positive arc $1 \to 1$, and if $b_1 = t_{11} = K_{1,\{1\}} = 1 > K_{1,\emptyset} = 0$, then the resulting asynchronous state graph has $2 = b_1 + 1$ attractors (that are stable states).

2.3.2.4 Formal Methods

SMBioNet. Given an interaction graph G, the number of asynchronous state graphs that we can build with the logical method is finite. For instance, 90 different asynchronous state graphs can be built from the interaction graph of Fig. 2.9.[2]

A natural question is then: *How to find, among this finite set of asynchronous state graphs, those that are coherent with biological observations (or hypothesis) on the dynamics of the system?*

Formal methods are useful to perform automatically such a selection. First, a *temporal logic* can be used to translate the dynamical observations into a temporal formula Φ, which can be handled by a computer. Then, one can use model-checking algorithms in order to check automatically if a given state graph (a model) satisfies or not formula Φ. Hence, to solve the question, the following (basic) approach can be used: *enumerate the different asynchronous state graphs, and select those that satisfy Φ using model-checking techniques.*

This enumerative approach has been implemented in a software called SMBIONET [8]. The temporal logic used is the well known Computational Tree Logic (CTL) [19], and the verification step is performed with the model-checker called NUSMV [11]. The Computational Tree Logic is briefly presented in the next paragraph. An illustration, on a real case, of this logic and the enumerative approach is then given in Sect. 2.3.3.

But before going further, let us briefly discuss the enumerative approach. The obvious limitation is that the number of state graphs to enumerate (which increases exponentially with the number of components) is often too huge to consider networks with more than ten genes or so. The obvious interest is that temporal logic and model checking allow us to handle automatically rather complex dynamical properties, and that the method is exhaustive: *all* the state graphs associated with G that are consistent with Φ are reported. (For other applications of formal methods in the context of gene regulatory networks, see for instance [6, 7, 13, 14] and the references therein.)

Computational Tree Logic. In an asynchronous state graph, a given state has generally several successors. So without additional information, all the successors are possible next states: the dynamical description is undeterministic. In other words, given an initial state x, the possible evolutions of the system are given by the set of paths starting from x, and these paths may be seen as a tree rooted at x.

---

[2]For the interaction graph of Fig. 2.9, we have $b_1 \in \{1, 2\}$, $t_{11}, t_{12} \in \{1, b_1\}$ and $b_2 = \theta_{21} = 1$. If $b_1 = 1$, then $t_{11} = t_{12} = 1$, and there are 18 possible instantiations of the parameters K, which lead to a set of 18 different asynchronous state graphs. If $b_1 = 2$ there are two cases. First, if $t_{11} = 1 < t_{21} = 2$, there are 60 possible instantiations of the parameters K, which lead to a set S of 42 different asynchronous state graphs. Second, if $t_{11} = 2 > t_{21} = 1$, there are 60 possible instantiations of the parameters K, which lead also to a set of 42 different asynchronous state graphs, but 12 of them are contained in S. Hence, the total number of asynchronous state graphs is $18 + 42 + 42 - 12 = 90$.

The Computational Tree Logic allows the formulation of properties on this tree. It is thus well adapted to formulate dynamical properties on undeterministic discrete dynamical systems. In particular, one can express possibilities in the future. For instance, the formula $\text{EF}(l_i = 0)$ expresses that "it is possible to reach a state in which the level of the ith component is 0", and the formula $\text{EG}(l_i = 0)$ expresses that "it is possible that the ith component stay for ever at the level 0".

Computational Tree Logic is defined in two steps. The first step consists in defining the *syntax* of the logic, i.e. rules for constructing formulas. The second step consists in defining the *semantic* of the logic, i.e. meaning of formulas.

The *syntax* of CTL is inductively defined by:

- For all genes i and integers k, $(l_i = k)$, $(l_i < k)$, $(l_i > k)$, $(l_i \leq k)$ and $(l_i \geq k)$ are (*atomic*) CTL formulas.
- If ϕ and ψ are two CTL formulas then $(\neg\phi)$, $(\phi \wedge \psi)$, $(\phi \vee \psi)$, $(\phi \Rightarrow \psi)$, $\text{EX}(\phi)$, $\text{EF}(\phi)$, $\text{EG}(\phi)$, $\text{E}(\phi \,\text{U}\, \psi)$, $\text{AX}(\phi)$, $(\text{AF}(\phi))$, $(\text{AG}(\phi))$, and $(\text{A}(\phi \,\text{U}\, \psi))$ are CTL formulas.

The *semantic* is given by the satisfaction relation $\models$ between the states x of a given asynchronous state graphs S and the CTL formulas ϕ. The semantic of atomic formulas is the following: $x \models (l_i = k)$ if and only if $x_i = k$; $x \models (l_i < k)$ if and only if $x_i < k$ and so on. The semantic of the classical logical connectives $\neg$ (negation), $\wedge$ (conjunction), $\vee$ (disjunction), and $\Rightarrow$ (implication) is usual: for instance, $x \models \phi \wedge \psi$ if and only if $x \models \phi$ and $x \models \psi$. The other connectives, called *temporal connectives*, are made with two letters and lead to formulas that are satisfied by a state x according to the set of infinite paths of S starting from x.[3] Intuitively, E and A correspond to existential and universal quantifiers respectively: E means "for at least one path" and A "for all paths". The other letters express properties along the paths: $\text{X}(\phi)$ means that ϕ is true at the neXt step, $\text{F}(\phi)$ means that ϕ is ture in the Future; $\text{G}(\phi)$ means that ϕ is Globally true, and $(\psi \,\text{U}\, \phi)$ means that ψ is always true Until ϕ becomes true. See Fig. 2.12 for an illustration. Formally, the semantic of temporal connectives is given by:

- $x \models \text{EX}(\phi) \iff$ there exists a successor of x satisfying ϕ.
- $x \models \text{AX}(\phi) \iff$ all the successors of x satisfy ϕ.
- $x \models \text{EF}(\phi) \iff$ there exists an infinite path starting from x which contains a state satisfying ϕ.
- $x \models \text{AF}(\phi) \iff$ all the infinite paths starting from x contain a state satisfying ϕ.
- $x \models \text{EG}(\phi) \iff$ there exists an infinite path starting from x which only contains states satisfyingϕ.

[3] An infinite path of S is an infinite sequence of states $x^0x^1x^2,\ldots$ such that, for all $k \in \mathbb{N}$: if x^k has a successor in S, then $x^k \to x^{k+1}$ is an arc of S, and $x^k = x^{k+1}$ otherwise.

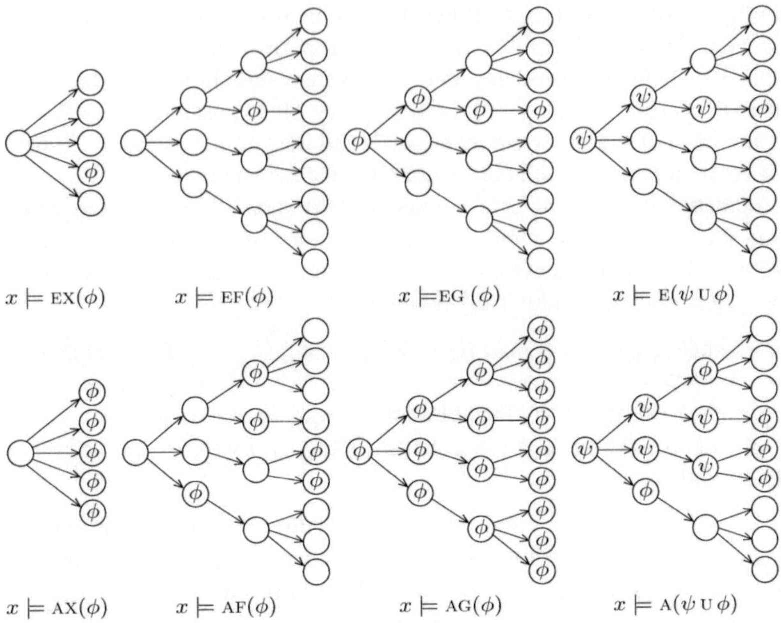

Fig. 2.12 Illustration of the semantic of temporal connectives. Paths starting from the root correspond to paths starting from x. The *circles* containing ϕ (resp. ψ) corresponds to states satisfying ϕ (resp. ψ)

- $x \models \mathrm{AG}(\phi) \iff$ all the infinite paths starting from x only contain states satisfying ϕ.
- $x \models \mathrm{E}(\psi\,\mathrm{U}\,\phi) \iff$ there exists both an infinite path $x^0x^1x^2\ldots$ with $x = x^0$ and $i \in \mathbb{N}$ such that $x^i \models \phi$ and $x^j \models \psi$ for all $j < i$.
- $x \models \mathrm{A}(\psi\,\mathrm{U}\,\phi) \iff$ for all infinite paths $x^0x^1x^2\ldots$ with $x = x^0$, there exists $i \in \mathbb{N}$ such that $x^i \models \phi$ and $x^j \models \psi$ for all $j < i$.

If all the states of a state graph S satisfy a given formula, we say that S satisfies this formula. For instance, the formula

$$(l_1 \geq 1) \Rightarrow \mathrm{AX}(\mathrm{AF}(l_1 \geq 1)),$$

is satisfied by S if and only if for every path $x^0, x^1, \ldots$ starting from an initial state x^0, with $x_1^0 \geq 1$, there exists $t > 0$ such that $x_1^t \geq 1$. In other words, the formula means that when the level of the first component is at least one at time t, then it will be at least one at another time $t' > t$. The asynchronous state graph of Fig. 2.11 satisfies this property. The formula

$$(l_1 = 0) \Rightarrow \mathrm{AG}(l_1 < 2)$$

is satisfied by S if and only if every path starting from a state x, with $x_1 = 0$, only contains states y such that $y_1 < 2$. In other words, the formula means that when the level of the first component is zero, then it will be always less than two. The asynchronous state graph of Fig. 2.11 does not satisfy this property, because of the presence of the path $(0,0) \to (1,0) \to (2,0)$.

To summarize, the interest of CTL is twofold. Firstly, it allows to express, in a simple way, rather complex dynamical properties on undeterministic transition systems. Secondly, every CTL formula ϕ can be checked on a transition system S in a polynomial time with respect to the size of ϕ and S. (More precisely, the complexity of the verification is in $\mathcal{O}(|\phi| \cdot |S|)$ where $|\phi|$ is the number of symbols in ϕ seen as a string, and $|S|$ is the sum of the number of vertices and the number of transitions of the state graph S.) Notice though, as discussed in Sect. 2.3.2.4, that the number of state graphs grows exponentially with the number of components.

2.3.3 *Success Story: Pseudomonas aeruginosa and Cystic Fibrosis*

The bacteria *Pseudomonas aeruginosa* [25] are commonly present in the environment and secrete mucus only in lungs affected by cystic fibrosis. As it increases the respiratory deficiency of the patient, it is the major cause of mortality. Bacteria isolated from cystic fibrosis lungs continue to grow in laboratory as mucous colonies for numerous generations (mucoid phenotype). A majority of these bacteria present a mutation. Does it mean that the mutation is the cause of the passage to the mucoid state?

A majority of biologists tend to follow this hypothesis. However, the regulatory network that controls the mucus production has been elucidated (Fig. 2.13a) and the regulatory graph contains two feedback circuits among which one is a positive one (Fig. 2.13b). This positive circuit makes possible a dynamic with two attractors that would allow, from a biological point of view, an epigenetic change (stable change of phenotype without mutation) from the non-mucoid state to the mucoid one.

From a biological point of view, it is very important to determine whether the mucoidy can be induced by an epigenetic phenomenon or not. In such a case, the elimination of the anti-AlgU gene (*via* a mutation) could be favored later on because an inhibitor complex is produced, which is toxic for the bacteria.

From a modeling point of view, and because the mathematical model of mucus production system is not yet determined, this question becomes: Can we exhibit, from the interaction graph of 2.13, a dynamical model (an asynchronous state graph) presenting at least two attractors, one in which mucus is regularly produced and one in which mucus is not produced?

Assuming that AlgU activates the mucus production at its maximal level b_{AlgU}, to state that a model which *regularly produces mucus* is equivalent to the fact that

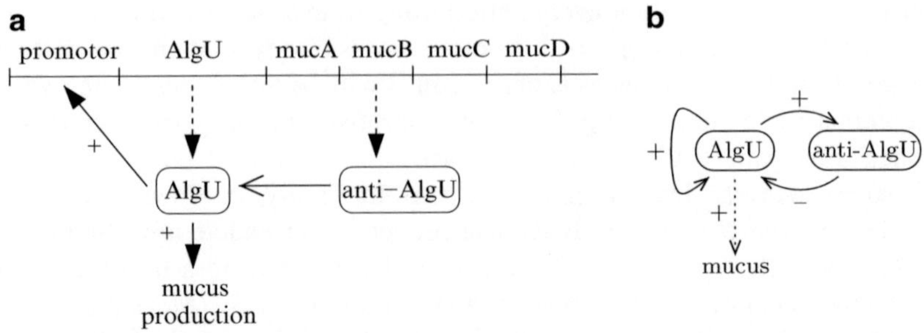

Fig. 2.13 Mucus production in *Pseudomonas aeruginosa* (**a**) The main regulatory genes (**b**) A possible interaction graph (identical to the one of Fig. 2.9)

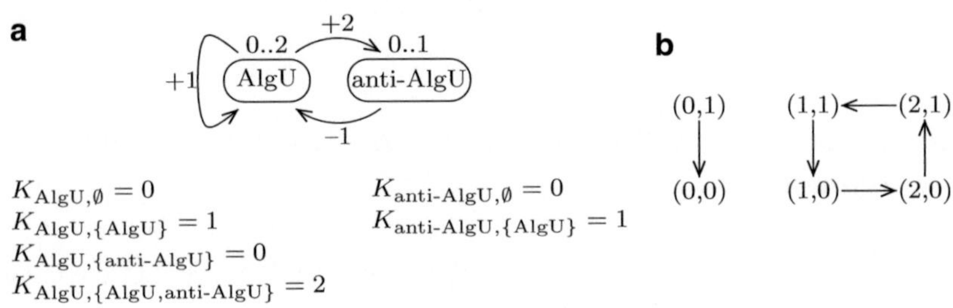

Fig. 2.14 Modeling the mucus production in *Pseudomonas aeruginosa* (**a**) The interaction graph of Fig. 2.13 together bounds, logical thresholds and logical parameters. (**b**) The resulting asynchronous state graph. It satisfies the formulas (2.28) and (2.29). It has two attractors: a cyclic attractor $\{(1,0),(2,0),(2,1),(1,1)\}$ in which mucus is regularly produced, and a stable state $(0,0)$ in which mucus is not produced

the concentration level of AlgU is repeatedly equal to b_{AlgU}. Thus this information can be expressed in CTL as:

$$(l_{\text{AlgU}} = b_{\text{AlgU}}) \Rightarrow \text{AX}(\text{AF}(l_{\text{AlgU}} = b_{\text{AlgU}})). \tag{2.28}$$

Moreover we know that the wild bacteria never produces mucus by themselves when starting from a basal state (second attractor):

$$(l_{\text{AlgU}} = 0) \Rightarrow \text{AG}(l_{\text{AlgU}} < b_{\text{AlgU}}). \tag{2.29}$$

Using SMBioNet, one shows that, among the 73 asynchronous state graphs that can be built using the logical method, from the interaction graph of Fig. 2.13, there are 17 asynchronous state graphs verifying the two previous formulas (one of them is display in Fig. 2.14). Consequently, because the set of remaining models is not empty, the epigenetic question receives a positive answer from the modeling standpoint. This epigenetic question has not only an academic interest because this prediction has been validated experimentally that could lead to new therapeutic strategies.

2.4 Outlook

A large amount of data on gene regulation is currently available. This created an upsurge of interest in modeling regulatory networks. Here, three usual approaches to modeling were presented: continuous differential systems (based on sigmoidal functions), piecewise affine systems (based on step functions), and discrete systems. On the one hand, continuous systems present a higher level of details and offer the ability to model dynamics. But on the other hand, discrete (or logical) systems are easier to analyze, need a smaller amount of data, can be deduced from qualitative observations, and thus allow modeling of larger systems. The choice then depends on the nature of input data and of the biological question under consideration. Furthermore, the well understood relationships between the continuous and discrete approaches presented here, allow one to follow a classical and simple strategy to model a new biological system: the discrete approach can be taken as a useful first step as long as the input data are qualitative, then more accurate descriptions can be achieved through continuous models based on the discrete ones when more precise input data are available.

Research on gene regulatory networks is rather active, and many research directions are relevant. From a computational point of view, it becomes crucial to develop techniques that allow the modeling of large systems, for instance using sensible model reductions or modular decompositions. Another direction consists in using experimental design approaches to select sets of experiments that are efficient to validate or to refute a model. From a biological point of view, a number of outstanding questions are open. The stochasticity and robustness of regulatory networks are not well understood. The dynamical influence of network architectures, and the evolutionary processes that produce them, are also far from being understood. Furthermore, models for the interplay between gene network and other processes, such as metabolism and cell signaling, have to be developed.

2.5 Online Resources

Several computer tools are available to help modeling and analyze genetic regulatory networks. A few examples are:

GNA (Genetic Network Analyzer)

http://www.genostar.com/en/genostar-software/gnasim.html

Modeling and simulation of GRN, using piecewise linear models.The user specifies the equations, the parameters (synthesis and degradation rates, thresholds), and inequality constraints between them.

GINsim (Gene Interaction Network simulation)

http://gin.univ-mrs.fr

Modeling and simulation of GRN, based on a discrete, logical formalism. The user may specify a model of a GRN in terms of asynchronous, multivalued logical functions.

SMBioNet (Selection of Models of Biological Networks)

http://www.i3s.unice.fr/~richard/smbionet

Modeling and model-checking analysis of GRN, based on a discrete, logical formalism (see Sect. 2.3.2.4).

The Biochemical Abstract Machine BIOCHAM

http://contraintes.inria.fr/BioCham

A modeling environment for systems biology, with features for inferring unknown model parameters from temporal logic constraints.

SBML (Systems Biology Markup Language)

http://sbml.org/Main_Page

A standard formalism for the construction and exchange of computer models of biological processes.

Acknowledgements It is a pleasure for GB, JPC and AR to thank the biologist Janine Guespin-Michel, who has actively participated to the definition of our formal logic methodology in such a way that our techniques from computer science and the SMBioNet software become truly useful for biologists. She has also been at the origin of the *Pseudomonas aeruginosa* hypothesis. The authors would also like to thank F. Cazals for his remarks and careful reading of the chapter.

List of Acronyms

DNA	Deoxyribose Nucleic Acid
mRNA	messenger Ribonucleic Acid
GRN	Gene Regulatory Networks
PWA	Piecewise affine

References

1. http://en.wikipedia.org/wiki/File:Ribosome_mRNA_translation_en.svg
2. http://http://en.wikipedia.org/wiki/File:Simple_transcription_elongation1.svg
3. U. Alon. *An Introduction to Systems Biology: Design Principles of Biological Circuits*. Chapman & Hall/CRC, Boca Raton, 2006.
4. J. Aracena. On the number of fixed points in regulatory boolean networks. *Bulletin of Mathematical Biology*, 70(5):1398–1409, 2008.
5. J. Aracena, J. Demongeot, and E. Goles. Positive and negative circuits in discrete neural networks. *IEEE Transactions of Neural Networks*, 15:77–83, 2004.
6. J. Barnat, L. Brim, I. Černá, S. Dražan, J. Fabrikozá, and D. Šafránek. On algorithmic analysis of transcriptional regulation by ltl model checking. *Theoretical Computer Science*, 2009.
7. G. Batt, M. Page, I. Cantone, G. Goessler, P. Monteiro, and H. de Jong. Efficient parameter search for qualitative models of regulatory networks using symbolic model checking. *Bioinformatics*, 26(18):i603–i610, 2010.
8. G. Bernot, J.-P. Comet, A. Richard, and J. Guespin. A fruitful application of formal methods to biological regulatory networks: Extending Thomas' asynchronous logical approach with temporal logic. *J. Theor. Biol.*, 229(3):339–347, 2004.

9. R. Casey, H. de Jong, and J.L. Gouzé. Piecewise-linear models of genetic regulatory networks: equilibria and their stability. *J. Math. Biol.*, 52:27–56, 2006.
10. M. Chaves and J.L. Gouzé. Exact control of genetic networks in a qualitative framework: the bistable switch example. *Automat.*, 47:1105–1112, 2011.
11. A. Cimatti, E. Clarke, E. Giunchiglia, F. Giunchiglia, M. Pistore, and M. Roven. NuSMV2: An Open Source Tool for Symbolic Model Checking. In *International Conference on Computer-Aided Verification (CAV 2002)*, 2002.
12. O. Cinquin and J. Demongeot. Roles of positive and negative feedback in biological systems. *C.R.Biol.*, 325(11):1085–1095, 2002.
13. F. Corblin, E. Fanchon, and L. Trilling. Applications of a formal approach to decipher discrete genetic networks. *BMC Bioinformatics*, 11(385), 2010.
14. F. Corblin, S. Tripodi, E. Fanchon, D. Ropers, and L. Trilling. A declarative constraint-based method for analyzing discrete genetic regulatory networks. *Biosystems*, 98(2):91–104, 2009.
15. F. Dardel and F. Képès. *Bioinformatics: genomics and post-genomics*. Wiley, Chichester, 2005.
16. H. De Jong. Modeling and simulation of genetic regulatory systems: a literature review. *Journal of computational biology*, 9(1):67–103, 2002.
17. H. de Jong, J.L. Gouzé, C. Hernandez, M. Page, T. Sari, and J. Geiselmann. Qualitative simulation of genetic regulatory networks using piecewise linear models. *Bull. Math. Biol.*, 66:301–340, 2004.
18. L. Edelstein-Keshet. *Mathematical models in Biology*. SIAM classics in applied mathematics, Philadelphia, 2005.
19. E.A. Emerson. *Handbook of theoretical computer science, Volume B : formal models and semantics*, chapter Temporal and modal logic, pages 995–1072. MIT Press, 1990.
20. S. Gama-Castro, H. Salgado, M. Peralta-Gil, A. Santos-Zavaleta, L. Muniz-Rascado, H. Solano-Lira, V. Jimenez-Jacinto, V. Weiss, J. S. Garcia-Sotelo, A. Lopez-Fuentes, L. Porron-Sotelo, S. Alquicira-Hernandez, A. Medina-Rivera, I. Martinez-Flores, K. Alquicira-Hernandez, R. Martinez-Adame, C. Bonavides-Martinez, J. Miranda-Rios, A. M. Huerta, A. Mendoza-Vargas, L. Collado-Torres, B. Taboada, L. Vega-Alvarado, M. Olvera, L. Olvera, R. Grande, E. Morett, and J. Collado-Vides. RegulonDB version 7.0: transcriptional regulation of Escherichia coli K-12 integrated within genetic sensory response units (Gensor Units). *Nucleic Acids Research*, 2010.
21. T.S. Gardner, C.R. Cantor, and J.J. Collins. Construction of a genetic toggle switch in *Escherichia coli*. *Nature*, 403:339–342, 2000.
22. L. Glass and S.A. Kauffman. The logical analysis of continuous, nonlinear biochemical control networks. *J. Theor. Biol.*, 39:103–129, 1973.
23. J.L. Gouzé. Positive and negative circuits in dynamical systems. *Journal of Biological Systems*, 6:11–15, 1998.
24. F. Grognard, J.-L. Gouzé, and H. de Jong. Piecewise-linear models of genetic regulatory networks: theory and example. In I. Queinnec, S. Tarbouriech, G. Garcia, and S. Niculescu, editors, *Biology and control theory: current challenges*, Lecture Notes in Control and Information Sciences (LNCIS) 357, pages 137–159. Springer-Verlag, 2007.
25. J. Guespin-Michel and M. Kaufman. Positive feedback circuits and adaptive regulations in bacteria. *Acta Biotheor.*, 49(4):207–18, 2001.
26. M. Kaufman, C. Soulé, and R. Thomas. A new necessary condition on interaction graphs for multistationarity. *Journal of Theoretical Biology*, 248:675–685, 2007.
27. H.K. Khalil. *Nonlinear systems*. Prentice Hall, New Jersey, 2002.
28. E. Klipp, R. Herwig, A. Howald, C. Wierling, and H. Lehrach. *Systems Biology in Practice*. Wiley-VCH, Weinheim, 2005.
29. E. Plahte, T. Mestl, and S.W. Omholt. Feedback loops, stability and multistationarity in dynamical systems. *Journal of Biological Systems*, 3:569–577, 1995.
30. E. Remy, P. Ruet, and D. Thieffry. Graphic requirement for multistability and attractive cycles in a boolean dynamical framework. *Advances in Applied Mathematics*, 41(3):335–350, 2008.
31. A. Richard. Positive circuits and maximal number of fixed points in discrete dynamical systems. *Discrete Applied Mathematics*, 157(15):3281–3288, 2009.

32. A. Richard. Negative circuits and sustained oscillations in asynchronous automata networks. *Advances in Applied Mathematics*, 44(4):378–392, 2010.
33. A. Richard and J.-P. Comet. Necessary conditions for multistationarity in discrete dynamical systems. *Discrete Applied Mathematics*, 155(18):2403–2413, 2007.
34. D. Ropers, H. de Jong, M. Page, D. Schneider, and J. Geiselmann. Qualitative simulation of the carbon starvation response in *Escherichia coli*. *Biosystems*, 84(2):124–152, 2006.
35. E.H. Snoussi. Qualitative dynamics of a piecewise-linear differential equations : a discrete mapping approach. *Dynamics and stability of Systems*, 4:189–207, 1989.
36. E.H. Snoussi. Necessary conditions for multistationarity and stable periodicity. *Journal of Biological Systems*, 6:3–9, 1998.
37. C. Soulé. Graphical requirements for multistationarity. *ComPlexUs*, 1:123–133, 2003.
38. C. Soulé. Mathematical approaches to differentiation and gene regulation. *C.R. Biologies*, 329:13–20, 2006.
39. R. Thomas. On the relation between the logical structure of systems and their ability to generate multiple steady states and sustained oscillations. In *Series in Synergetics*, volume 9, pages 180–193. Springer, 1981.
40. G. Yagil and E. Yagil. On the relation between effector concentration and the rate of induced enzyme synthesis. *Biophys. J.*, 11:11–27, 1971.

Part II
Biomedical Signal and Image Analysis

Chapter 3
Noninvasive Cardiac Signal Analysis Using Data Decomposition Techniques

Vicente Zarzoso, Olivier Meste, Pierre Comon, Decebal Gabriel Latcu, and Nadir Saoudi

3.1 Preliminaries and Motivation

3.1.1 Cardiac Electrophysiology and Genesis of the ECG Signal

The heart is one of the most important bodily organs, as it is responsible for pumping blood to all cells of the body. Anatomically, it is divided into four chambers, two atria and two ventricles, which work in a well coordinated fashion as an ingenious double pump system. The left-hand side of the heart is in charge of oxygenated blood. This comes from the lungs and enters the left atrium through the pulmonary veins. The left ventricle then pumps it to the rest of the body through the aorta. The right-hand side of the heart deals with deoxygenated blood, which enters the right atrium through the caval veins and is then pumped by the right ventricle through the pulmonary arteries towards the lungs, where it is re-oxygenated again. These two pumping functions operate simultaneously during a cardiac beat. A schematic diagram of the heart and the blood circulation system is shown in Fig. 3.1.

V. Zarzoso (✉) · O. Meste
I3S – UMR 7271 CNRS/UNS, Algorithmes-Euclide-B, 2000 Route des Lucioles, B.P. 121, 06903, Sophia Antipolis, France
e-mail: zarzoso@i3s.unice.fr; meste@i3s.unice.fr

P. Comon
GIPSA-Lab - UMR 5216, Grenoble Campus, B.P. 46, 38402 St Martin d'Hères, France
e-mail: pierre.comon@grenoble-inp.fr

D.G. Latcu · N. Saoudi
Cardiology Department, Princess Grace Hospital, 1 avenue Pasteur, B.P. 489, 98012, Monaco Cedex, France
e-mail: dg.latcu@chpg.mc; nsaoudi@chpg.mc

F. Cazals and P. Kornprobst (eds.), *Modeling in Computational Biology and Biomedicine*, DOI 10.1007/978-3-642-31208-3_3, © Springer-Verlag Berlin Heidelberg 2013

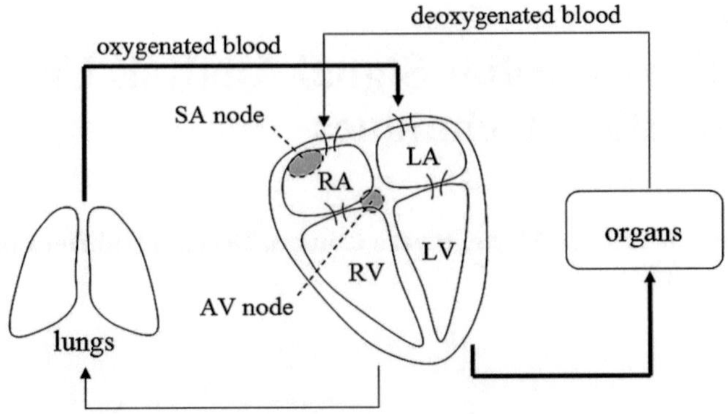

Fig. 3.1 Schematic depiction of the heart and the blood circulation system. *Thick* and *thin solid lines* represent oxygenated and deoxygenated blood, respectively. *RA* right atrium, *LA* left atrium, *RV* right ventricle, *LV* left ventricle, *SA* sino-atrial, *AV* atrio-ventricular

In resting conditions, myocardial cells are polarized, i.e., a potential difference exists across the cellular membrane due to a difference in concentrations of positive and negative ions. A beat is initiated by an electrical excitation generated at the sino-atrial node, the natural pacemaker of the heart, located on the top of the right atrium. As a result of this excitation, a depolarization wavefront propagates from the top to the bottom of the atria, causing a synchronized atrial contraction and the effective passage of blood towards the ventricles through the mitral and tricuspid valves. The depolarization wavefront is delayed at the atrio-ventricular node to allow an effective blood flow from atria to ventricles, and then rapidly propagates through the bundle of His, the bundle branches and the Purkinje fibers towards the bottom of the ventricular myocardium. Finally, the electrical activation propagates from the apex to the base of the ventricles, making them contract upwards and eject the blood through the pulmonary and aortic valves towards the lungs and the rest of the body. The beating process just described corresponds to a healthy electrophysiological state and is referred to as *normal sinus rhythm.*

The electrical activity generated in the heart is observable on the body surface. The electrocardiogram (ECG) records the cardiac electric potentials as measured on standard locations of the skin resulting in 12 leads: leads I, II and III are potential differences between electrodes on the wrists and the left foot; the augmented bipolar leads aVR, aVL and aVF use the same electrodes but take a virtual electrode (the so-called Wilson's central terminal) as a reference; and the precordial leads V1 to V6 are derived from electrodes across the chest (Fig. 3.2). In physiological conditions (sinus rhythm), the coordinated atrial depolarization and contraction is reflected as the P wave in the surface ECG. The atrio-ventricular pause is reflected as a an isoelectric line. Ventricular depolarization (linked to mechanical contraction) and repolarization (linked to mechanical relaxation) are observed as the QRS complex and the T-wave, respectively, which is followed by another isoelectric line until the next cardiac beat. Figure 3.3 illustrates the main inflections of the ECG signal in

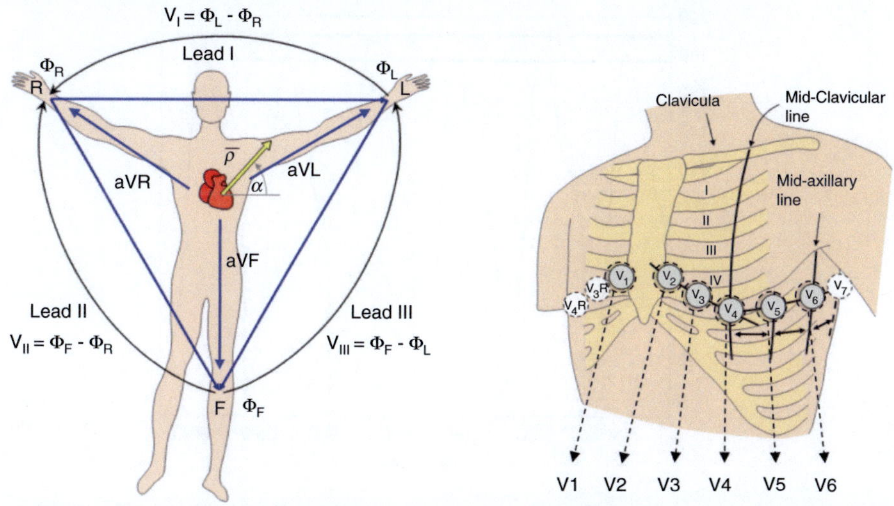

Fig. 3.2 Electrode placement in the standard 12-lead ECG (modified from [14] with permission). Symbol Φ denotes the electric potential measured on the body surface

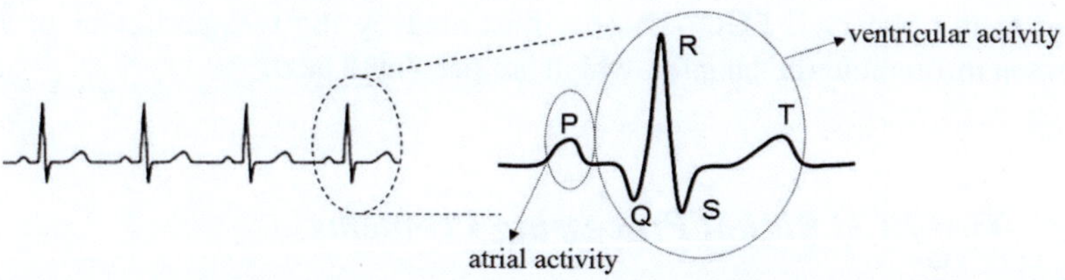

Fig. 3.3 The quasi-repetitive P-QRST complex observed in the ECG of a healthy subject (normal sinus rhythm). The P wave, the QRS complex and the T wave are associated with atrial depolarization, ventricular depolarization and ventricular repolarization, respectively

sinus rhythm conditions. In a given healthy subject, the shape and timing of these waves and complexes vary across leads.

Since its invention in the Nineteenth century, the ECG has become a major tool for the diagnosis and analysis of cardiac disturbances. The ECG conveys information about the patient's heart in a noninvasive manner, with the consequent benefits in procedural time, cost and risk of complications relative to invasive diagnostic modalities such as endocardial recordings. Deviations from the typical shape and timing of its waves and complexes are linked to specific cardiac conditions. While some of these variations can be perceived by the naked eye, others are more subtle and not always apparent even to the specialist, especially when several leads may be providing apparently inconsistent information. In such cases, one can no longer rely on human perception alone, and signal processing techniques become necessary to reveal and quantify underlying structures that remain otherwise

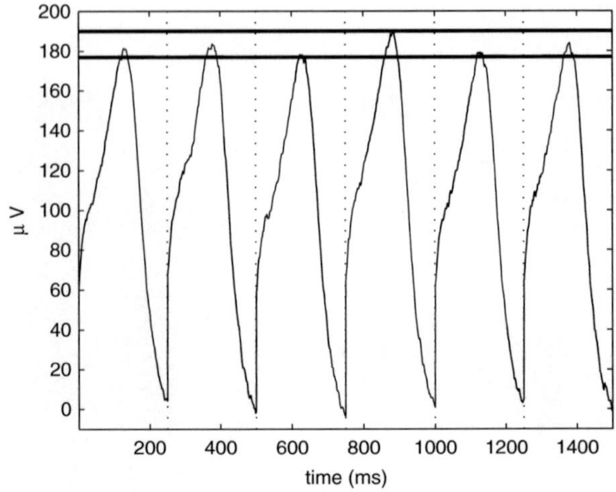

Fig. 3.4 Six consecutive segmented T waves, concatenated to aid visualization. The alternans phenomenon is visible in the horizontal slab delimited by the two *horizontal bold lines*

hidden in the observed ECG. This is illustrated by the two particular problems addressed throughout the chapter, which are presented next.

3.1.2 Two ECG Signal Processing Problems

3.1.2.1 T-Wave Alternans Detection

As introduced in the previous section, the T wave corresponds to the ventricular repolarization stage of the cardiac cycle. The study of this cardiac wave is often included in the global analysis for risk stratification of cardiac patients. The reason is that deviations from a normal T-wave template is often linked to future arrhythmia such as ventricular fibrillation, a serious heart condition leading to sudden cardiac death. In the T-wave alternans (TWA) phenomenon, the amplitude of consecutive T waves can vary as little as a few μV. As an example, six successive T waves observed in a particular ECG lead (data from the STAFF III database [15]) are concatenated in Fig. 3.4, where the alternans effect is visible between the two horizontal lines: the alternans amplitude is of the order of 10 μV here. Since these subtle variations are not always apparent to the specialist, specific ECG signal processing techniques are required for a reliable automatic detection and quantitative estimation of the alternans phenomenon.

3.1.2.2 Atrial Activity Extraction in Atrial Fibrillation

Atrial fibrillation (AF) is the sustained cardiac arrhythmia most often encountered in clinical practice. Its incidence increases with age, and about 10 % of the population over 70 years old is affected by the condition; this amounts to more than six million people in Europe and the United States alone. AF accounts for about two thirds of hospitalizations related to cardiac disturbances, and its cost to health-care providers is estimated to around 3,000 euros per patient per year in the Western world. AF patients usually complain of relatively minor symptoms like palpitations, tiredness, dizziness or lack of breath, but the condition can also entail more serious complications such as thrombo-embolic disorders. Indeed, AF is held responsible for up to 20 % of cerebral vascular accidents (brain strokes), which often lead to major impairment and even death [11].

Despite its incidence and risks of serious complications, the causes of AF are not yet fully understood. Ectopic foci located around the pulmonary veins and reentrant electrical pathways in the atrial myocardium due to the progressive shortening of the atrial myocytes' refractory period are two of the most generally acknowledged mechanisms behind the genesis and perpetuation of the disease. These mechanisms result in several electrical depolarization wavefronts propagating in a disorganized manner across the atria. This phenomenon is illustrated in Fig. 3.5a, which shows a 5-s segment recorded by a catheter-tip electrode on the atrial wall of a persistent AF patient (data recorded at the Cardiology Department of Princess Grace Hospital, Monaco). Such a catheter is typically inserted under sedation or general anesthesia through the patient's femoral vein up to the right atrium prior to the ablation of atrial tissue with radiofrequency energy [12]. Whereas endocardial bipolar recordings in sinus rhythm are characterized by very short impulses synchronized with the ventricular rate (Sect. 3.1.1), local activations during AF spread out in time and are no longer synchronized with the heartbeat, represented by the QRS complexes in the ECG plot of Fig. 3.5b. This disorganized electrical activation causes an inefficient atrial contraction, so that the atria are said to quiver or fibrillate rather than beat. As a result, blood ejection towards the ventricles is incomplete and stagnation in the atria can easily generate thrombi.

Compared with endocardial recordings, the ECG presents the obvious advantage of not requiring a catheter, and is thus a much more efficient procedure in terms of time and cost, with practically no complications for the patient. Figure 3.5b shows the surface ECG signal recorded simultaneously in lead V1 from our persistent AF patient. With an electrode very close to the right atrium (Fig. 3.2), this lead records atrial activity quite clearly. AF is characterized by the absence of P wave and the presence of fibrillatory waves, or f waves, at about 200–500 cycles per minute (around 3–9 Hz). The fibrillatory waves are visible in the TQ segments between consecutive beats, as illustrated by the plot. This atrial activity signal is of particular interest to cardiologists as it can provide useful noninvasive information about the state and evolution of the disease. For instance, the dominant atrial frequency (the inverse of the atrial cycle length) has been shown to correlate with the atrial myocytes' refractory period and the probability of spontaneous cardioversion [2]. Unfortunately, the atrial signal is masked in both time and frequency domains by the

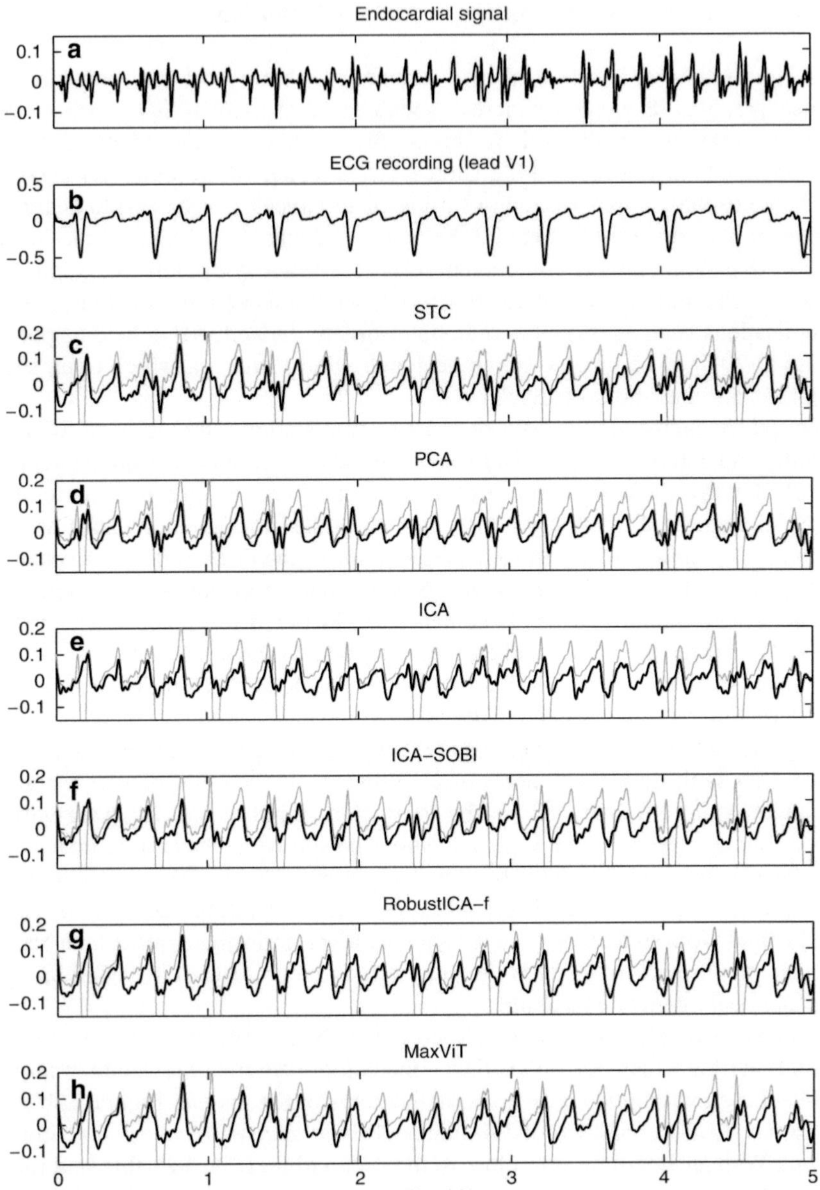

Fig. 3.5 Estimation of the atrial activity signal in the surface ECG during AF. (**a**) Endocardial signal recorded from the atrial wall (*left atrial appendage*) of a persistent AF patient. (**b**) Surface ECG signal recorded simultaneously in lead V1 from the same patient. (**c**) Atrial activity signal estimated in lead V1 by the spatio-temporal QRST cancellation (STC) method of [17, 27]. (**d**)–(**h**) Atrial activity signal estimated in lead V1 by the BSS-based methods described in Sects. 3.3.3–3.3.4. For reference, *light grey lines* show the signal recorded in lead V1. Signal amplitudes are in mV. The corresponding frequency spectra are shown in Fig. 3.6

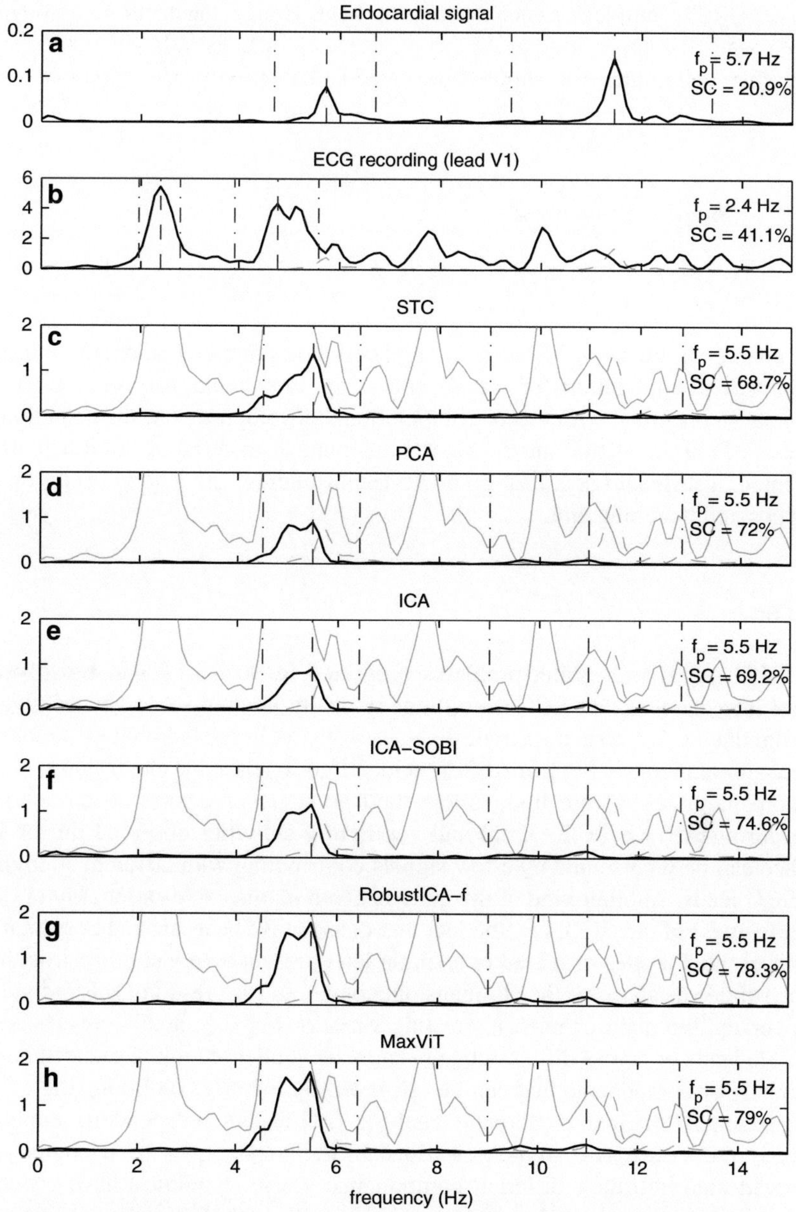

Fig. 3.6 Estimation of the atrial activity signal in the surface ECG during AF. The plots show the frequency spectra of the signals whose time courses are displayed in Fig. 3.5. Improved atrial signal estimates are typically associated with higher spectral concentration (SC) values around the dominant frequency (f_p) and its harmonics. For reference, the endocardial and lead V1 signal spectra are plotted in *light-grey dashed* and *solid lines*, respectively. Endocardial spectra in the background have been magnified ($\times 10$) along the vertical axis to aid visualization. Otherwise, power spectral densities are represented in the same linear scale and only relative amplitudes are important

stronger QRST complex at each ventricular beat. Hence, the accurate noninvasive analysis and characterization of AF from surface recordings calls for suitable signal processing techniques for suppressing ventricular activity and enhancing atrial activity.

3.1.3 Chapter Overview

Public

This chapter is mainly addressed to engineers and computer scientists interested in biomedical applications of signal processing, but should also be of interest to physicians concerned with recent computational developments in the detection and analysis of cardiac disturbances. Most mathematical material covered hereafter is presented in sufficient detail for graduate-level students with a good background in statistics and linear algebra.

Outline

The ECG signal processing problems presented in Sect. 3.1.2 can be solved by taking into account that ECG recordings typically contain some sort of diversity or redundancy. A classical form of diversity consists in the fact that different leads record simultaneously but from different locations in space the electrophysiological phenomena under examination, so that the lead signals are correlated. Such is the case during AF, where the atrial and ventricular activities observed on the body surface can be represented by a few signals contributing with different strengths to the ECG leads. Another kind of diversity is given by the cyclostationarity or quasi-repetitiveness of the ECG, i.e., the fact that consecutive beats are rather similar. For instance, the T waves observed in a single lead present a very similar structure up to the subtle variations of the alternans phenomenon. As a result of this redundancy or diversity, the main or most interesting variations of the signal—as observed in multiple leads or across different beats—may be condensed into a reduced number of signals or components that can be estimated by linearly combining the original observations. Principal component analysis (PCA) and independent component analysis (ICA) are two popular well-established techniques to decompose the observed data into their different components. These techniques have been used with success in the processing of real-world signals, including electrophysiological data [27], [9, Chap. 18]. The present chapter shows how the two ECG signal analysis problems introduced in Sect. 3.1.2 can be effectively modeled and tackled by such data decomposition techniques. The chapter is structured as follows. Section 3.2 deals with the problem of TWA detection and develops a solution using PCA. The characterization of AF from the surface ECG is the topic of Sect. 3.3, which also presents ICA as a tool to overcome the limitations of PCA in this context. The summary and concluding remarks of Sect. 3.4 bring the chapter to an end.

Notations

All quantities considered in this chapter belong to the field of real numbers, which is denoted by $\mathcal{R}$. Scalars, vectors and matrices are represented by lightface (x, X), boldface lowercase ($\mathbf{x}$) and boldface uppercase ($\mathbf{X}$) characters, respectively; $[\mathbf{X}]_{ij}$ denotes the (i, j)-entry of matrix $\mathbf{X}$. Symbol $\mathrm{E}\{\cdot\}$ stands for the mathematical expectation, and $(\cdot)^{\mathrm{T}}$ for the matrix transpose operator.

3.2 T-Wave Alternans Detection via Principal Component Analysis

This section analyzes the TWA phenomenon using PCA. The model definition proposed in Sect. 3.2.1 points to PCA as a good candidate for the estimation of the model parameters. The basic concepts behind this statistical technique are recalled in Sect. 3.2.2, allowing us to develop PCA-based solutions to TWA detection in Sect. 3.2.3. Finally, Sect. 3.2.4 reports some illustrative experimental results.

3.2.1 Mathematical Modeling of T-Wave Alternans

First, we assume that an ECG is recorded at the body surface and that the best lead for the observation of TWA is selected. An initial detection stage provides the time epochs of the R wave locations and consequently allows the T-wave segmentation. We assume each T wave appears at a fixed delay after its corresponding R wave. Let t_i denote the time index of the ith T wave and L the number of samples composing each T wave, assumed to be constant. If we refer to the signal recorded at the lead of interest as $x(t)$, the L samples of the ith T wave can be stored in vector $\mathbf{x}_i$ as follows:

$$\mathbf{x}_i = [x(t_i), x(t_i + T_s), \ldots, x(t_i + (L-1)T_s)]^{\mathrm{T}} \in \mathcal{R}^L,$$

where T_s denotes the sampling period. The concatenated T waves plotted in Fig. 3.4 correspond to indices $330 \leq i \leq 335$ in the dataset used as an example. As introduced in Sect. 3.1.2.1, the alternans effect is visible between the two horizontal bold lines displayed for reference.

Now, we can put forward the following model:

$$\mathbf{x}_i = \alpha_i \big(\mathbf{v}_1 + a(-1)^i \mathbf{v}_2\big) + \beta_i \mathbb{1} + \mathbf{n}_i. \tag{3.1}$$

The key idea underlying this model is that each observed T wave can be decomposed as the sum of three components, namely, the average T wave, the alternans wave and a constant wave, represented by the L-dimensional vectors $\mathbf{v}_1$, $\mathbf{v}_2$ and $\mathbb{1}$, respectively. These components are linearly weighted by parameters α_i and β_i, which stand for a scaling coefficient and the vertical offset (mean value), respectively, before adding up to yield the ith observed T wave, corrupted by the additive noise represented by vector $\mathbf{n}_i$. The binary value 0 or 1 for the a variable will allow us to detect and classify alternans episodes. This model accounts for a baseline component [4] that is assumed to be constant in the T-wave interval. The scaling coefficient α_i represents the modulation of the ECG signal during the recording, which can be due to respiration movements or recording conditions. According to model (3.1) the alternans amplitude lies in the range 0–10 μV for the particular example of Fig. 3.4. The alternans phenomenon can be characterized by estimating the model parameters in Eq. (3.1) from the observed T-wave data. We will see next how to do so by using PCA.

3.2.2 *Principal Component Analysis*

Principal component analysis (PCA) is a classical statistical technique for decomposing multivariate observations into uncorrelated variables known as *principal components*. A recent review of its application to ECG signal processing can be found in [5]. Chapters 5 and 7 of this book apply PCA to other biomedical data. The concepts recalled in this section will be useful not only in the design of PCA-based TWA detectors, but also in the derivation of ICA techniques later in Sect. 3.3. We will assume throughout this section that the data to be analyzed are represented by an L-dimensional random vector $\mathbf{x} \in \mathcal{R}^L$. The following mathematical formulations of this statistical technique focus on the ensemble data covariance matrix, defined as

$$\mathbf{R_x} = \mathrm{E}\{\mathbf{x}\mathbf{x}^{\mathrm{T}}\}, \tag{3.2}$$

where the data are assumed to have zero mean, $\mathrm{E}\{\mathbf{x}\} = \mathbf{0}$. The rank of this matrix yields the dimension of the subspace of $\mathcal{R}^L$ actually spanned by the observed data. As will be seen next, the data covariance matrix plays a central role in PCA. The practical setting where several realizations (typically time samples) of $\mathbf{x}$ are observed will be discussed in Sect. 3.2.2.4.

3.2.2.1 PCA as a Variance Maximization Decomposition

Among the various mathematical formulations of PCA, one of the most intuitive is perhaps in terms of linear combinations or projections with maximum variance. Let us consider a one-dimensional (scalar) signal or component $z \in \mathcal{R}$ derived as a linear combination of the observed data $\mathbf{x} \in \mathcal{R}^L$:

$$z = \mathbf{w}^{\mathrm{T}}\mathbf{x}. \tag{3.3}$$

Then, PCA can be defined as follows:

1. The first principal direction $\mathbf{w}_1$ is the unit-norm vector maximizing the variance or power of (3.3) as measured by the function:

$$\Psi_{\mathrm{PCA}}(\mathbf{w}) = \mathrm{E}\{z^2\} = \mathbf{w}^{\mathrm{T}}\mathbf{R}_{\mathbf{x}}\mathbf{w}. \tag{3.4}$$

2. The second principal direction $\mathbf{w}_2$ is the unit-norm maximizer of criterion (3.4) lying orthogonal to the first principal direction $\mathbf{w}_1$, i.e., $\mathbf{w}_2^{\mathrm{T}}\mathbf{w}_1 = 0$.

$\vdots$

k In general, the kth principal direction $\mathbf{w}_k \in \mathcal{R}^L$ is the unit-norm maximizer of criterion (3.4) lying orthogonal to the previous principal directions $\{\mathbf{w}_j\}_{j=1}^{k-1}$, i.e., $\mathbf{w}_k^{\mathrm{T}}\mathbf{w}_j = 0$, for $j < k$.

The principal components $\{z_1, z_2, \ldots, z_k\}$, are obtained by replacing $\mathbf{w}$ with the corresponding principal directions in Eq. (3.3). Simple algebraic manipulations show that the kth principal direction $\mathbf{w}_k$ is the kth dominant eigenvector of the data covariance matrix $\mathbf{R}_{\mathbf{x}}$ defined in Eq. (3.2). Let

$$\mathbf{R}_{\mathbf{x}} = \mathbf{U}\mathbf{D}\mathbf{U}^{\mathrm{T}} \tag{3.5}$$

denote its eigenvalue decomposition (EVD), where the columns of unitary matrix $\mathbf{U} = [\mathbf{u}_1, \mathbf{u}_2, \ldots, \mathbf{u}_L] \in \mathcal{R}^{L\times L}$ contain the eigenvectors and diagonal matrix $\mathbf{D} = \mathrm{diag}(\lambda_1, \lambda_2, \ldots, \lambda_L) \in \mathcal{R}^{L\times L}$ stores the eigenvalues arranged in decreasing order. Then the principal directions are found in the columns of $\mathbf{U}$ and the principal components $\mathbf{z} = [z_1, z_2, \ldots, z_L]^{\mathrm{T}} \in \mathcal{R}^L$ can be computed as

$$\mathbf{z} = \mathbf{U}^{\mathrm{T}}\mathbf{x}. \tag{3.6}$$

Thus, according to this decomposition, the original data are expressed as the product of unitary matrix $\mathbf{U}$ of principal directions and vector $\mathbf{z}$ of principal components with decreasing variance:

$$\mathbf{x} = \mathbf{U}\mathbf{z}.$$

3.2.2.2 PCA as a Decorrelating Transform

Eqs. (3.5)–(3.6) prove that the covariance matrix of the principal components is diagonal:

$$\mathbf{R}_{\mathbf{z}} = \mathrm{E}\{\mathbf{z}\mathbf{z}^{\mathrm{T}}\} = \mathbf{U}^{\mathrm{T}}\mathbf{R}_{\mathbf{x}}\mathbf{U} = \mathbf{D}.$$

As a result, the principal components are uncorrelated:

$$\mathrm{E}\{z_i z_j\} = [\mathbf{R}_{\mathbf{z}}]_{ij} = 0, \qquad \text{for } i \neq j.$$

Hence, PCA can be considered as a decorrelating transform, whereby the original data are transformed into uncorrelated components. Because of their decorrelation, the principal components are less redundant than the original observations. In Sect. 3.3 we will see that ICA extends this statistical independence property to orders higher than two.

3.2.2.3 PCA as a Data Compression Technique

Another interesting property of PCA, closely related to the previous one, is its data reduction or approximation capabilities., i.e., its ability to construct simple representations of the available data with reduced dimensionality. Indeed, PCA solves an important problem in signal processing that can be stated as follows. Let us consider the linear component of Eq. (3.3), where $\mathbf{w} \in \mathcal{R}^L$ is an unknown vector of linear combination coefficients. The best approximation of this signal to the original data can be computed by minimizing the mean square error (MSE)

$$\Psi_{\mathrm{MSE}}(\mathbf{w}, \mathbf{h}) = \mathrm{E}\{\|\mathbf{x} - \mathbf{h}z\|^2\}, \tag{3.7}$$

where $\mathbf{h} \in \mathcal{R}^L$ is an unknown vector allowing the projection of z back onto the original L-dimensional observation space and $\|\cdot\|$ denotes the ℓ_2-norm of its vector argument. Note that Ψ_{MSE} is also a function of vector $\mathbf{w}$ through relationship (3.3). To find the optimal values of $\mathbf{w}$ and $\mathbf{h}$, we must cancel the gradient of Eq. (3.7) with respect to both vectors, leading to the equalities

$$\mathbf{w} = \mathbf{h}/\|\mathbf{h}\|^2, \tag{3.8}$$

$$\mathbf{h} = \mathbf{R_x w}/(\mathbf{w}^{\mathrm{T}}\mathbf{R_x w}), \tag{3.9}$$

where we have assumed that $\mathbf{R_x}$ is full rank. We set $\|\mathbf{w}\| = 1$ to fix the scale ambiguity in Eq. (3.7), since a scale factor can be exchanged between $\mathbf{w}$ and $\mathbf{h}$ without altering the MSE. Combining Eqs. (3.8) and (3.9) proves that the optimal $\mathbf{w}$ and $\mathbf{h}$ are identical, and equal to the dominant eigenvector of $\mathbf{R_x}$, i.e., $\mathbf{w} = \mathbf{h} = \mathbf{u}_1$. As we have seen in Sect. 3.2.2.1, this eigenvector is also the dominant principal direction of the observed data, $\mathbf{w}_1$, so that signal z in Eq. (3.3) turns out to be the dominant principal component of $\mathbf{x}$, i.e., the entry z_1 of vector $\mathbf{z}$ in Eq. (3.6). Additional algebraic manipulations show that the MSE of this optimal rank-1 approximation is $\sum_{i=2}^{L} \lambda_i$, where λ_i denotes the ith eigenvalue of $\mathbf{R_x}$. The second principal direction $\mathbf{w}_2$ is computed by repeating the above procedure on the error vector or residual data $(\mathbf{x} - \mathbf{w}_1 z_1)$, and so forth. This minimum-error derivation of PCA proves, as a by-product, that the second principal direction $\mathbf{w}_2$ must lie orthogonal to $\mathbf{w}_1$, since the error vector is indeed orthogonal to $\mathbf{w}_1$:

$$\mathbf{w}_1^{\mathrm{T}}(\mathbf{x} - \mathbf{w}_1 z_1) = \mathbf{w}_1^{\mathrm{T}}(\mathbf{x} - \mathbf{w}_1\mathbf{w}_1^{\mathrm{T}}\mathbf{x}) = \mathbf{w}_1^{\mathrm{T}}\mathbf{x} - \|\mathbf{w}_1\|^2\mathbf{w}_1^{\mathrm{T}}\mathbf{x} = 0.$$

The minimum-variance derivation of the previous section imposed this orthogonality property to avoid extracting the same principal component twice.

We have just proved an important result that can be generalized as follows: the orthogonal projection of the data onto their first k principal directions ($k < L$), given by $\mathbf{U}_k\mathbf{U}_k^{\mathrm{T}}\mathbf{x}$, with $\mathbf{U}_k = [\mathbf{u}_1, \mathbf{u}_2, \ldots, \mathbf{u}_k]$, is the rank-$k$ approximation best resembling the data in the MSE sense; its minimum MSE is given by $\sum_{i=k+1}^{L} \lambda_i$. The original data are thus approximated by the k principal directions and the realizations of the corresponding principal components. If N data realizations are available, this approximation reduces the storage requirements from $O(LN)$ to $O(kN)$ in order of magnitude. The popularity of PCA as a data compression technique hinges upon this property. In particular, discarding principal components associated with negligible eigenvalues allows us to reduce the dimensionality of the data with little loss of information. No information loss is actually incurred when the covariance matrix is rank-deficient with $(L - k)$ null eigenvalues, as the L-dimensional data lie in a subspace of dimension $k < L$ in that case.

3.2.2.4 PCA in Practice

As we have seen in the preceding sections, PCA relies on the computation of the data covariance matrix $\mathbf{R}_{\mathbf{x}}$. In practice, N data samples or realizations $\{\mathbf{x}_n\}_{n=1}^{N}$ are observed, which can be stored in the observation matrix $\mathbf{X} = [\mathbf{x}_1, \mathbf{x}_2, \ldots, \mathbf{x}_N] \in \mathcal{R}^{L\times N}$. Then, the covariance matrix can be estimated from the available data by sample averaging:

$$\hat{\mathbf{R}}_{\mathbf{x}} = \frac{1}{N}\sum_{n=1}^{N} \mathbf{x}_n\mathbf{x}_n^{\mathrm{T}} = \frac{1}{N}\mathbf{X}\mathbf{X}^{\mathrm{T}}. \tag{3.10}$$

PCA can then be obtained from the EVD of this covariance matrix estimate. Nevertheless, a computationally more robust yet theoretically equivalent alternative to compute PCA is the singular value decomposition (SVD) of the observed data matrix

$$\mathbf{X} = \hat{\mathbf{U}}\hat{\mathbf{D}}\mathbf{V}^{\mathrm{T}}, \tag{3.11}$$

where the columns of unitary matrices $\hat{\mathbf{U}}$ and $\mathbf{V}$ contain, respectively, the left and right singular vectors, and diagonal matrix $\hat{\mathbf{D}}$ contains the singular values of $\mathbf{X}$. Plugging Eq. (3.11) into Eq. (3.10) and comparing the result with Eq. (3.5), it turns out that the left singular matrix $\hat{\mathbf{U}}$ provides an estimate of the principal directions whereas the singular-value matrix $\hat{\mathbf{D}}$ is an estimate of $\sqrt{N}\mathbf{D}^{\frac{1}{2}}$. According to Eq. (3.6) the realizations of the principal components are stored in the columns matrix $\mathbf{Z} = \hat{\mathbf{D}}\mathbf{V}^{\mathrm{T}}$. Algorithm 1 summarizes how to compute the SVD-based PCA.

3.2.3 PCA-Based Solution to T-Wave Alternans Detection

Now let us come back to the TWA detection problem. Although estimating the parameters in model (3.1) is directly feasible [16], we can relax some constraints by

Algorithm 1 PCA based on the SVD of the observed data matrix

1: Store the N observed data samples into matrix $\mathbf{X} = [\mathbf{x}_1, \mathbf{x}_2, \ldots, \mathbf{x}_N]$.
2: Compute the SVD of the data matrix as $\mathbf{X} = \hat{\mathbf{U}}\hat{\mathbf{D}}\mathbf{V}^{\mathrm{T}}$.
3: Recover the principal directions in the columns of matrix $\hat{\mathbf{U}}$.
4: Recover the principal component samples in the columns of matrix $\mathbf{Z} = \hat{\mathbf{D}}\mathbf{V}^{\mathrm{T}}$.

Dimensionality reduction can be performed by discarding the singular vectors associated with negligible singular values in Steps 3–4 (see Sect. 3.2.2.3 for details).

using an observation model well adapted to the PCA approach. Indeed, model (3.1) can be written as

$$\mathbf{x}_i = \mathbf{M}\boldsymbol{\theta}_i + \mathbf{n}_i, \tag{3.12}$$

where $\mathbf{M} = [\mathbf{v}_1, \mathbf{v}_2, \mathbb{1}]$ and $\boldsymbol{\theta}_i = [\alpha_i, \alpha_i a(-1)^i, \beta_i]^{\mathrm{T}}$. Thus, the information contained in the current T wave is summarized by a few parameters, represented by vector $\boldsymbol{\theta}_i$, using some global knowledge, condensed into matrix $\mathbf{M}$, over the total amount of data. In general, the columns of $\mathbf{M}$ are not orthogonal, i.e., $\mathbf{M}^{\mathrm{T}}\mathbf{M}$ is not a diagonal matrix, because there is no evidence that vectors $\mathbf{v}_1$, $\mathbf{v}_2$ and $\mathbb{1}$ are mutually orthogonal. As a result, the principal directions of $\mathbf{x}_i$ are unlikely to coincide with these vectors.

To overcome this problem, we will first remove the contribution of the offset vector from the original T-wave data by making use of the dimensionality reduction capabilities of PCA recalled in Sect. 3.2.2.3. To this end, we minimize the MSE (3.7) but fixing the projection vector $\mathbf{h} = \mathbb{1}$. This leads to the minimization of function $\mathrm{E}\{\|\mathbf{x}_i - \mathbb{1}z\|^2\}$ with respect to the linear combination vector $\mathbf{w}$, with $z = \mathbf{w}^{\mathrm{T}}\mathbf{x}_i$. From Eqs. (3.7) and (3.8), it turns out that $\mathbf{w} = \mathbb{1}/\|\mathbb{1}\|^2$ and the offset-corrected data are thus given by

$$\tilde{\mathbf{x}}_i = \mathbf{x}_i - \frac{\mathbb{1}\mathbb{1}^{\mathrm{T}}}{\|\mathbb{1}\|^2}\mathbf{x}_i = \left(\mathbf{I} - \frac{\mathbb{1}\mathbb{1}^{\mathrm{T}}}{\|\mathbb{1}\|^2}\right)\mathbf{x}_i. \tag{3.13}$$

This is the orthogonal projection of the original T-wave data $\mathbf{x}_i$ onto the orthogonal complement of vector $\mathbb{1}$, as could be intuitively expected. Now the overall variance of the offset-corrected observations is mainly due to the scaled T and alternans waves only, so we can write

$$\tilde{\mathbf{x}}_i = \tilde{\mathbf{M}}\tilde{\boldsymbol{\theta}}_i + \tilde{\mathbf{n}}_i, \tag{3.14}$$

where $\tilde{\mathbf{M}} = [\mathbf{v}_1, \mathbf{v}_2]$, $\tilde{\boldsymbol{\theta}}_i = [\alpha_i, \alpha_i a(-1)^i]^{\mathrm{T}}$, and $\tilde{\mathbf{n}}_i$ is the noise term $\mathbf{n}_i$ in Eq. (3.12) after projection (3.13). As a result, PCA should now produce dominant principal directions $\mathbf{u}_1$ and $\mathbf{u}_2$ related to $\mathbf{v}_1$ and $\mathbf{v}_2$, so that vector $\tilde{\boldsymbol{\theta}}_i$ can be estimated in the least square sense by

$$\tilde{\boldsymbol{\theta}}_i = \tilde{\mathbf{M}}^{\mathrm{T}}\tilde{\mathbf{x}}_i. \tag{3.15}$$

Assuming that the alternans effect is now condensed in $\tilde{\boldsymbol{\theta}}_i$, the detection is performed over this set of data. An alternative but equivalent development is given

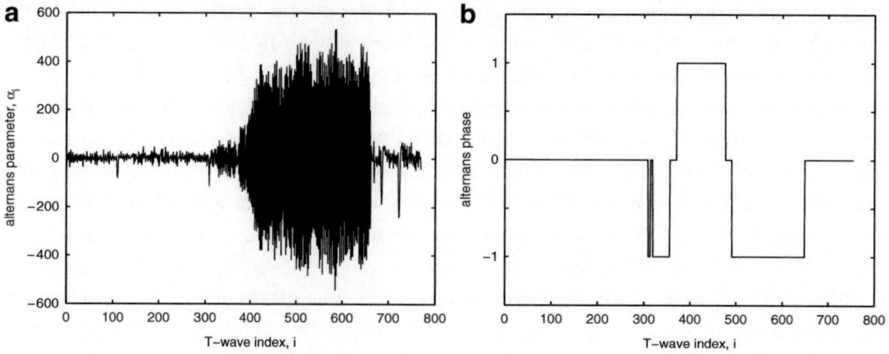

Fig. 3.7 TWA detection during angioplasty. (**a**) First component of vector $\boldsymbol{\theta}_i$ in model (3.12) estimated by the PCA-based method of Sect. 3.2.3. The alternans of the parameter appears in the range $300 \le i \le 650$. (**b**) Welch's t test output for $p < 0.01$ from the PCA results displayed in (**a**). Changes in alternans phase $(0, 1, -1)$ are clearly visible

in [16] by considering the matrix made up of the N available T-wave realizations: $\mathbf{X} = [\mathbf{x}_1, \mathbf{x}_2, \ldots, \mathbf{x}_N] \in \mathcal{R}^{L \times N}$. The rank-$k$ subspace best describing in the least square sense the observation space, i.e., the subspace spanned by the columns of $\mathbf{X}$, is spanned by the columns of any given rank-ℓ matrix $\mathbf{A}$ and the $(k-\ell)$ dominant left singular vectors of matrix $\left(\mathbf{I} - \mathbf{A}(\mathbf{A}^{\mathrm{T}}\mathbf{A})^{-1}\mathbf{A}^{\mathrm{T}}\right)\mathbf{X}$. Fixing $k = 3$ and $\mathbf{A} = \mathbb{1}$, this result yields the sample version of the above derivations. To enhance the varying nature of the TWA phenomenon, the data can be transformed according to the derivative operator $\delta\mathbf{x}_i = (\mathbf{x}_i - \mathbf{x}_{i-1})$ before applying the detection method.

The PCA-based approach described in this section is valid for any extended model of observation disturbances. For instance, a detrending step can also be applied as in [17], where a more refined model accounting for an affine baseline wandering has also been proposed in the context of noninvasive atrial activity extraction during AF using a similar framework.

3.2.4 *Success Story: T-Wave Alternans Detection During Angioplasty*

The TWA detection approach described in the previous section has been applied to real data acquired during angioplasty (STAFF III database) [15]. Angioplasty consists in the inflation of a balloon in the coronary artery to treat arterial stenosis. This procedure temporarily provokes ischemia and induces changes in cardiac cell properties that may cause TWA. Figure 3.7a displays the first entry of vector $\boldsymbol{\theta}_i$ in model (3.12) (or of vector $\tilde{\boldsymbol{\theta}}_i$ in Eqs. (3.14) and (3.15)) when using $\delta\mathbf{x}_i$ from the T-wave dataset.

The actual detection of the alternans effect can be carried out in different manners. In any case, a sliding window of fixed length is applied over the output of the PCA-based detector, i.e., the data displayed in Fig. 3.7a. For each position of the window, two ensembles are formed by splitting the odd and even samples. Their means are compared assuming normal distribution laws with unequal variances. The statistical significance (p value) of their difference is verified by means of a Welch's t test. For a window length of 16 beats, results are shown in Fig. 3.7b where the values 0, 1 and -1 refer, respectively, to no detection, detection of sequence "$+-+-+-$" and detection of sequence "$-+-+-+$" (phase change). After a short delay, the angioplasty procedure causes the alternans of the ventricular repolarization, as clearly detected by the PCA-based method. Interestingly, phase changes of the alternans sequence are present in this example. An alternative approach applies PCA over the detection window instead of on the entire set of T waves.

3.3 Atrial Activity Extraction via Independent Component Analysis

As presented in Sect. 3.1.2.2, the problem of noninvasive atrial activity extraction aims at suppressing the ventricular interference that masks the atrial signal observed in the ECG during AF episodes. Classical methods for solving this problem compute an average ventricular beat and subtract it from the recordings after suitable time alignment and amplitude scaling [17, 22, 27]. This average beat subtraction technique relies on the quasi-repetitiveness of the QRST complex and its uncorrelation with the atrial signal. The spatio-temporal QRST cancellation (STC) method of [17] (also described in [27]) belongs to this family. Using a signal model reminiscent of Eq. (3.12), each segmented QT interval is modeled as a linear combination of the average ventricular beats of every lead (or their dominant subspace computed via PCA) plus an additive noise term including atrial activity. An optimal linear combination of the average beats is then subtracted from each QT interval and each lead in the recording, canceling out the ventricular activity. Average beat subtraction requires a previous beat detection and classification stage, and thus its performance severely degrades in the presence of ectopic beats. By construction, this approach is unable to cancel interference other than the ventricular activity itself. Other classical techniques for atrial signal extraction are summarized in [25].

3.3.1 *Linear Mixture Model*

A recent alternative approach to atrial activity enhancement models the ECG lead signals, $\{x_i(t)\}_{i=1}^{L}$, as a linear superposition of contributions from signal components or *sources* originating in the atria, $\{s_j^{\mathrm{A}}(t)\}_{j=1}^{M_{\mathrm{A}}}$, and the ventricles, $\{s_j^{\mathrm{V}}(t)\}_{j=1}^{M_{\mathrm{V}}}$, as well as other sources of noise, interference and artifacts, $\{s_j^{\mathrm{N}}(t)\}_{j=1}^{M_{\mathrm{N}}}$, [20, 21]:

$$x_i(t) = \sum_{j=1}^{M_{\mathrm{A}}} h_{ij}^{\mathrm{A}} s_j^{\mathrm{A}}(t) + \sum_{j=1}^{M_{\mathrm{V}}} h_{ij}^{\mathrm{V}} s_j^{\mathrm{V}}(t) + \sum_{j=1}^{M_{\mathrm{N}}} h_{ij}^{\mathrm{N}} s_j^{\mathrm{N}}(t) \qquad i = 1, 2, \ldots, L. \tag{3.16}$$

With the aid of some additional notations, this model accepts a convenient matrix formulation. Let vectors

$$\mathbf{s}_{\mathrm{A}}(t) = [s_1^{\mathrm{A}}(t), s_2^{\mathrm{A}}(t), \ldots, s_{M_{\mathrm{A}}}^{\mathrm{A}}(t)]^{\mathrm{T}},$$
$$\mathbf{s}_{\mathrm{V}}(t) = [s_1^{\mathrm{V}}(t), s_2^{\mathrm{V}}(t), \ldots, s_{M_{\mathrm{V}}}^{\mathrm{V}}(t)]^{\mathrm{T}},$$
$$\mathbf{s}_{\mathrm{N}}(t) = [s_1^{\mathrm{N}}(t), s_2^{\mathrm{N}}(t), \ldots, s_{M_{\mathrm{N}}}^{\mathrm{N}}(t)]^{\mathrm{T}}$$

contain, respectively, the atrial, ventricular and noise sources. Let the linear superposition coefficients be stored in matrices $[\mathbf{H}_{\mathrm{A}}]_{ij} = h_{ij}^{\mathrm{A}}$, $1 \le i \le L$, $1 \le j \le M_{\mathrm{A}}$, $[\mathbf{H}_{\mathrm{V}}]_{ij} = h_{ij}^{\mathrm{V}}$, $1 \le i \le L$, $1 \le j \le M_{\mathrm{V}}$, and $[\mathbf{H}_{\mathrm{N}}]_{ij} = h_{ij}^{\mathrm{N}}$, $1 \le i \le L$, $1 \le j \le M_{\mathrm{N}}$. Finally, let the lead outputs be stacked in vector $\mathbf{x}(t) = [x_1(t), x_2(t), \ldots, x_L(t)]^{\mathrm{T}}$. According to these notations, model (3.16) can be compactly expressed as

$$\mathbf{x}(t) = [\mathbf{H}_{\mathrm{A}}, \mathbf{H}_{\mathrm{V}}, \mathbf{H}_{\mathrm{N}}] \begin{bmatrix} \mathbf{s}^{\mathrm{A}}(t) \\ \mathbf{s}^{\mathrm{V}}(t) \\ \mathbf{s}^{\mathrm{N}}(t) \end{bmatrix} = \mathbf{H}\mathbf{s}(t), \tag{3.17}$$

where $\mathbf{H} = [\mathbf{H}_{\mathrm{A}}, \mathbf{H}_{\mathrm{V}}, \mathbf{H}_{\mathrm{N}}] \in \mathcal{R}^{L \times M}$ and $\mathbf{s}(t) = [\mathbf{s}_{\mathrm{A}}(t)^{\mathrm{T}}, \mathbf{s}_{\mathrm{V}}(t)^{\mathrm{T}}, \mathbf{s}_{\mathrm{N}}(t)^{\mathrm{T}}]^{\mathrm{T}} \in \mathcal{R}^{M}$, with $M = M_{\mathrm{A}} + M_{\mathrm{V}} + M_{\mathrm{N}}$. Since most often vertical offsets do not convey any physiological information, the source signals are assumed to have zero mean. The mixing coefficients are determined by the relative location between sources and electrodes, and the propagation characteristics of the body as a conductive medium, which can be considered as purely resistive in the frequency range of interest [14]. Each mixing-matrix column represents the contribution of the corresponding source to the different electrodes and can be associated with the electric potential spatial distribution of that source on the body surface; hence, a mixing-matrix column is also known as the *spatial topography* of its respective source [3].

Now, if the sources $\mathbf{s}(t)$ and the mixing matrix $\mathbf{H}$ were available in model (3.17), the atrial contributions to the recordings could easily be computed free from ventricular activity and other disturbances by isolating $\mathbf{H}^{\mathrm{A}}$ and $\mathbf{s}^{\mathrm{A}}(t)$:

$$\hat{\mathbf{x}}_{\mathrm{A}}(t) = \mathbf{H}_{\mathrm{A}}\mathbf{s}_{\mathrm{A}}(t).$$

Unfortunately, neither the sources nor the mixing matrix are known in practice, and Eq. (3.17) defines an *inverse problem* referred to as blind source separation (BSS) in instantaneous linear mixtures [9]. Given the observed signals, BSS aims at estimating the source signals and the mixing matrix in model (3.17). The separation is achieved by transforming the observations via a separating matrix $\mathbf{W}$, ideally the inverse of the mixing matrix, $\mathbf{W} = \mathbf{H}^{-1}$. Inverse problems arising in confocal microscopy imaging and brain signal analysis are addressed in Chaps. 4 and 7, respectively, of this book.

The BSS model can be considered as a generalization of classical approaches such as adaptive noise canceling [24]. This earlier approach estimates the interference by suitably filtering some of the observations called reference signals. A crucial assumption for the success of the noise canceling technique is that the reference signals must be correlated with the interference but uncorrelated with the signal of interest. In the atrial activity extraction problem, this assumption constrains the electrode location, since the reference leads must be free of atrial contributions. By contrast, the BSS model is more flexible in that the contribution of each source to each observation (i.e., the coefficients h_{ij}) can be practically arbitrary provided the mixing matrix remains left invertible or full column rank, i.e., its columns are linearly independent. A necessary condition for the left invertibility of $\mathbf{H}$ is that $M \leq L$.

Apart from this requirement on the mixing matrix, additional assumptions are necessary to solve the BSS problem (3.17). These assumptions concern certain source properties that are exploited to perform the separation. During AF, atrial and ventricular activities can be assumed to arise from relatively uncoupled electrophysiological phenomena, since the atrial electrical wavefronts impinge on the atrio-ventricular node, thus generating a ventricular beat, in a rather random fashion (see Sect. 3.1.2.2; cf. the normal sinus activation described in Sect. 3.1.1). Hence, the atrial and ventricular sources present a certain degree of statistical independence. Depending on the degree of source independence assumed, BSS can be carried out by different approaches. The PCA technique reviewed in Sect. 3.2.2 can be considered as a BSS technique exploiting the source independence assumption up to second order (uncorrelation), whereas ICA exploits independence at orders higher than two. The remaining of the chapter will summarize these techniques as well as their advantages and limitations in the context of BSS.

Before continuing the exposition, remark that a scale factor can be interchanged between a source and its mixing-matrix column without modifying the observations nor the source independence. To fix this *scale ambiguity*, we can assume, without loss of generality, that the sources have unit variance, $\mathrm{E}\{s_i^2(t)\} = 1$, for $1 \leq i \leq M$, leading to an identity source covariance matrix: $\mathbf{R_s} = \mathrm{E}\{\mathbf{s}\mathbf{s}^{\mathrm{T}}\} = \mathbf{I}$. For analogous reasons, the exact source ordering cannot be determined without further information about the sources or the mixture; this is the so-called *permutation ambiguity*. The permutation ambiguity makes it difficult to target a specific source without separating all sources first, unless additional information about the source of interest is introduced in the separation method; this issue will be further discussed in Sect. 3.3.4. In the sequel, we will assume for simplicity that $L = M$ (square mixing) and the time index t will be dropped for convenience.

3.3.2 PCA Solution to BSS

Let us first examine whether the PCA method introduced in Sect. 3.2.2 is suitable for BSS and under which conditions it can recover the source and the mixing matrix in model (3.17). To guarantee the source identity covariance assumption justified by the scale ambiguity, the principal components in Eq. (3.6) must be scaled by their standard deviation to provide the PCA source estimate:

$$\hat{\mathbf{s}}_{\mathrm{PCA}} = \mathbf{D}^{-\frac{1}{2}}\mathbf{z} = \mathbf{D}^{-\frac{1}{2}}\mathbf{U}^{\mathrm{T}}\mathbf{x}. \tag{3.18}$$

Hence, the separating matrix yielding the estimated sources from the observations as $\hat{\mathbf{s}}_{\mathrm{PCA}} = \mathbf{W}_{\mathrm{PCA}}\mathbf{x}$ is given by $\mathbf{W}_{\mathrm{PCA}} = \mathbf{D}^{-\frac{1}{2}}\mathbf{U}^{\mathrm{T}}$, which amounts to the mixing matrix estimate:

$$\hat{\mathbf{H}}_{\mathrm{PCA}} = \mathbf{U}\mathbf{D}^{\frac{1}{2}}. \tag{3.19}$$

According to Sect. 3.2.2.4, the PCA estimates can also be computed from the SVD of the observed data matrix as

$$\hat{\mathbf{H}}_{\mathrm{PCA}} = \frac{1}{\sqrt{N}}\hat{\mathbf{U}}\hat{\mathbf{D}}, \qquad \hat{\mathbf{S}}_{\mathrm{PCA}} = \sqrt{N}\mathbf{V}^{\mathrm{T}}, \tag{3.20}$$

where the columns of $\hat{\mathbf{S}}_{\mathrm{PCA}}$ contain the N samples of $\hat{\mathbf{s}}_{\mathrm{PCA}}$ corresponding to the observations in $\mathbf{X}$.

In any case, it can be remarked that the columns of the estimated mixing matrix in Eqs. (3.19) and (3.20) are always orthogonal due to the orthogonality of the principal directions. As a result, PCA will be unable to perform the separation whenever the actual structure of $\mathbf{H}$ in model Eq. (3.17) violates this orthogonality constraint. In the problem of atrial activity extraction, forcing an orthogonal structure for $\mathbf{H}$ seems a difficult task due to the spatial proximity of the atrial and ventricular sources, and would most probably require a specific patient-dependent electrode placement. In general, since both $\hat{\mathbf{s}}_{\mathrm{PCA}}$ and $\mathbf{s}$ have an identity covariance matrix, they will be related through an unknown $(M \times M)$ orthonormal transformation:

$$\hat{\mathbf{s}}_{\mathrm{PCA}} = \mathbf{Q}\mathbf{s}. \tag{3.21}$$

The separation can thus be completed by finding $\mathbf{Q}$ and applying its transpose to $\hat{\mathbf{s}}_{\mathrm{PCA}}$. The covariance of $\hat{\mathbf{s}}_{\mathrm{PCA}}$ does not depend on $\mathbf{Q}$. Therefore, finding this matrix requires information other than the second-order statistics exploited by PCA.

As recalled in Sect. 3.2.2.2, PCA transforms the original data into uncorrelated components, i.e., into a random vector with diagonal covariance matrix. Hence, in the context of BSS, PCA tries to recover the sources by diagonalizing the observation covariance matrix, thus recovering the source diagonal covariance structure and the source independence at order 2 (uncorrelation). By analogy with whitening filters that whiten or flatten the frequency spectrum of a given signal by diagonalizing its correlation matrix, PCA can be seen as a *spatial whitening* operation, and the principal components or, rather, their scaled versions in $\hat{\mathbf{s}}_{\mathrm{PCA}}$, are often called *whitened signals*. As explained above, the mixing matrix linking the whitened signals with the sources reduces to the unitary transformation in Eq. (3.21). In consequence, even if PCA does not generally do the job, it does at least half of it in a computationally affordable manner, as it is based on second-order statistics and standard matrix decompositions such as the EVD or the SVD (Sects. 3.2.2.1 and 3.2.2.4).

3.3.3 Beyond PCA: ICA

We have just seen that if the mixing matrix reduces to an orthogonal matrix $\mathbf{Q}$, then the covariance of $\mathbf{x} = \mathbf{Qs}$ does not depend on the mixing matrix at all, and PCA fails to perform its identification. By contrast, independent component analysis (ICA), a statistical tool for transforming multivariate data into *independent* random variables [7], is able to identify any full column rank mixing matrix under rather general conditions summarized later in this section. Second-order statistics are not sufficient to account for statistical independence, as illustrated by the inability of PCA to perform the separation in the general case. Through the use of second-order statistics, PCA implicitly assumes that the principal components have Gaussian distributions and it yields indeed the maximum-likelihood estimate of the separating matrix for uncorrelated Gaussian sources in noiseless scenarios. Hence, ICA exploits, either explicitly or implicitly, deviations from Gaussianity. This can be done with the help of optimization criteria based on statistical tools such as entropy, mutual information, or cumulants, as described next.

3.3.3.1 Statistical Tools

A Gaussian probability density function is entirely characterized by its mean and variance, i.e., its moments of order 1 and 2 only. Hence, a simple intuitive way to measure deviations from Gaussianity is via moments of order higher than two. The rth-order moment of a real random variable z is defined as $\mu_{(r)} = \mathrm{E}\{z^r\}$. In the multivariate case, the set of second-order moments of random vector $\mathbf{z} \in \mathcal{R}^M$ can be stored in its covariance matrix, with elements $[\mathbf{R_z}]_{ij} = \mathrm{E}\{z_i z_j\}$, as defined in matrix form by Eq. (3.2). Similarly, the $(M \times M \times M \times M)$ array containing all fourth-order moments can be defined as: $\mu_{ijk\ell} = \mathrm{E}\{z_i z_j z_k z_\ell\}$. Yet if vector $\mathbf{z}$ is Gaussian then this moment can be expressed as a function of moments of order 1 and 2 only. If we assume for simplicity that $\mathbf{z}$ is zero-mean, then it can be shown that

$$\mu_{ijk\ell} = R_{ij}R_{k\ell} + R_{ik}R_{j\ell} + R_{i\ell}R_{jk},$$

which reduces in the scalar case to the well known relation $\mu_{(4)} = 3\mu_{(2)}^2$. It follows that a natural way to measure deviation from Gaussianity of a random vector $\mathbf{z}$ consists of computing the so-called *fourth-order cumulant*:

$$\gamma_{ijk\ell} = \mathrm{Cum}(z_i, z_j, z_k, z_\ell) = \mu_{ijk\ell} - R_{ij}R_{k\ell} - R_{ik}R_{j\ell} - R_{i\ell}R_{jk}. \quad (3.22)$$

This cumulant indeed cancels out for all indices (i, j, k, ℓ) when $\mathbf{z}$ is made up of Gaussian components.

Now under linear transforms $\mathbf{z}' = \mathbf{W}\mathbf{z}$, moments and cumulants enjoy a multilinearity property. For instance, the covariance matrix of $\mathbf{z}'$ is related to that of $\mathbf{z}$ by: $R'_{ij} = \sum_{mn} W_{im}W_{jn}R_{mn}$, with $W_{im} = [\mathbf{W}]_{im}$, or in compact form by $\mathbf{R}' = \mathbf{W}\mathbf{R}\mathbf{W}^{\mathrm{T}}$. Similarly, the cumulant of order 4 of $\mathbf{z}'$ is related to that of $\mathbf{z}$ by

$$\gamma'_{ijk\ell} = \sum_{mnpq} W_{im}W_{jn}W_{kp}W_{\ell q}\gamma_{mnpq}.$$

In particular, if we estimate one source from the observation $\mathbf{x}$ as $\hat{s} = \mathbf{w}^{\mathrm{T}}\mathbf{x}$, then the cumulant $\gamma = \mathrm{Cum}(\hat{s}, \hat{s}, \hat{s}, \hat{s})$ is related to the cumulants $C_{mnpq} = \mathrm{Cum}(x_m, x_n, x_p, x_q)$ by

$$\gamma = \sum_{mnpq} w_m w_n w_p w_q C_{mnpq}, \quad (3.23)$$

where $w_m = [\mathbf{w}]_m$. With the help of these statistical tools, let us see first how we can extract one source estimate $\hat{s}$ from the mixture $\mathbf{x}$ in Eq. (3.17).

3.3.3.2 Axiomatic Derivation

Our goal is to find a vector $\mathbf{w}$ such that $\hat{s} = \mathbf{w}^{\mathrm{T}}\mathbf{x}$ is close to one of the sources s_i. In other words, the row vector $\mathbf{g}^{\mathrm{T}} = \mathbf{w}^{\mathrm{T}}\mathbf{H}$ should contain a single nonzero entry; such vectors are called *trivial filters*. To this end, we will maximize an objective function $\Psi(\mathbf{w})$, depending on $\mathbf{w}$ through $\hat{s}$. It may hence be seen as a function $\Psi[\hat{s}]$ of some feature of $\hat{s}$, typically its probability distribution, which we distinguish with brackets. According to Sect. 3.2.2.1, PCA uses the output power (3.4) as an objective function, which is maximized subject to the constraint $\|\mathbf{w}\| = 1$. But we have seen in Sect. 3.3.2 that, depending on the mixing matrix structure, the maximization of this function is not always successful in extracting one of the sources. To perform a successful source extraction in the noiseless case, the optimization criterion $\Psi(\mathbf{w}) = \Psi[\mathbf{w}^{\mathrm{T}}\mathbf{x}] = \Psi[\mathbf{g}^{\mathrm{T}}\mathbf{s}]$ should satisfy the following properties, which may be viewed as axioms:

1. For any trivial filter $\mathbf{t}$, there should exist one source s_i such that

$$\Psi[\mathbf{t}^{\mathrm{T}}\mathbf{s}] = \Psi[s_i].$$

   This means that the objective function Ψ should be insensitive to trivial filters. We can refer to this property as *invariance*.

2. Any linear combination of independent sources should decrease the objective. Put in mathematical terms, given a trivial filter $\mathbf{t}$, then for any spatial filter $\mathbf{g}$ we have:
$$\Psi[\mathbf{g}^{\mathrm{T}}\mathbf{s}] \leq \Psi[\mathbf{t}^{\mathrm{T}}\mathbf{s}].$$
This property is called *domination*.
3. Finally, the maximum of Ψ should be reached only for the expected sources. In other words, the equality should occur in the inequality above only for trivial mixtures. This can be written under the form of the *discrimination* property below:
$$\text{If } \exists\, \mathbf{g} \,:\, \Psi[\mathbf{g}^{\mathrm{T}}\mathbf{s}] = \max_{\mathbf{t}} \Psi[\mathbf{t}^{\mathrm{T}}\mathbf{s}] \;\Rightarrow\; \mathbf{g} \text{ trivial}.$$

The discrimination property avoids the existence of spurious maxima. Optimization criteria satisfying these properties are referred to as *contrast criteria*. In particular, the above properties define the so-called multiple-input single-output contrasts, since the extracting system generates a single output signal (an estimate of one of the sources) from several mixtures of the sources acting as multiple inputs to the system. These quite natural properties have already been put forward in [7] for the multiple-input multiple-output BSS problem, and in [10] for the single-input single-output blind channel equalization problem.

3.3.3.3 Kurtosis Contrast Criterion

A variety of optimization criteria can be devised depending on the assumptions available on the sources [8, 9]. In the remaining of this chapter, we will solely assume that the M sources are mutually statistically independent and that at least $(M-1)$ are non Gaussian. Moreover, contrary to [7, 27], we will concentrate here on the extraction of sources one by one, a separation procedure known as *deflation*. Contrast criteria are in fact not the same in both cases [9, Chap. 3].

The first idea is to search for extremal values of γ, the fourth-order cumulant of $\hat{s}$, linked to the cumulants of the observations through Eq. (3.23). According to Sect. 3.3.3.1, this search will indeed maximize the gap to Gaussianity. But one can notice that $|\gamma|$ is unbounded above if $\|\mathbf{w}\|$ is unbounded, which is undesirable. This problem can be fixed by normalizing $\hat{s}$ by its standard deviation, leading to the *kurtosis maximization (KM)* criterion:

$$\Psi_{\mathrm{KM}}[\hat{s}] = \frac{\gamma}{\mathrm{var}^2\{\hat{s}\}} = \frac{\mathrm{E}\{\hat{s}^4\} - 3\mathrm{E}^2\{\hat{s}^2\}}{\mathrm{E}^2\{\hat{s}^2\}}. \tag{3.24}$$

This criterion can be expressed in terms of vector $\mathbf{w}$ by exploiting the multilinearity property (3.23):
$$\Psi_{\mathrm{KM}}(\mathbf{w}) = \frac{\sum_{mnpq} w_m w_n w_p w_q C_{mnpq}}{(\sum_{ij} w_i w_j R_{ij})^2}, \tag{3.25}$$

where C_{mnpq} and $R_{ij} = [\mathbf{R}_\mathbf{x}]_{ij}$ denote the fourth-order cumulant and the covariance of the observation $\mathbf{x}$, respectively. These quantities can easily be estimated from sufficiently long observation samples. Then it can be shown [9, Chap. 3] that $|\Psi_{\mathrm{KM}}(\mathbf{w})|$ is indeed a multiple-input single-output contrast as defined in Sect. 3.3.3.2.

In the absence of noise, i.e., if $\mathbf{x} = \mathbf{H}\mathbf{s}$ holds exactly, where s_i are statistically independent, the fact that the absolute kurtosis can serve as a contrast criterion accepts an intuitive interpretation based on the Central Limit Theorem. This well-known theorem in Statistics states that the Gaussianity of a linear mixture increases as the number of independent variables contributing to the mixture increases. Hence, looking for the matrix $\mathbf{W}$ maximizing the independence among the components of $\hat{\mathbf{s}} = \mathbf{W}\mathbf{x}$ is equivalent to maximizing the non-Gaussianity of every component $\hat{s}_i$. It follows that maximizing the absolute kurtosis can be seen, at order 4, as a sensible criterion to act in the opposite direction of the Central Limit Theorem: reduce the mixing by decreasing Gaussianity.

When the entries of a random vector $\mathbf{z}$ are statistically independent, all its cumulants $\gamma_{ijk\ell}$ are null except for $i = j = k = \ell$, so that the fourth-order cumulant array will show a diagonal structure. As we have just seen, non-Gaussianity and independence are closely related, and thus maximizing non-Gaussianity as described above will implicitly diagonalize the cumulant array of the observations. Hence, in the same way as PCA finds uncorrelated components by diagonalizing the observation covariance matrix (Sect. 3.2.2.2), ICA looks for independent components by diagonalizing the observation cumulant array at orders higher than two.

It is important to stress that if there exist more than one Gaussian source, then the Gaussian sources cannot be recovered. In fact, as pointed out in Sect. 3.3.3.1, the kurtosis of Gaussian variables is null, and hence cannot be used as an objective function to estimate this kind of sources. On the other hand, all non-Gaussian sources can be extracted regardless of the number of Gaussian sources present in the mixture, as long as the full rank observation assumption holds (Sect. 3.3.1).

Before closing this section, note that criterion $\Psi_{\mathrm{KM}}(\mathbf{w})$ is insensitive to scale, which makes sense because the actual amplitude of the original sources cannot be recovered by resorting solely to their statistical independence; this is the scale ambiguity noted at the end of Sect. 3.3.1. In order to avoid divergence or instabilities of vector $\mathbf{w}$ during the optimization process, it is hence also desirable to fix its norm, e.g., to impose $\|\mathbf{w}\| = 1$.

3.3.3.4 Extraction of One Source

An efficient algorithm, the so-called *RobustICA* [26], can be employed to maximize the kurtosis contrast. The key to this algorithm lies in the fact that the contrast criterion (3.24)–(3.25) is a rational function in vector $\mathbf{w}$. As a result, once a search direction $\mathbf{g}$ has been fixed (for instance the gradient), then the *global* optimum of the contrast along the search direction can easily be computed. To see this, it is

Algorithm 2 RobustICA algorithm for extracting an independent source from a linear mixture

```
Compute the fourth-order cumulants C_mnpq = Cum(x_m, x_n, x_p, x_q) and the covariance
  matrix R_x = E{xx^T} of the observation x.
Initialize the extracting vector as w = w^(0).
i = 0
repeat
    Compute a suitable search direction g^(i) (e.g., the gradient of the kurtosis contrast) at
      point w^(i).
    Compute the coefficients of the fourth-degree polynomial P(ν, w^(i), g^(i)) (as detailed
      in [26]).
    Select the root ν_i yielding the largest value of the objective |Ψ_KM(w^(i) + νg^(i))|.
    Update the extracting vector as w^(i+1) = w^(i) + ν_i g^(i).
    Normalize to unit norm: w^(i+1) ← w^(i+1)/||w^(i+1)||.
    Check convergence using a suitable stopping criterion.
    i ← i + 1
until convergence
Compute ŝ = w^T x, where w is the extracting vector after convergence.
```

sufficient to observe that the one-dimensional cost $\Psi_{\text{KM}}(\mathbf{w} + \nu\mathbf{g})$ is also a rational function in the *step size* ν for any given vectors $\mathbf{w}$ and $\mathbf{g}$. Hence, all its stationary points could be calculated by using standard numerical routines from a computer library.

But we can do even better than that. In fact, it turns out that the stationary points of $\Psi_{\text{KM}}(\mathbf{w} + \nu\mathbf{g})$ with respect to ν are the roots of a polynomial $P(\nu, \mathbf{w}, \mathbf{g})$ of degree four only (a quartic), as defined in [26]. Hence, they can be computed within a finite number of operations, following Ferrari's formula for quartics. By plugging back these roots into the original objective $\Psi_{\text{KM}}(\mathbf{w} + \nu\mathbf{g})$, the globally optimum step size ν is easily selected for the current iteration. Algorithm 2 summarizes the off-line version of the RobustICA algorithm for extracting one source.

As as search direction $\mathbf{g}$, we can use the gradient of criterion (3.24) and (3.25), with components

$$g_i(\mathbf{w}) = [\nabla\Psi_{\text{KM}}(\mathbf{w})]_i = 4\frac{\sum_{mnp} w_m w_n w_p C_{imnp}}{(\mathbf{w}^{\text{T}}\mathbf{R}_{\mathbf{x}}\mathbf{w})^2} - 4\gamma\frac{\sum_j R_{ij} w_j}{(\mathbf{w}^{\text{T}}\mathbf{R}_{\mathbf{x}}\mathbf{w})^3},$$

where γ is the output fourth-order cumulant given by Eq. (3.23). On-line versions operating on a sample-by-sample basis can also be devised by observing that the gradient can be rewritten as

$$\nabla\Psi_{\text{KM}}(\mathbf{w}) = 4\frac{\text{Cum}(\hat{s}, \hat{s}, \hat{s}, \mathbf{x})}{\text{var}^2\{\hat{s}\}} - 4\gamma\frac{\text{E}\{\hat{s}\mathbf{x}\}}{\text{var}^3\{\hat{s}\}}, \tag{3.26}$$

where $\text{Cum}(\hat{s}, \hat{s}, \hat{s}, \mathbf{x}) = \text{E}\{\hat{s}^3\mathbf{x}\} - 3\text{E}\{\hat{s}^2\}\text{E}\{\hat{s}\mathbf{x}\}$.

3.3.3.5 Deflation Algorithm

If more than one source are to be extracted, the above procedure may be run several times. To extract each source only once, an idea is to remove from the observation the contribution of the sources already extracted. More specifically, denote $\hat{s}_1 = \mathbf{w}_1^{\mathrm{T}}\mathbf{x}$ the first source extracted by the algorithm described in Sect. 3.3.3.4. Then, a new observation vector $\mathbf{x}_1$ can be built by removing the contribution of $\hat{s}_1$ as

$$\mathbf{x}_1' \stackrel{\text{def}}{=} \mathbf{x} - \mathbf{h}_1\hat{s}_1, \tag{3.27}$$

where $\mathbf{h}_1$ is the regression vector minimizing the MSE (3.7) with respect to $\mathbf{h}$, whose solution is given by Eq. (3.9):

$$\mathbf{h}_1 = \frac{\mathbf{R}_{\mathbf{x}}\mathbf{w}_1}{\mathbf{w}_1^{\mathrm{T}}\mathbf{R}_{\mathbf{x}}\mathbf{w}_1} = \frac{\mathrm{E}\{\hat{s}_1\mathbf{x}\}}{\mathrm{E}\{\hat{s}_1^2\}}. \tag{3.28}$$

Now the rank of the covariance of $\mathbf{x}_1'$ is not full anymore, because the subtraction necessarily decreases it by one. Hence, the size of vector $\mathbf{x}_1'$ can be diminished by one without losing information. As explained in Sect. 3.2.2.3, this dimensionality reduction may be performed with the help of the PCA of $\mathbf{x}_1$ by retaining its $(M-1)$ dominant principal components and neglecting the eigenvectors associated with null eigenvalues. Including variance normalization as in Eq. (3.18), this operation can be expressed as

$$\mathbf{x}_1 \stackrel{\text{def}}{=} \mathbf{P}_1\mathbf{x}_1' \quad \text{with} \quad \mathbf{P}_1 = \mathbf{D}_1^{-\frac{1}{2}}\mathbf{U}_1^{\mathrm{T}}, \tag{3.29}$$

where $\mathbf{D}_1$ and $\mathbf{U}_1$ contain the $(M-1)$ nonzero eigenvalues and their corresponding eigenvectors, respectively, of the covariance matrix of $\mathbf{x}_1'$. Vector $\mathbf{x}_1$ is now of dimension $(M-1)$, is uncorrelated with the first extracted source $\hat{s}_1$, and has a full rank covariance matrix. The deflation procedure can thus continue until a single source is left, when vector $\mathbf{x}_M$ is of dimension one. The complete deflation algorithm is summarized in Algorithm 3. In practical implementations of this algorithm, mathematical expectations need to be replaced by, e.g., sample estimates as defined in Eq. (3.10).

Improved deflation algorithm. When a large number of sources are extracted, it may happen that the same source is extracted several times despite the use of regression. This is due to rounding errors that accumulate through successive deflation stages, especially when processing short observation windows. In order to cope with this problem, one solution consists of minimizing the MSE function $\mathrm{E}\{\|\mathbf{x} - \mathbf{H}_k\hat{\mathbf{s}}_k\|^2\}$ with respect to $\mathbf{H}_k \in \mathcal{R}^{L\times k}$, where $\hat{\mathbf{s}}_k = [\hat{s}_1, \hat{s}_2, \ldots, \hat{s}_k]^{\mathrm{T}}$ are the first k extracted sources. This minimization leads to replacing the regression step (3.27) and (3.28), or line 4 in Algorithm 3, by the alternative regression equation:

$$\mathbf{x}_k' \stackrel{\text{def}}{=} \mathbf{x} - \mathrm{E}\{\mathbf{x}\hat{\mathbf{s}}_k^{\mathrm{T}}\}\mathrm{E}\{\hat{\mathbf{s}}_k\hat{\mathbf{s}}_k^{\mathrm{T}}\}^{-1}\hat{\mathbf{s}}_k. \tag{3.30}$$

Algorithm 3 RobustICA algorithm for extracting all independent sources from a linear mixture

1: Compute the PCA of $\mathbf{x}$ and let $\mathbf{x}_0 = \hat{\mathbf{s}}_{\text{PCA}}$ be the whitened observation vector of dimension M, where $\hat{\mathbf{s}}_{\text{PCA}}$ is given by Eq. (3.18).
2: **for** $k = 1 \to (M-1)$ **do**
3: Extract $\hat{s}_k$ from $\mathbf{x}_{k-1}$ with the help of Algorithm 2 (Sect. 3.3.3.4).
4: Subtract the contribution of $\hat{s}_k$ from $\mathbf{x}_{k-1}$ by using the regression formulas (3.27)–(3.28): $\mathbf{x}'_k = \mathbf{x}_{k-1} - \frac{\mathrm{E}\{\hat{s}_k \mathbf{x}_{k-1}\}}{\mathrm{E}\{\hat{s}_k^2\}}$.
5: Compute the standardized observation vector $\mathbf{x}_k$ with reduced dimensionality as defined in (3.29): $\mathbf{x}_k = \mathbf{P}_k \mathbf{x}'_k$.
6: **end for**
7: Set $\hat{s}_M = x_M$.

The deflation algorithm explained earlier in this section corresponds, up to dimensionality reduction, to the approximation $\mathrm{E}\{\hat{\mathbf{s}}_k \hat{\mathbf{s}}_k^{\mathrm{T}}\} \approx \mathrm{diag}(\mathrm{E}\{\hat{s}_1^2\}, \mathrm{E}\{\hat{s}_2^2\}, \ldots, \mathrm{E}\{\hat{s}_k^2\})$ in Eq. (3.30), that is, to the case where the estimated source covariance matrix is diagonal. By taking into account the whole covariance matrix of the estimated sources, this alternative deflation scheme achieves improved robustness to residual source correlations that may remain when dealing with short sample sizes.

3.3.3.6 Remarks

An interesting advantage of the RobustICA algorithm of Sect. 3.3.3.4 is that it does not require prewhitening, which improves its asymptotic (large sample) performance [26]. Hence, the whitening stage (Algorithm 3, line 1) can just be omitted, and $\mathbf{x}_0$ can simply be initialized with the observed data $\mathbf{x}$. Nevertheless, running PCA before extracting the sources improves numerical conditioning due to the diagonal covariance matrix and unit variance estimates. It can hence still be used as a preprocessing step before deflation, even if it limits asymptotic performance. All the results stated in the present framework for the full separation of M sources hold in two situations: (1) in the presence of at least $(M-1)$ non-Gaussian sources without noise, or (2) in the presence of M non-Gaussian sources with additive Gaussian noise. In the presence of non-Gaussian noise, results become approximate and can become erroneous if the non-Gaussian noise is too strong. Finally, although ICA performs independent source separation under linear mixing model (3.17), it can also be employed to find interesting alternative data representations (useful, e.g., for classification) even if this observation model is not fulfilled.

3.3.4 Refining ICA for Improved Atrial Signal Extraction

ICA was first applied to atrial signal estimation in AF ECG episodes in [20, 21], yielding satisfactory results when the amplitude of the atrial sources presents non-Gaussian probability distributions. This is usually the case in early, often more organized, stages of AF or in related conditions such as atrial flutter. As the disease evolves, however, the atrial activity becomes more disorganized and tends to present quasi-Gaussian distributions, making it difficult for ICA techniques to perform its extraction from other sources of Gaussian noise and interference. Hence, refinements of the classical ICA approach are necessary for a successful atrial signal estimation in the general case.

The most successful refinements capitalize on the time coherence or narrowband spectrum of the atrial signal in the surface ECG. Indeed, the atrial frequency spectrum is typically concentrated around a dominant peak located in the 3–9 Hz band and its harmonics. To benefit from this property, a two-stage approach is adopted in [6]. In the first stage, classical ICA estimates the ventricular activity sources, which are strongly non-Gaussian signals. The remaining ICA sources contain a mixture of atrial components and noise, and are further processed by another separation technique known as second-order blind identification (SOBI) [1]. Like PCA (Sect. 3.3.2), SOBI is based on the diagonalization of correlation matrices, but also considers time lags different from zero:

$$\mathbf{R}_{\mathbf{x}}(\tau) = \mathrm{E}\{\mathbf{x}(t)\mathbf{x}(t-\tau)^{\mathrm{T}}\},$$

(cf. Eq. (3.2)). Hence, this second stage is particularly suited to the separation of sources with long correlation functions or, equivalently, narrowband frequency spectra. The improvement brought about by the second processing stage is more beneficial in cases where the atrial source distribution is close to Gaussian.

The spectral concentration, or relative power contained in a narrow band around the fundamental frequency, is a quantitative measure of the time coherence of the atrial signal [6]. The spectral concentration is explicitly exploited in [18] by assuming that the atrial source dominates, in terms of power, the other sources in the narrow frequency band, denoted $[f_1, f_2]$ Hz, where it appears; we call this band (contained within the 3–9 Hz AF band) the significant *spectral support* of the atrial activity signal. According to this assumption, the optimal separating filter can be found by maximizing the filter output relative power in the atrial spectral support. After whitening the observations, e.g., by means of PCA as in Sect. 3.3.2, the atrial activity extracting filter can be computed algebraically as the dominant eigenvector of the frequency-constrained spectral covariance matrix

$$\tilde{\mathbf{R}}_{\mathbf{x}}^{(f_1,f_2)} = \int_{f_1}^{f_2} \mathcal{R}\mathrm{e}\left\{\tilde{\mathbf{x}}(f)\tilde{\mathbf{x}}(f)^{\mathrm{H}}\right\} \mathrm{d}f,$$

where $\tilde{\mathbf{x}}(f)$ represents the Fourier transform of $\mathbf{x}(t)$, symbol $(\cdot)^{\mathrm{H}}$ denotes the Hermitian (conjugate-transpose) operator and $\mathcal{R}\mathrm{e}\{\cdot\}$ yields the real part of its complex argument. As opposed to classical ICA, this second stage is not based on higher-order statistics but on *conditional* second-order statistics computed on the significant spectral support of the desired signal and, as a result, it can also deal with near-Gaussian atrial signals. This atrial signal extraction technique, referred to as *maximum variance in tails (MaxViT)*, can be cast in a more general source separation framework based on a conditional maximum likelihood principle [18,19].

The narrowband character of the atrial signal during AF can also be exploited by using higher-order statistics, even if the amplitude of the atrial time course is near Gaussian [26]. The trick consists of working in the frequency domain, where the marked sparsity of the atrial signal due to its reduced spectral support is linked with strong non-Gaussianity. As a result, the atrial source can typically be found among the most kurtic frequency-domain sources extracted by a BSS technique based on kurtosis maximization such as RobustICA [26], which is able to handle complex-valued sources with noncircular distributions. We refer to this frequency-domain ICA method as *RobustICA-f*. A common feature of the RobustICA-f and MaxViT methods is that, by incorporating some prior information about the frequency content of the desired source, they mitigate the permutation ambiguity of the BSS model (Sect. 3.3.1), thus sparing the need to separate all sources to recover the atrial signal. Further details on these refined BSS-based techniques for atrial activity extraction can be found in [25].

3.3.5 Success Stories

3.3.5.1 Atrial Activity Extraction in Persistent Atrial Fibrillation

Recall that the first two plots of Figs. 3.5 and 3.6 show the endocardial and surface ECG recordings of a persistent AF patient, as introduced in Sect. 3.1.2.2. The whole 12-lead ECG over 10 s is available for processing, yet only lead V1 is plotted over the last 5 s to ease visualization. In plots (c)–(h), the results of different atrial signal estimation methods are compared with the lead V1 signal plotted in the background. In Fig. 3.6, the endocardial signal spectrum is also plotted on a magnified amplitude scale ($\times 10$) using light-grey dashed lines. The spectral concentration (SC in the plots) is computed as the relative power around the dominant or fundamental frequency (f_p) and the second harmonic. The vertical dashed and dash-dotted lines mark the location of f_p and the frequency bounds used in the computation of spectral concentration.

Automatic beat detection in the surface ECG reveals an average R-R period of 418 ms, linked to the fundamental frequency of 2.4 Hz, i.e., 144 beats per minute, displayed in Fig. 3.6b. The fundamental frequency of 5.7 Hz in Fig. 3.6a corresponds to an average atrial cycle length of 175 ms. These values illustrate the lack of synchronization between the atrial and ventricular activities, supporting the appropriateness of the independence assumption exploited by the BSS approach to atrial activity extraction (Sect. 3.3.1). Remark that the endocardial signal is not known to the atrial signal estimation methods evaluated next.

The STC approach of [17] (see also [27]) is employed as a benchmark. As briefly summarized at the beginning of Sect. 3.3, this approach mainly differs from BSS in that it does not aim at the atrial sources, but directly at the atrial contribution to the leads under study. In lead V1, STC produces the estimated atrial signal shown in Figs. 3.5c and 3.6c. The method does a good job, but some residual ventricular interference still remains around the R-peak locations.

Results by BSS-based methods are shown in plots (d)–(h) of Figs. 3.5 and 3.6. PCA (Sect. 3.3.2) is able to condense into just four principal components over 99.5 % of the variance of the 12-lead ECG recording, which illustrates the data compression capabilities of this decomposition technique (Sect. 3.2.2.3). Although not shown here due to space limitations, the first two components are linked to ventricular activity, while the third and fourth components mainly contain atrial activity. These produce the reconstructed atrial signal shown in Figs. 3.5d and 3.6d. Again, ventricular residuals are still visible, but seems a little less noticeable than for STC in this example. The spectral concentration increases accordingly.

The kurtosis-based RobustICA of [26] (see also Sect. 3.3.3) yields the atrial signal estimate shown in Figs. 3.5e and 3.6e, labeled 'ICA'. The ICA approach is able to concentrate the atrial activity into a single independent source, while PCA required two components to describe this activity in this particular dataset. The spectral concentration slightly decreases as compared to PCA, but the time course shows reduced ventricular interference. With a kurtosis of -0.6, the estimated atrial source is relatively close to Gaussian, as may be expected in persistent forms of AF. ICA is thus expected to benefit from a processing refinement based on the time coherence of the atrial signal, as explained in the previous section.

To carry out this refinement, we note that the last six sources obtained by RobustICA have a kurtosis value below 1.5. These quasi-Gaussian sources are passed on to the SOBI algorithm aiming to diagonalize 17 correlation matrices equally spaced at 20 ms time lags, as proposed in [6]. These lags could have been optimized by taking into account a preliminary estimation of the AF dominant frequency. Nevertheless, the atrial signal obtained with this simple lag selection improves on the spectral concentration of PCA and ICA, as shown in the 'ICA-SOBI' plot of Figs. 3.5f and 3.6f.

Section 3.3.4 recalled that the RobustICA-f [26] and MaxViT [18] methods both operate in the frequency domain. The former is based on higher-order statistics whereas the latter only exploits second-order statistics. Despite this key difference, both methods produce very similar atrial signal estimates and yield the highest spectral concentration values, as can be observed in the last two plots of Figs. 3.5 and 3.6.

The surface atrial signal estimated by all tested methods presents a harmonic structure reminiscent of that of the endocardial recording, with just a 0.2 Hz offset in the fundamental frequency value. This yields a noninvasive AF cycle length estimate of 182 ms, quite close to the 175 ms measured invasively on the atrial endocardium.

3.3.5.2 Measuring Atrial Fibrillation Organization with PCA

Although PCA yields satisfactory results in the illustrative example of the previous section, we have seen in Sect. 3.3.2 that it generally fails to perform the separation under the general form of model (3.17). Yet this classical technique proves useful when the focus is on the signal subspaces of the sources, i.e., the linear span of their spatial topographies (or mixing matrix columns, as defined in Sect. 3.3.1), rather than the sources themselves. A good example in the context of AF analysis is the noninvasive measurement of atrial signal organization [3]. The working hypothesis can simply be put as follows: the more complex or disorganized the atrial activations during AF, the higher the number of spatial topographies required to explain or reconstruct the observations with certain accuracy (spatial complexity) and the higher the time variability or nonstationarity of the spatial topographies in the surface recording (temporal complexity).

To quantify this notion of spatio-temporal complexity, the TQ intervals (containing atrial activity only) can be concatenated and divided into several consecutive segments. A BSS model like that in Eq. (3.17) is assumed for each segment c:

$$\mathbf{x}^{(c)} = \mathbf{H}^{(c)}\mathbf{s}^{(c)}.$$

As explained in Sects. 3.2.2 and 3.3.2, PCA obtains such a decomposition in which the sources are uncorrelated and arranged in increasing order of variance (principal components). We compute the PCA mixing-matrix estimate $\hat{\mathbf{H}}^{(1)}$ from the first segment, and then project the data of the following segments on its first k columns or spatial topographies, denoted $\hat{\mathbf{H}}_k^{(1)}$. Such a projection can be computed as in Sect. 3.2.2.3:

$$\hat{\mathbf{x}}_k^{(c)} = \hat{\mathbf{H}}_k^{(1)}[(\hat{\mathbf{H}}_k^{(1)})^{\mathrm{T}}\hat{\mathbf{H}}_k^{(1)}]^{-1}(\hat{\mathbf{H}}_k^{(1)})^{\mathrm{T}}\mathbf{x}^{(c)} = \mathbf{U}_k^{(1)}(\mathbf{U}_k^{(1)})^{\mathrm{T}}\mathbf{x}^{(c)},$$

where $\mathbf{U}_k^{(1)}$ contains the orthonormal k principal directions of the first segment; see also Eq. (3.19). The normalized MSE between $\hat{\mathbf{x}}^{(c)}$ and $\mathbf{x}^{(c)}$ can then be computed and averaged to quantify how well the dominant principal directions of the first segment are able to explain the observed data in the remaining segments. This parameter is computed for $k = 3$ on the basis of the classical dipole model, which assumes that cardiac activity in physiological conditions can be explained by three components only [14]. It turns out [3] that this noninvasive index is able to clearly distinguish two populations of patients that appear to be related, respectively, to type I (well-organized) and type II/III (disorganized) AF according to Konings' criteria for invasive atrial recordings [13].

3.4 Conclusion and Outlook

This chapter has examined two problems in cardiology involving the analysis of the surface ECG, namely, TWA detection and atrial activity estimation during AF. The diversity or redundancy provided by the spatially separated electrodes and quasi-periodic waveforms of this noninvasive recording modality can sometimes

Table 3.1 Comparative summary of the main features of PCA and ICA

Features	PCA	ICA
Source assumptions	Uncorrelated (second-order independent) Gaussian or non-Gaussian sources (principal components)	Independent (at orders higher than two) non-Gaussian sources (independent components)
Mixing matrix assumptions	Full column rank with orthogonal columns (scaled principal directions)	Full column rank with arbitrary structure
Statistics	Second-order statistics (covariance matrix)	Higher-order statistics (typically, fourth-order cumulants)
Solutions	Matrix decompositions (EVD, SVD)	Iterative algorithms (e.g., RobustICA)
Computational cost	Lower than ICA's	Higher than PCA's
Additional features	Optimal compression in the MSE sense	Insensitive to additive Gaussian noise

be difficult to analyze by the cardiologist, especially when different leads or time intervals provide seemingly conflicting information. Yet, as we have seen throughout the chapter, this diversity can effectively be exploited by signal processing techniques decomposing the observed data into latent components or sources that are often easier to interpret than the observed ECG. Such components yield alternative representations of the original data according to specific features of interest. While PCA explains the data in terms of second-order statistics (variance and covariance) and results in uncorrelated sources, ICA can sometimes provide deeper insights by searching for independence through the use of higher-order statistics (cumulants). The main features of PCA and ICA are summarized in Table 3.1. These linear data decomposition techniques are capable of revealing underlying structures of the ECG signal that remain otherwise hidden to the naked eye. As a result, such methods prove useful in noninvasively detecting and estimating cardiac electrophysiological phenomena of interest, such as TWA and atrial activity during AF, thus aiding the cardiologist to make subsequent clinical decisions. Indeed, an accurate TWA detection allows a more precise assessment of the risks of sudden cardiac death, whereas a clean atrial activity signal simplifies the estimation and improves the statistical significance of clinically pertinent parameters such as dominant atrial frequency or atrial cycle length.

Although the chapter has focused on ECG signals, these decomposition techniques have also shown their success in processing other biomedical data such as functional magnetic resonance images, electroencephalograms and electromyograms, to name but a few [27], [9, Chap. 18]. Chapters 5 and 7 of this book apply PCA to cardiac imaging and brain signal analysis, respectively. Other linear data decomposition approaches including nonnegative matrix factorization, sparse component analysis and tensor factorizations have drawn intense research attention in recent years [9], and so have nonlinear dimensionality reduction techniques [23] (see also Chap. 7). Their application to ECG signal processing and other biomedical problems is a promising avenue of ongoing research.

3.5 Online Resources

The RobustICA algorithm

http://www.i3s.unice.fr/~zarzoso/robustica.html

This website proposes a Matlab implementation of the *RobustICA algorithm* described in Sect. 3.3.3

The LVA Central website

http://lvacentral.inria.fr

This website contains a variety of algorithms for latent variable analysis and signal separation, including ICA, as well as datasets for evaluating their performance

The PhysioNet database

http://www.physionet.org

This website, maintained by the MIT, features a large collection of recorded physiological signals and related open-source software, including but not limited to ECG signal processing

Inria's CardioSense3D project

http://www-sop.inria.fr/CardioSense3D

This project aims at modeling cardiac electrical and mechanical activity simultaneously

Acknowledgements Part of the work summarized in this chapter is supported by the French National Research Agency under contract ANR 2010 JCJC 0303 01 "PERSIST".

List of Acronyms

AF	Atrial fibrillation
BSS	Blind source separation
ECG	Electrocardiogram
EVD	Eigenvalue decomposition
ICA	Independent component analysis
KM	Kurtosis maximization
MaxViT	Maximum variance in tails
MSE	Mean square error
PCA	Principal component analysis
SOBI	Second-order blind identification
STC	Spatio-temporal QRST cancellation
SVD	Singular value decomposition
TWA	T-wave alternans

References

1. A. Belouchrani, K. Abed-Meraim, J.-F. Cardoso, and E. Moulines. A blind source separation technique using second-order statistics. *IEEE Transactions on Signal Processing*, 45(2):434–444, February 1997.
2. A. Bollmann and F. Lombardi. Electrocardiology of atrial fibrillation. *IEEE Engineering in Medicine and Biology Magazine*, 25(6):15–23, November/December 2006.
3. P. Bonizzi, M. S. Guillem, A. M. Climent, J. Millet, V. Zarzoso, F. Castells, and O. Meste. Noninvasive assessment of the complexity and stationarity of the atrial wavefront patterns during atrial fibrillation. *IEEE Transactions on Biomedical Engineering*, 57(9):2147–2157, September 2010.
4. L. Burattini, W. Zareba, and R. Burattini. The effect of baseline wandering in automatic T-wave alternans detection from Holter recordings. In *Proc. Computers in Cardiology*, volume 33, pages 257–260, Valencia, Spain, Sept. 17–20, 2006.
5. F. Castells, P. Laguna, L. Sörnmo, A. Bollmann, and J. Millet Roig. Principal component analysis in ECG signal processing. *EURASIP Journal on Advances in Signal Processing*, 2007:21 pages, 2007.
6. F. Castells, J. J. Rieta, J. Millet, and V. Zarzoso. Spatiotemporal blind source separation approach to atrial activity estimation in atrial tachyarrhythmias. *IEEE Transactions on Biomedical Engineering*, 52(2):258–267, February 2005.
7. P. Comon. Independent Component Analysis. In J-L. Lacoume, editor, *Higher Order Statistics*, pages 29–38. Elsevier, Amsterdam, London, 1992.
8. P. Comon. Contrasts, independent component analysis, and blind deconvolution. *Int. Journal Adapt. Control Sig. Proc.*, 18(3):225–243, April 2004.
9. P. Comon and C. Jutten, editors. *Handbook of Blind Source Separation, Independent Component Analysis and Applications*. Academic Press, Oxford, UK, 2010.
10. D. Donoho. On minimum entropy deconvolution. In *Applied Time-Series Analysis II*, pages 565–609. Academic Press, 1981.
11. V. Fuster, L. E. Rydén, D. S. Cannom, H. J. Crijns, A. B. Curtis, et al. ACC/AHA/ESC guidelines for the management of patients with atrial fibrillation – executive summary. *Circulation*, 114(7):700–752, 2006.
12. P. Jaïs, D. C. Shah, M. Hocini, L. Macle, K.-J. Choi, et al. Radiofrequency ablation for atrial fibrillation. *European Heart Journal Supplements*, 5(Supplement H):H34–H39, 2003.
13. K. T. Konings, C. J. Kirchhof, J. R. Smeets, H. J. Wellens, O. C. Penn, and M. A. Allessie. High-density mapping of electrically induced atrial fibrillation in humans. *Circulation*, 89(4):1665–1680, April 1994.
14. J. Malmivuo and R. Plonsey. *Bioelectromagnetism: Principles and Applications*. Oxford University Press, New York, 1995.
15. J. P. Martínez and S. Olmos. Methodological principles of T wave alternans analysis: a unified framework. *IEEE Transactions on Biomedical Engineering*, 52(4):599–613, April 2005.
16. O. Meste, D. Janusek, and R. Maniewski. Analysis of the T wave alternans phenomenon with ECG amplitude modulation and baseline wander. In *Proc. Computers in Cardiology*, volume 34, pages 565–568, Durham, NC, Sept. 30–Oct. 3, 2007.
17. O. Meste and N. Serfaty. QRST cancellation using Bayesian estimation for the auricular fibrillation analysis. In *Proc. 27th Annual International Conference of the IEEE Engineering in Medicine and Biology Society*, pages 7083–7086, Shanghai, China, Sept. 1–4, 2005.
18. R. Phlypo, V. Zarzoso, and I. Lemahieu. Atrial activity estimation from atrial fibrillation ECGs by blind source extraction based on a conditional maximum likelihood approach. *Medical & Biological Engineering & Computing*, 48(5):483–488, May 2010.
19. R. Phlypo, V. Zarzoso, and I. Lemahieu. Source extraction by maximizing the variance in the conditional distribution tails. *IEEE Transactions on Signal Processing*, 58(1):305–316, January 2010.

20. J. J. Rieta, F. Castells, C. Sánchez, V. Zarzoso, and J. Millet. Atrial activity extraction for atrial fibrillation analysis using blind source separation. *IEEE Transactions on Biomedical Engineering*, 51(7):1176–1186, July 2004.
21. J. J. Rieta, V. Zarzoso, J. Millet-Roig, R. García-Civera, and R. Ruiz-Granell. Atrial activity extraction based on blind source separation as an alternative to QRST cancellation for atrial fibrillation analysis. In *Proc. Computers in Cardiology*, volume 27, pages 69–72, Boston, MA, Sept. 24–27, 2000.
22. M. Stridh and L. Sörnmo. Spatiotemporal QRST cancellation techniques for analysis of atrial fibrillation. *IEEE Transactions on Biomedical Engineering*, 48(1):105–111, January 2001.
23. J. B. Tenenbaum, V. de Silva, and J. C. Langford. A global geometric framework for nonlinear dimensionality reduction. *Science*, 290(5500):2319–2323, December 2000.
24. B. Widrow, J. R. Glover, J. M. McCool, et al. Adaptive noise cancelling: principles and applications. *Proceedings of the IEEE*, 63(12):1692–1716, December 1975.
25. V. Zarzoso. Extraction of ECG characteristics using source separation techniques: exploiting statistical independence and beyond. In A. Naït-Ali, editor, *Advanced Biosignal Processing*, chapter 2, pages 15–47. Springer, Berlin, 2009.
26. V. Zarzoso and P. Comon. Robust independent component analysis by iterative maximization of the kurtosis contrast with algebraic optimal step size. *IEEE Transactions on Neural Networks*, 21(2):248–261, February 2010.
27. V. Zarzoso, R. Phlypo, O. Meste, and P. Comon. Signal extraction in multisensor biomedical recordings. In P. Verdonck, editor, *Advances in Biomedical Engineering*, chapter 3, pages 95–143. Elsevier BV, Oxford, UK, 2009.

Chapter 4
Deconvolution and Denoising for Confocal Microscopy

Praveen Pankajakshan, Gilbert Engler, Laure Blanc-Féraud, and Josiane Zerubia

4.1 Introduction

4.1.1 The World of Microscopy and Beyond

The appearance of the first microscopes in the late sixteenth century can be attributed to the fascination and curiosity of man towards the miniature world of the unseen. The idea of using a lens (originally made to examine fabrics) to look at small living objects not only revolutionized our view of the living world but also created the field of biology. Robert Hooke discovered, with his earliest microscope, that most living organisms are build up of small compartments that he called "Cells" (because they reminded him of the small rooms where monks lived). This discovery led to the study of biology and completely changed the way people looked at the world. Even several generations after this earliest invention, the aesthetic appeal of the details visible in this tiny microscopic world continues to enamor many.

P. Pankajakshan (✉)
Institut Pasteur, 25 rue du Docteur Roux, 75724 Paris, France
e-mail: praveen.pankaj@gmail.com

G. Engler
INRA, 400 route des Chappes, 06903 Sophia Antipolis, France
e-mail: gilbert.engler@sophia.inra.fr

L. Blanc-Féraud
I3S (CNRS/UNS) UMR 7271 CNRS/UNSA and Inria, Algorithmes/Euclide-B,
2000 Route des Lucioles, 06903 Sophia Antipolis, France
e-mail: blancf@i3s.unice.fr

J. Zerubia
Inria Sophia Antipolis Méditerranée, Ariana project-team, 2004 Route des Lucioles,
06902 Sophia Antipolis, France
e-mail: josiane.zerubia@inria.fr

F. Cazals and P. Kornprobst (eds.), *Modeling in Computational Biology and Biomedicine*,
DOI 10.1007/978-3-642-31208-3_4,

In the end of the nineteenth century, microscopists began to notice that it was impossible to resolve features of less than half a micrometer in size. This realization later led to the invention of an *electron microscope* that uses a beam of electrons, instead of light, to create an image of the specimen. An electron microscope is capable of much higher magnification and has much greater resolving power than a light microscope. Modern electron microscopes can reach a resolution (resolution can be defined as the minimal distance of separation between two point sources so that they can be resolved as separate binary objects) of less than 0.1 nm, allowing one to image individual atoms. However, successful examination of biological samples is only possible after it has been chemically treated, to immobilize the macromolecules, and then stained with heavy metals, to provide electron contrast to the cell components. Moreover, the sample must be sliced very thin to allow the electrons to penetrate into the sample. As a result, imaging living cells and tissues with an electron microscope is simply impossible and it is mainly the specimen preparation procedure that limits the resolving power to be at its best around 1 nm. Light and electron microscopy have made it possible to unravel details in the millimeter to nanometer scale for most living beings, from viruses to bacteria, from unicellular to multicellular organisms belonging to the animal and plant kingdom. However, imaging individual molecules within "alive" biological samples with a precision to "nanometer spatial and millisecond time resolution" remains a challenge.

With every decade, the microscope has improved, propelled by the technological advancements of each epoch. While In the field of optical microscope, opticians devised new ways of perfecting the lens system, and thereby increasing the quality of the images obtained. Computers began to be used in microscopy primarily only as a tool for improved image acquisition, storage, retrieval and display. It is only in the last decade that the processing power of computers became much readily available for image analysis and to improve the quality of acquired images. Since then, computational methods such as *deconvolution* have been very successfully applied to reduce out-of-focus light in biological samples, and to extract sharp three-dimensional (3-D) reconstructions of the raw data from 3-D wide-field microscopy. We will see that the progression from *macro* [61] to *nano* [34] scales, and beyond can be partly credited to these early advancements.

4.1.2 Imaging by Fluorescence

In life sciences, living or chemically fixed cells are usually partially transparent (optically). In general, the microscopes use properties of the sample such as absorption and refraction index variations to look at cellular structures. Nevertheless, one of the most important methods to generate contrast in biological samples comes from the use of fluorescent molecules. These molecules can occur naturally in cells. Otherwise, the proteins in the sample can be labeled with specific fluorescent proteins (*fluorophores*) or specific amino-acid moieties that can be labelled with arsenical dyes that *fluoresce* under light illumination. The ability to specifically label virtually all cell components with chemically synthesized fluorophores lead to a large variety of fluorescence imaging strategies. Furthermore, the discovery of inherently fluorescent gene products such as the Green Fluorescent Protein (GFP), opened the way for biologists to genetically tag a specific protein in a living organism. Fluorescence imaging, being essentially a dark-field method, exhibits an exceptionally high image contrast compared to other wide field methods and therefore became the method of choice to localize molecules in cells. It is no wonder that Osamu Shimomura, Martin Chalfie and Roger Y. Tsien [81] were jointly awarded the 2008 Nobel Prize in Chemistry for the discovery of the GFP, demonstrating it's value and contributing to our understanding of how they fluoresce.

4.1.2.1 Biological Context

Nowadays, a major interest for cell biologists is to identify and precisely localize macromolecules, i.e., the building blocks of cellular organelles and supra-molecular complexes. A first breakthrough to allow such molecular imaging was due to the development of methods based on the use of antibodies and in-situ hybridization techniques. These approaches combined with fluorescence microscopy made it possible to image with high contrast individual cellular components using fluorescent dyes. A second breakthrough resulted from the discovery of the so called "fluorescent proteins". These auto-fluorescent molecules can be physically attached to any protein, for which the gene is known, and re-introduced into a living cell. It has the ability to non-destructively image tagged-molecules in real-time with optical resolution. It was mainly this tool that boosted microscopy in recent years. In animal cells, the cellular structures can be labeled with fluorescence dyes, using histochemical techniques, and efficiently imaged, if little or no background autofluorescence is present. A confocal microscope can generate several optical sections, at different depths, inside a chemically fixed or living specimen. With the help of these optical sections, a 3-D data set of the specimen, representing the spatial distribution of the fluorescent labeled cell components, can be constructed. Due to this optical sectioning capacity, the confocal microscope has become a major workhorse instrument in many biomedical research laboratories. Variants of the confocal microscope have been developed to increase imaging speed (spinning disk confocal) and tissue penetration depth (multi-photon) [37]. For an overview of these confocal variants, see the handbook of Pawley [57].

In life-cell imaging, questions on the functioning of macromolecular machines remain largely unanswered since many cell components have dimensions below the 200 nm x-y resolution limit (and 400 nm axial resolution limit) obtainable from a standard light microscope. Recently introduced super resolution optical imaging techniques in combination with adapted deconvolution strategies will be crucial to address some of these questions.

However, imaging biological samples, especially living specimens, with a microscope is difficult and tricky. Although a multitude of parameters will influence the final quality and faithfulness of the image, we can simplify and state that imaging chemically-fixed biological samples follows a straightforward goal to preserve structure while maximizing contrast and resolution. Due to the action of chemicals like acetone, alcohols or cross-linking aldehydes, cell components are immobilized and the biological sample is killed. Such a fixation step is optimized to preserve the fine-morphology of the cells and tissues of the biological samples. Chemical fixation makes biological samples and its constituents immobile, and therefore removes any time constraint during image recording. By clearing biological samples before imaging, most bulk components like water and various soluble components are extracted from the tissues. These components are then replaced by a homogeneous medium of refraction index close to 1.515, a process that will greately improve image quality and allow one to obtain near ideal optical imaging conditions. Interestingly, live cell imaging must seek a continuous balance between image quality and information content (which requires more signal), and the need to preserve on cell viability and unaltered biological processes. Simply maximizing contrast and resolution by extending data collection mostly leads to cell damage and permanent loss of signal intensities (bleaching). As such, preserving the viability often implies producing noisy data.

In addition, most botanical samples contain highly refractile cell walls that surround an aqueous cell content often filled with highly auto-fluorescent and light-scattering components. This can lead to extreme refraction index heterogeneities within the sample and seriously compromise deep imaging for 3-D microscopy.

4.1.2.2 Fluorescence

Fluorescence is the phenomenon whereby light is first absorbed by a crystal or molecule and then rapidly (of the order of nanoseconds) re-emitted at a slightly longer wavelength (Fig. 4.1a). The Jablonski fluorescence diagram in Fig. 4.1a, was named after the Polish physicist *Aleksander Jablonski*, and it illustrates the energy of the electronic states of a molecule and the transitions between them. The states are arranged vertically by energy arrows and the transitions between them are given by straight arrows. The event time progresses from left to right. This process can be explained as follows. The fluorescent molecule creates an excited electronic singlet state S_2 by absorbing some energy E. When it finally relaxes back to its native ground state, S_0 it emits a photon having a wavelength longer than that of the excitation beam. This shift in wavelength towards the red spectrum occurs because the energy of the emitted beam is on an average lower than that of the illumination. We recall that the energy and the wavelength are related by the expression $\mathrm{Energy} \propto 1/\mathrm{Wavelength}$. This relative shift is known as the *Stokes shift* and the emitted beam is said to be *red-shifted* (Fig. 4.1b).

4.1.2.3 Fluorescence Microscopes

Fluorescence microscopes are optical instruments capable of imaging a specimen in 3-D. Under ideal conditions, the number of photons emitted are proportional

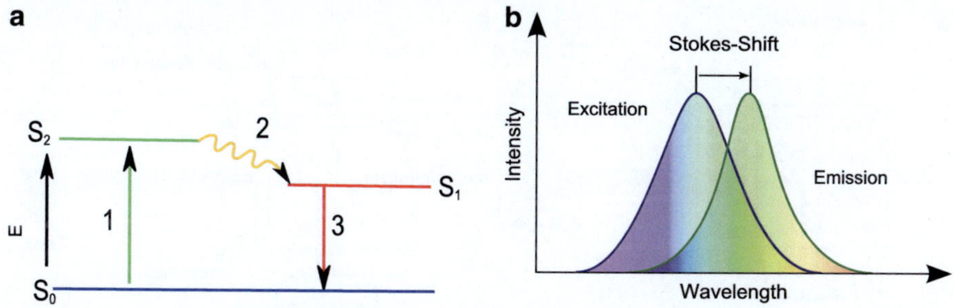

Fig. 4.1 (**a**) Jablonski diagram of a fluorescent event. S_0 is the ground state or the steady state of the fluorescent molecule and the molecule is excited to the singlet state S_2 by absorbing the energy E in a specific wavelength. The molecule undergoes some internal conversion or vibrational relaxation and transits to the relaxed state S_1 by releasing some of the energy. The molecule finally returns to its ground state by releasing the remaining energy as fluorescence. (**b**) The excitation and emission spectra showing the Stokes shift towards the *red* light (Adapted from Wikimedia Commons)

to the excitation intensities. Fluorescence microscopes are equipped with a set of filter block, each consisting of an excitation filter, a dichroic mirror and an emission filter. The excitation filter, as the name implies, is inserted along the path of the illumination light so that an excitation beam of the desired wavelength can be chosen. On the other hand, the emission filter is designed to permit only the emitted light generated by the fluorescent objects, and to proceed to the detector (Fig. 4.2a). Photons emitted from the fluorescent sample are collected by a photon sensing device (a PMT or the Charge coupled device (CCD) of a camera). The contrast obtained by this kind of optical method is far superior than classical transmission microscopy methods that are based on light absorption by counter-stained specimens. Two dimensional fluorescence micrographs can then be reconstructed, to form a 3-D representation, by computationally combining the image data from a stack of images.

For the sake of completeness, we should mention, very briefly, a number of fluorescent microscopes that have been more recently developed and commercialized to overcome the diffraction barrier described in Sect. 4.1. These far-field[1] super-resolution techniques include Stimulated emission depletion (STED) , Photon activated light microscope (PALM) (and similar), and the Structured-illumination microscope (SIM). PALM, is based on the principle of photo-activation of a very small fraction of switchable fluorophores.[2] First a small fraction of the

[1]The near field (or near-field), far field (or far-field), and the transition zone are regions of the electromagnetic radiation field scattering off an object. Certain characteristics of electromagnetic fields dominate at a large distance (or zone) from the scattering object, while a different characteristic can dominate at a shorter distance.

[2]Molecules having two states, one fluorescent and the other non-fluorescent, and the ability to be switched from one state to the other by excitation with a shortwave light.

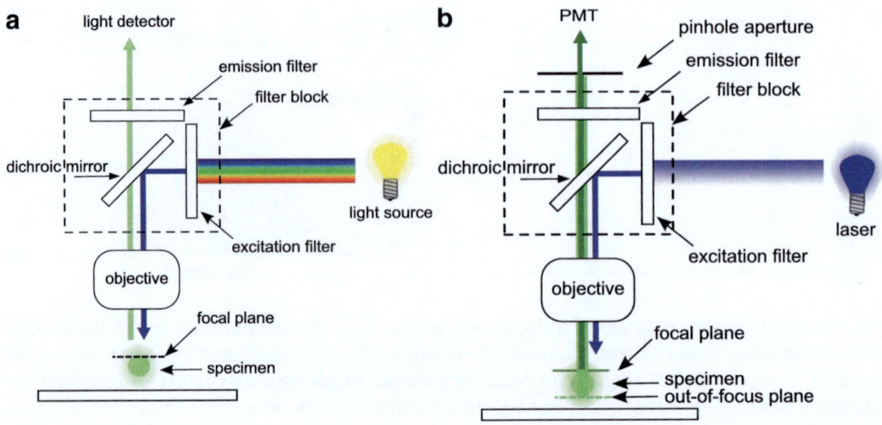

Fig. 4.2 Comparison of the ray paths in (**a**) wide-field, and (**b**) confocal microscopes (Adapted from Wikimedia Commons). The two schematics show the filter block consisting of the excitation filter, emission filter and the dichroic mirror. In (**b**), the light from the in-focus plane reaches the PMT while the emissions from the out-of-focus plane are blocked by the pinhole aperture

available molecules ($>1\,\%$) are activated and imaged. A two-dimensional (2-D) Gaussian model is fitted to the detected intensities from each individual photo-activated molecule. The image is built up slowly by summing the locations of molecules imaged in successive activations. In wide-field SIM, the resolution is improved by imposing and modulating a grid pattern in the illumination/excitation light. A grid is projected onto the focus plane of the objective and then shifted laterally with respect to the sample, and also oriented in three directions, 60° apart. The sequence of images obtained, as a result of this lateral shifting and reorienting, are de-interlaced in the Fourier space [32], and the inverse Fourier transform yields the super-resolved reconstructed image in real space. We recommend the guide [66] for the recent techniques in super-resolution microscopy (both near and far-field) and also the article in [55].

4.1.2.4 Mathematical Context

The process of deblurring can be mathematically achieved by deconvolution. Although deconvolution is discussed here in the context of microscopy, this technique is neither new nor restricted to this field alone. Blur could be very commonly encountered in almost all imaging applications such as satellite imaging, remote sensing, and astronomical observations [16]. In general they fall under one of the following three categories:

- Motion blurring caused due to a significant 'relative' motion between the observed object and the imaging device;
- Interference by media. For example, in ground-based telescopes, when imaging celestial objects, there can be image quality degradation due to atmospheric

turbulence. This can be also the case for microscopes when the specimen medium is turbid;
- Out-of-focus blur where part of the object is not in focus.

The mathematical description of each of the above cases is different principally because the underlying physical phenomenon causing the deterioration are different. In fluorescence microscopy, we often deal with the out-of-focus blur. The difference between this and the other categories is that in fluorescence microscopy the blurring phenomenon is in 3-D (i.e., both radial and axial).

In order to solve this image processing problem, it is necessary to break it down into the following three sub-problems:

- A forward problem (also called direct problem), where knowing the object and the observation mechanism, we establish a mathematical description of the object observed. This model will be a compromise between exact description of the observed physical phenomenon and a simple one for processing;
- An instrumentation problem, in which a complete description of the imaging properties of the instrument must be acquired and modeled;
- An inverse problem, where the object must be estimated from the preceding models and the data.

The above three sub-problems are unique for the optics and the imaging sensor used. We will discuss these sub-problems in this chapter, with respect to fluorescence microscopy, although not necessarily in the above order.

4.1.3 Bettering the Resolution

Better resolution is often a desired feature to inspect the specimen in detail. However, the quality of the image produced and its resolution depend on the lens, its Numerical aperture (NA),[3] and the wavelength of excitation light (λ). Ernst Abbe is credited with showing that light, with a wavelength of λ, when traveling through a lens of NA, will make a spot of radius governed by the following law:

$$d = \frac{0.61\lambda}{\mathrm{NA}}. \tag{4.1}$$

Most commercial microscope objective lenses in the range of $40\times$ to $100\times$ magnification, have an effective working NA of about 1.4 (in immersion oil medium). In such a case, the resolution limit, from Eq. (4.1), is a little less than

[3]The numerical aperture of a lens measures its maximum light collection angle. It can be calculated as $\mathrm{NA} = n \sin\alpha$, where n is the refractive index of the imaging medium between the objective lens and the coverglass, and α is the maximum semi-angle subtended by the incident light cone accepted by the lens.

half the wavelength, λ. This means, if the light used has a wavelength of about 500 nm, the observable structures in the specimen should be larger than 250 nm. With low Signal to Noise Ratio (SNR), even the Abbe resolution limit (Eq. 4.1) is not realized. However, fluorescent proteins like Green Fluorescent Protein (GFP) are approximately 2.5 nm in diameter and classical dyes like 4',6-diamidino-2-phenylindole (DAPI), a DNA specific dye, are even lesser in size, and at least 100 times smaller than the resolution limit! The situation can be compared to a hypothetical case of trying to observe a rice grain but being limited to the size of a rugby ball! We will discuss more on this limit in Sect. 4.2.1.2. Increasing the NA can improve the resolution, but this is not practical. Higher NA means larger objective lenses that translates into additional cost. This also lowers the working distance between the objective lens and the specimen. The main limit on the NA is imposed by the refractive index of the specimen. For example, specimens immersed in water (refractive index 1.33) can only really support objectives of NAs up to 1.25. On the other hand oil at a refractive index of 1.515, can still support objectives of 1.45. If we use a shorter wavelength, it can induce more light scattering and it will also damage the viability of the biological specimen.

Since Abbe, the resolution of a microscope was considered to be limited by this barrier of half a wavelength. Once suitable image sensors and fast computers became available, it was clear that the resolution of the acquired image could be improved by twofold in the radial direction and sometimes fourfold in the axial direction by using computational methods such as deconvolution [67]. Computers can thus be considered as a secondary lens system, and when combined with a microscope optical system, the system is known as *deconvolution microscope* [45, 72].

Although there are many kinds of microscopes in the market, we will restrict our discussion to the resolution improvement of the Wide-field microscope (WFM) and the Confocal laser scanning microscope (CLSM). In the process, it is our intention to take our readers through the fascinating and beautiful world of fluorescence microscopy, and also walk them through the development of this *auxiliary computational lens*.

4.1.4 Chapter Overview

Public

While writing this chapter, we aimed to reach a wide range of readers, graduate students to cell biologists, interested in an introduction to deconvolution microscopy. We have assumed that the reader has some prior exposure to linear system theory and digital image processing, as are found in [4, 11]. For the basics of digital image processing in microscopy we refer to the articles [29, 58, 69]. For the basic principles of fluorescence, we suggest [21, 43], and to know more about fluorescent compounds (or *fluorophores*) that are used for the specific labeling of components in biological systems to [77, 81]. We will place more emphasis on developing the "auxiliary computational supplement to the optics", which is also the basis of many of the super-resolution methods, and leave the classical microscope optics to the huge number of books and articles devoted to this subject [10, 57]. In addition, the resources provided at the end may serve as a starting reference for those who are interested in developing and researching these techniques further.

Overview

Section 4.1.2 is written by keeping in mind those readers who wish to have a first-hand understanding of the phenomenon of fluorescence, and its application to fluorescence microscopy. The CLSM is introduced as a special case of fluorescence microscopy, and for a better understanding, it is compared with the WFM. Image resolution is limited primarily by noise, out-of-focus blurs and aberrations. In simple terms, blurring can be described as a non-random dispersal of light that occurs when it passes through the entire imaging system including the sample. The image acquisition process can be roughly divided into two parts: an optical part consisting of the lens system and the detection part consisting of either a photodetector or a camera. To create a faithful representation of the biological specimen, ideally, the image acquisition process should not introduce any distortions. This of course is almost never the case under practical conditions. When we will discuss the origin of these distortions and make an analysis of the limiting factors in Sect. 4.2.1.2, it will become clear why these distortions cannot be physically eliminated. In Sect. 4.2.1.4, we model this distortion process at the acquisition level, which is known as the forward problem. One can say that the inverse problem, which is the computational restoration of the specimen from the images, to be half-solved if the forward problem is well modeled. We present in Sect. 4.2.2 the different challenges that appear during a direct restoration approach and the different methodologies that can solve this reverse problem of estimating the specimen of interest. As the estimation process is in itself under-determined, a unique solution can be obtained only by introducing some a priori knowledge of the system and/or the specimen. We recall that a linear system is said to be under-determined if the number of unknowns in the system is larger than the number of known entities. Noise can be reduced by *denoising*, but in Sect. 4.2.3 we will show how it can also be contained with the restoration process, simultaneously, by introducing some constraints on the solution space through *regularization*. Case studies on the application of deconvolution are discussed, from the literature, in Sect. 4.2.4. Some recent ideas and developments in this field are discussed in Sect. 4.3 along with future research challenges. Finally, we end this chapter in Sect. 4.4 with a list of online resources on interactive tutorials, tools, and relevant open-source/commercial softwares that we feel are most likely to be of use. A list of the acronyms used and their expansions are provided at the end of this chapter.

Notations

All the scalar variables in this chapter are denoted by the lowercase (x), while the vectors by boldface lowercase alphabets ($\mathbf{x}$), and the matrices by the boldface uppercase alphabets ($\mathbf{X}$). As the images are discretized, their spatial support is $\Omega_s = \{(x, y, z) : 0 \leq x \leq N_x - 1, 0 \leq y \leq N_y - 1, 0 \leq z \leq N_z - 1\}$. By $\mathcal{O}(\Omega_s) = \{o = (o_{xyz}) : \Omega_s \subset \mathbb{N}^3 \mapsto \mathbb{R}\}$, we refer to the possible observable objects, and we assign the function $h : \Omega_s \mapsto \mathbb{R}$ as the microscope Point-spread function (PSF). The observed intensities are denoted by $i(\mathbf{x}) : \mathbf{x} \in \Omega_s$ (bounded and positive), while a 3-D convolution operation between two functions is denoted by '$*$'. However, when the same symbol is used as a superscript over a given function ($h^*(\mathbf{x})$), it represents the Hermitian adjoint operation. $\mathcal{F}$ is the forward Fourier transform and $\mathcal{F}^{-1}$ the inverse Fourier transform operations respectively. As the images are in 3-D, we present them in the 2-D format in this chapter by taking the maximum of the intensity along the z-direction for each pixel in the 2-D XY plane (or the Maximum Intensity Projection (MIP)) or by taking the maximum along the y-direction for each pixel in the 2-D XZ plane.

The objective lenses of a microscope are defined by their magnification (M), NA, and the medium in between the lens and the cover slip. For example, a lens of $63\times$ magnification, 1.2NA, and water as medium between the lens and cover slip is written as '$63\times/1.2$ water immersion'.

4.2 Development of the Auxiliary Computational Lens

4.2.1 Confocality Improves Resolving Power

4.2.1.1 Background Fluorescence Rejection

The euphoria with the developments in the WFM was short lived, because biologists quickly realized that the illumination excites almost the entire depth of the sample instead of just the focus plane. Therefore, the intensities detected are not localized from a single focal plane. For example, in Fig. 4.3a, we show a *Convallaria* sample that illustrates the inevitable presence of out-of-focus blur in a WFM image, that perturbs contrast and image resolution. When imaging biological cells, the amount of radiation must be limited. When exposed to irradiation, fluorescent molecules tend to react with molecular oxygen to produce free radicals that can damage subcellular components and compromise the entire cell. The eventual photochemical destruction of the fluorophore (or *photobleaching*) usually occurs after extensive exposure to light [25]. This can especially be the case in time-lapse microscopy. Moreover, live samples are sensitive to the amount of light that they are exposed to. Even plant cells that are especially attuned to light collection do not

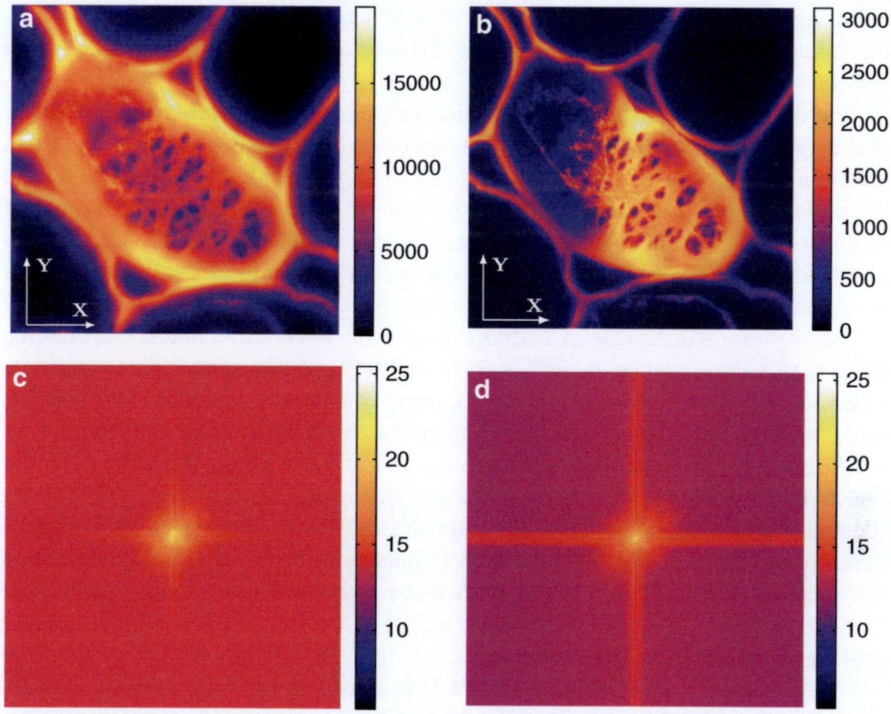

Fig. 4.3 Comparison between two fluorescence microscopes: Cross section through a *Convallaria* rhizome showing a blow up of a cortical parenchyma cell and its highly fluorescent wall (Courtesy INRA). This sample was imaged on a (**a**) WFM (maximum intensity is 20,000 IU) and (**b**) CLSM (maximum intensity is 3,000 IU) The intensities are linearly proportional to the number of photons collected. (**c**) 2-D Fourier transform of the WFM image. (**d**) 2-D Fourier transform of the CLSM image

have any protection against flux levels of $100–10,000$ times higher than the normal levels [69].

Although a CLSM is in principle a standard upright or inverted fluorescence microscope equipped with a high quality objective lens, there are several fundamental differences between it and a WFM. CLSM has a smaller Depth-of-field (DOF) [70], higher contrast, reduction of out-of-focus light, 'background rejection', and full three-dimensional image scanning ability. In a conventional WFM, the entire image is recorded onto the CCD camera. In contrast, in a CLSM, the specimen is irradiated sequentially point-by-point using a laser beam as excitation source and a pinhole that is "confocal" with this source. To be detected, the emitted light must pass through this confocal aperture before being detected by a PMT. Because fluorescence from out-of-focus planes will be out-of-focus at the plane of the pinhole, most of it will not pass to the PMT with the result that the light that is recorded comes from the optical section defined by the focus plane. The entire specimen is scanned in 2-D or 3-D, generating a 2-D or 3-D image. There are many online interactive and elegant virtual tutorials, from several microscope vendors, that compares WFM and CLSM imaging. These are available in the resources given in Sect. 4.4.

The development of the confocal microscope allowed the efficient removal of blur by physical means by filtering out the emission light that does not contribute to a well focused and blur-free image. The PMT can only detect light that passes through the pinhole. Since the diameter of the pinhole aperture can be adjusted, more or less out-of-focus light can be eliminated from the detected light. The physical elimination of the out of-focus light by the aperture placed in front of the detector is explained in Fig. 4.2b (a schema illustrating the confocal ray path). It is basically this configuration that gives a CLSM the possibility to create an image representing the emission fluorescence intensities corresponding to a thin optical slice or a single plane out of a thick fluorescent specimen. Depending on the objective lens used, this so called "optical sectioning property" can generate slices as thin as 500 nm.

Unlike in Fig. 4.3a, in the confocal image in Fig. 4.3b, a large fraction of the blur is eliminated and image details inside the sample become visible. This can be further illustrated by looking at the 2-D Fourier transform [32] of a single section. We notice that, for the WFM in Fig. 4.3c, the high frequency information is unavailable. It is also impossible to image a single focal plane using a WFM, because in the 3-D Optical transfer function (OTF), there is a cone of frequencies that are missing (the missing cone problem [14]). By comparing Fig. 4.3c with Fig. 4.3d, it might seem like there is more information in the low frequencies in Fig. 4.3c than in Fig. 4.3d. However, the CLSM image in Fig. 4.3d retains much of the higher frequencies, shown by the rays along the horizontal and vertical axes, providing sharp details. The resolution of the CLSM in terms of cut-off frequency can be improved by a factor of $\sqrt{2}$ [70], and the Full-width at half maximum (FWHM) is improved by a factor of $1/\sqrt{2} \approx 0.707$ for a very small diameter. In practice, as the signal level is very low, the level decreases with the square of the diameter.

Although the original design of Minsky [49] scanned the object by moving the specimen stage, current commercial adaptation of a CLSM scans the specimen by using galvanometric mirrors to tilt the laser beam as it passes through the back focal plane of the objective.

4.2.1.2 Fundamental Imaging Challenges

When using very thin samples obtained after chemical fixation and histological processing, sharp in-focus images can be obtained with high NA objective lenses. Nevertheless, imaging intact living cells or tissues that largely surpass the thickness of the imaging plane (DOF) suffers from a greatly reduced contrast since most fluorescence observed is out-of-focus light blurring in-focus details and hence reducing image contrast [56, 74].

Diffraction Barrier. Apart from fluorescence from nearby planes, the most important source of blur is *diffraction*. When light from a point source passes through a

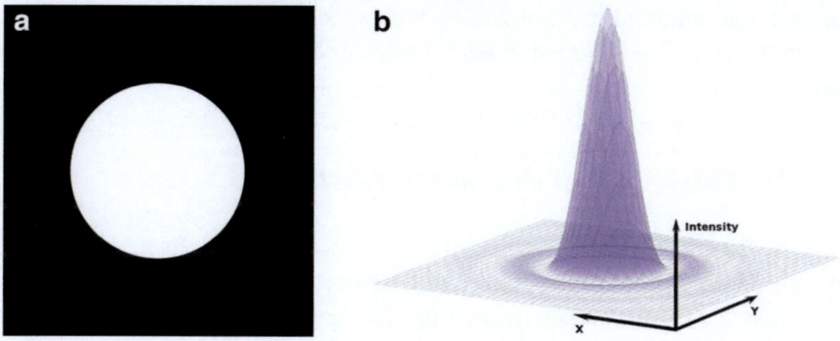

Fig. 4.4 (**a**) A diffracting aperture represented here as a uniform disc (Courtesy Inria), and (**b**) the 2-D Airy disk pattern after diffraction by the aperture (Adapted from Wikimedia Commons)

small circular aperture (Fig. 4.4a), such as an objective lens in a microscope, it does not produce a bright dot as a 2-D image, but rather a diffused circular disk, known as Airy disk, surrounded by much fainter concentric circular rings (Fig. 4.4b). This Airy disk is the system's response to an impulse (here a Dirac source) signal and it constitutes the best description of the imaging properties of the optical system (here a microscope) [10]. This non-random spreading of a point light source, or blurring, is a consequence of diffraction by a limited aperture (either the objective or the condenser lens or both). An image whose resolution is thus constrained is said to be "diffraction-limited" [32]. The experimental measurement of this barrier by Abbe (Eq. 4.1), which was discussed in Sect. 4.1, is exactly due to diffraction.

Effect of Pinhole Size. The size of the pinhole in a CLSM is calibrated against the diffraction rings. The sizes of the pinholes are annotated by their back-projected[4] values in Airy units (AU). We define 1 AU= $(1.22\lambda_{ex})/\mathrm{NA}$, where λ_{ex} is the excitation wavelength. As we reduce the pinhole size, the diffraction rings are blocked out. At 1 AU, the pinhole has about the size of the central principal maximum. From our experiments in [52], with a sample from the *Arabidopsis thaliana* plant immersed in water, it was observed that, with sizes $<$ 1 AU there might be more contrast and less blur, but there are very few photons detected at the PMT. This is on top of the low Quantum efficiency (QE)[5] of the PMTs (around 6 %). While, images obtained with pinhole sizes $>$ 3 AU have larger spot and more intense signal (signal from the region of interest and out-of-focus signal), but loss in contrast as well due to unwanted fluorescence. As far as the photon statistics are concerned, one could even say that the microscope almost behaves like a wide-field because of

[4]Back-projected diameter is the diameter of a pinhole in the object space. It is equal to the ratio between the real physical diameter of the pinhole and the total magnification of the system.

[5]Quantum efficiency for a photosensitive device measures the percentage of photons hitting the photoreactive surface that will produce an electron-hole pair. It is an accurate measurement of the device's electrical sensitivity to light.

the law of large numbers (Sect. 4.2.1.4). We can summarize that the usable pinhole range is between 1–3 AU, to establish a trade-off between the amount of noise and the blur.

4.2.1.3 The Physics of the Point-Spread Function

In order to mathematically reduce the non-focused blur from the image, the knowledge of the Point-spread function (PSF) is essential. By definition, the PSF is the response of the system to an impulse or a point source. In other words, an image of an infinitely small "point source" of light is spread out into a 3-D diffraction pattern. As every photon emitted from the sample has a different phase, the imaging is incoherent. In incoherent imaging, the PSF can be defined as the irradiance distribution in the image plane of a point source in the specimen plane. In a microscope, it reflects the fluorescence activity at every point in the specimen, and it constitutes the essential building block for restoring a specimen's intensities. Determining the PSF of an imaging system, as faithfully as possible, is a prerequisite to optimally eliminate blur. This is the instrumentation problem that was discussed in Sect. 4.1.2.4. This PSF can be either calculated theoretically using a mathematical model or determined empirically by acquiring 3-D image of a sub-resolution sized fluorescent microsphere [35, 46, 68].

PSF Model. Due to the difficulty in obtaining experimental PSFs and recognizing the variability in the data, theoretical PSF models are often used for deconvolution. As the diffraction effect is more noticeable in the far-field region, a point in the aperture would appear Fourier transformed in the far-field [10]. Thus, if we consider a converging spherical wave in the object space from the objective lens, the near-focus amplitude coherent PSF, denoted by h_A, can be written in terms of the inverse Fourier transform of the coherent OTF, denoted by OTF_A, as

$$h_\mathrm{A}(\mathbf{x}) = \int_{\mathbf{k}\in\Omega_f} \mathrm{OTF}_\mathrm{A}(\mathbf{k})\exp(i\mathbf{k}\cdot\mathbf{x})d\mathbf{k}, \tag{4.2}$$

where $i = \sqrt{-1}$, $\mathbf{x} \in \Omega_s$, $\mathbf{k} \in \Omega_f$ are the 3-D coordinates in the image and the inverse space respectively. We use the subscript A to denote the coherent nature of the light OTF and the PSF. The calculation in Eq. (4.2) involves a 3-D Fourier transform, which is often expensive to compute. Although, when we write the axial Fourier coordinate, denoted by k_z, as a function of the lateral coordinates $k_z = ((2\pi n/\lambda)^2 - (k_x^2 + k_y^2))^{1/2}$, where n is the refractive index of the medium between the lens and the specimen. The computation of the PSF is reduced to N_z (the number of 2-D sections) 2-D Fourier transforms:

$$h_\mathrm{A}(x, y, z; \lambda) = \int_{k_x}\int_{k_y} P(k_x, k_y, z; \lambda, \mathrm{NA})\exp(i(k_x x + k_y y))dk_y dk_x, \tag{4.3}$$

where $P(\cdot,\cdot,z)$ describes the overall complex field distribution in the pupil of a non-aberrated objective lens. The pupil function is a description of the magnitude and phase of the wavefront that a point source produces at the exit pupil of the imaging system. In simple terms, the Eq. (4.3) states that the field distribution at a point (x, y, z) in the image space can be obtained by applying Fourier transform on the overall pupil function. For a microscope with defocus, the pupil function can be written as [75],

$$P(k_x, k_y, z; \lambda, \mathrm{NA}) = \begin{cases} \exp\left(i2\pi z(\frac{n^2}{\lambda^2} - (k_x^2 + k_y^2))^{\frac{1}{2}}\right), & \text{if } (k_x^2 + k_y^2)^{\frac{1}{2}} < \frac{\mathrm{NA}}{\lambda}, \\ 0, & \text{otherwise,} \end{cases} \tag{4.4}$$

where the defocus z takes a value between $[-(N_z/2)\Delta_z, (N_z/2-1)\Delta_z]$ and Δ_z is the step size between two slices (or axial sampling). When the $z = 0$ (at the focal plane), the function in Eq. (4.4) is the disk in Fig. 4.4a. Just like the one-dimensional Fourier transform of a unit step function gives a *sinc* function, the 2-D Fourier transform of a disk should inductively give the Airy disk in Fig. 4.4b. In [52], the Debye's scalar diffraction model for a lens system was derived that serves as the basis for obtaining an analytical expression for the CLSM PSF. As standard detectors does not measure the coherent PSF but only the intensity PSF or the incoherent PSF, we write the model as:

$$h_{\mathrm{clsm}}(\mathbf{x}; \lambda_{\mathrm{ex}}, \lambda_{\mathrm{em}}, \mathrm{NA}) \propto |\Pi(\mathbf{x}) * h_{\mathrm{A}}(\mathbf{x}; \lambda_{\mathrm{ex}}, \mathrm{NA})|^2 \times |h_{\mathrm{A}}(\mathbf{x}; \lambda_{\mathrm{em}}, \mathrm{NA})|^2, \tag{4.5}$$

where $\Pi(\mathbf{x})$ is the pinhole model. If we assume the pinhole to be very small, it can be modeled as a Dirac, and the PSF can be simply calculated by multiplying the squared coherent excitation and the emission PSFs. The WFM PSF can also be calculated from Eq. (4.5) by simply taking the square-root of the ideal CLSM PSF (Algorithm 4). If the pinhole is larger, it is modeled as a simple uniform circular disc with its radius in nm or AU.

Algorithm 4 Theoretical confocal PSF calculation

Input: Voxel sizes, peak wavelengths λ_{ex} λ_{em}, numerical aperture NA, magnification M, refractive index n.
Output: PSF $h(\mathbf{x})$.
1: Calculate defocus z.
2: Pupil bandlimit: $k_{\max} \leftarrow (\mathrm{NA}/\lambda_{\mathrm{ex}})$.
3: **if** $(k_x^2 + k_y^2)^{1/2} < k_{\max}$ **then**
4: $\quad P(k_x, k_y, z; \lambda_{\mathrm{ex}}, \mathrm{NA}) \leftarrow \exp\left(i2\pi z(n^2 - {\lambda_{\mathrm{ex}}}^2(k_x^2 + k_y^2))^{1/2}/\lambda_{\mathrm{ex}}\right)$
5: **else**
6: $\quad P(k_x, k_y, z; \lambda_{\mathrm{ex}}, \mathrm{NA}) \leftarrow 0$.
7: **end if**
8: For every z, $h_{\mathrm{ex}}(x, y, z) \leftarrow \mathrm{IFFT}(P(k_x, k_y, z); \lambda_{\mathrm{ex}}, \mathrm{NA})$.
9: Repeat steps $2:7$ for λ_{em}, and for every z, $h_{\mathrm{em}}(x, y, z) \leftarrow \mathrm{IFFT}(P(k_x, k_y, z); \lambda_{\mathrm{em}}, \mathrm{NA})$.
10: For ideal pinhole, $h_{\mathrm{clsm}}(\mathbf{x}) \leftarrow |h_{\mathrm{ex}}(\mathbf{x})|^2 \times |h_{\mathrm{em}}(\mathbf{x})|^2$

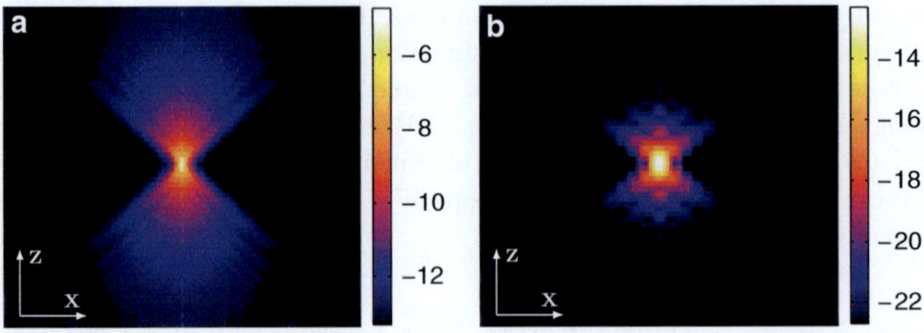

Fig. 4.5 The axial MIP PSF on a log-scale for a 40X/1.4 oil immersion lens as imaged from (**a**) WFM with a radial pixel size of 161 nm and a slice width of 300 nm; (**b**) CLSM with a radial pixel size of about 47 nm and a slice width of 166 nm

Under ideal conditions, the PSF of a CLSM or WFM is a fully symmetric 3-D diabolo (or hour-glass) shaped image of an ideal point source (Fig. 4.5a, b). Note how the spatial support of the kernel is larger for the WFM than the CLSM. Increase in the support in the spatial domain signifies a decrease in the support in the Fourier domain. This provides an additional weight to the argument of reduction in resolution for WFM. Much of the energy in the CLSM PSF is focussed in the central ellipsoidal region. For this reason, an 'ideal CLSM PSF' is also modeled as a 3-D Gaussian [87].

In the non-ideal case, the shape of the PSF is governed by the optical system used and the experimental settings. In [65], it was shown that the PSF is dependent on the lens, the wavelength and on the spherical aberration. In Fig. 4.6a–c, we illustrate this effect of the optical setup on the PSF. We note how the lowering of the NA from 1.4 (in (a)) to 0.5 (in (c)) increases the size of the spatial support (increased spreading of light) or decreases the frequency support.

Similarly, we can also show the dependence of the PSF on the fluorophore that is used for tagging the sample. For example, in Fig. 4.7, the radial and the axial plots of the different PSFs are shown using the same setup but by using four different chemical fluorophores. According to the excitation light that is used, the resolution decreases as we go from 405 to 633 nm. This is quite in conjunction with the Abbe limit (Eq. 4.1) we discussed earlier in Sect. 4.1.1. We conclude that a single PSF alone is insufficient for deconvolving the observed images under all settings. The practical way to handle this problem is to either generate the different PSFs before deconvolution or to store them off-line for the different objectives and settings. The effect of the specimen on the PSF is also important but a detailed discussion on this subject would take another chapter in itself!

Comments on Empirical PSF and It's Distillation. In a confocal, the PSF may be independently obtained by identifying point sources from the observation, and later used in combination with one of the image restoration algorithms. This identification procedure [22] requires manual intervention, because both the PSF and the point sources may not easily distinguished from noise. Due to this reason,

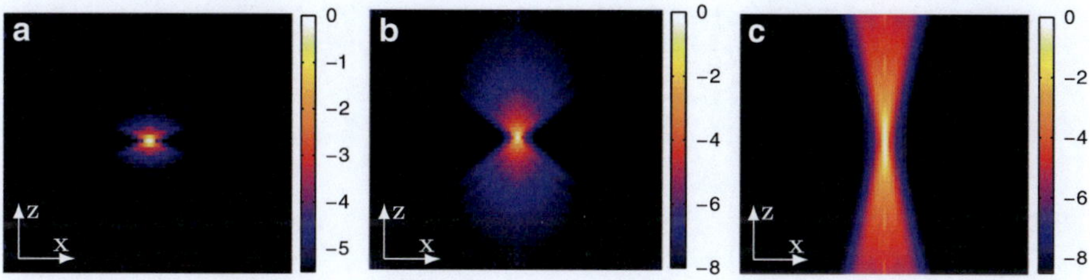

Fig. 4.6 The axial MIP of a WFM PSF on a log-scale for three different objectives: (a) 100X/1.4 oil immersion, (b) 40X/1.2 water immersion, (c) 20X/0.5 dry air

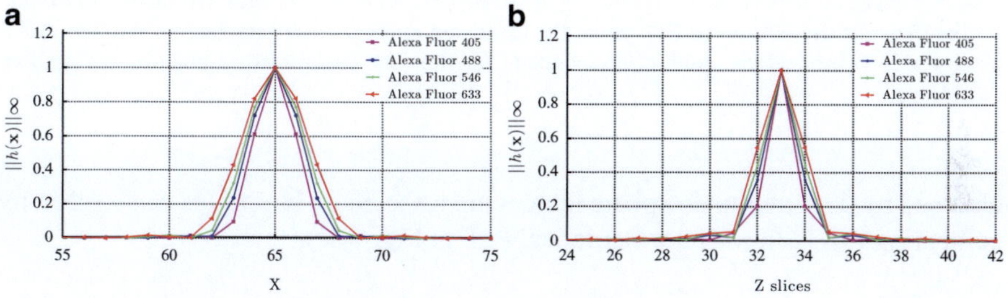

Fig. 4.7 Calculated ℓ_∞ normalized PSFs for a WFM with a 100X/1.4 oil immersion lens for four different fluorophores: AlexaFluor fluorophore with an excitation peak of 405 nm, 488 nm, 546 nm and 633 nm. (a) shows the radial plot, and (b) gives the axial plot

often sub-resolution fluorescent beads are imaged and used as PSFs after some processing. Due to the limited amount of light reaching the detector after the pinhole, such images have very low signal.

In both the above approaches, the PSFs are noisy, and can influence the results of a deconvolution algorithm [40]. Often, several images are collected from a single bead (Fig. 4.8a), and averaged to get the PSF. The SNR improvement as a result of averaging is roughly proportional to the square root of the number of volumes used for averaging. This process is called bead distillation. It involves extraction of a stack of single-bead images from the raw acquisition, aligning the selected images, and computing the averaged PSF [22, 24] sometimes assuming rotational symmetry around the z-axis.

In the first method of PSF distillation [53], the parameters of the theoretical model of the PSF in Eq. (4.5) are estimated from the acquired images by fitting. These parameters are re-inserted back into the model to distill the PSF that closest matches with the acquired image. In the second method, the phase of the back focal pupil is estimated from the intensity images and the PSF is generated from this retrieved phase [33]. Both these methods produce a noise-free and blur-free PSF for deconvolution. In Fig. 4.8b we show the axial MIP of one such distilled PSF from the bead image of Fig. 4.8a. As the PSF is radially symmetrical, we show here only the axial MIP. We used larger beads for this experiment, and we see that the PSFs

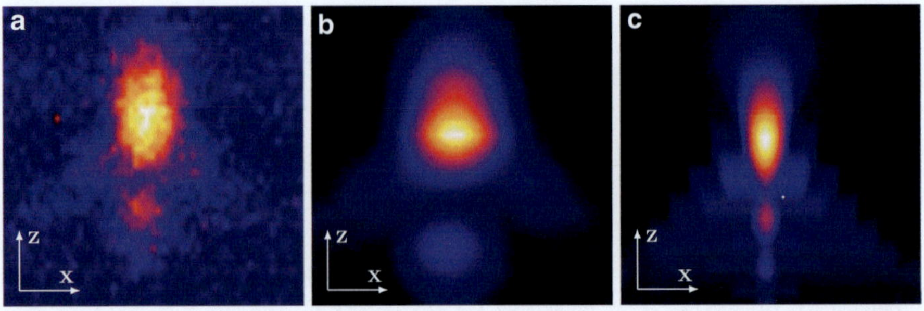

Fig. 4.8 170 nm fluorescent bead as imaged by a Zeiss LSM 510 confocal microscope fit with a 63X/1.2 water immersion lens. Axial MIP of the (**a**) observed bead images, (**b**) distilled PSF and (**c**) the theoretically calculated PSF using the estimated experimental parameters [53]. The radial pixel size is 37 nm and the width of each slice is 151 nm. The backprojected pinhole size is about 0.5 AU

obtained by distilling microsphere images are affected by its size (as can be seen by comparing Fig. 4.8b with the theoretical PSF in Fig. 4.8c).

4.2.1.4 Image Formation Model

Mathematical Blurring Model. In incoherent imaging, the distribution of intensity in the image volume is found by integrating the intensity distributions in the diffraction images of the PSF associated with each point in the specimen. Mathematically, the process of blurring is modeled as a convolution between the object intensity function, $o : \Omega_s \mapsto \mathbb{R}$ at the 3-D coordinate $\mathbf{x}' = (x', y', z') \in \Omega_s$ in the specimen volume, and the system PSF, $h : \Omega_s \mapsto \mathbb{R}$. By the scalar diffraction theory, the observed image $i : \Omega_s \mapsto \mathbb{R}$ (in the absence of any other degradation) can be written using a discrete framework as

$$i(\mathbf{x}) = \sum_{\mathbf{x}' \in \Omega_s} h(\mathbf{x} - \mathbf{x}')o(\mathbf{x}'), \quad (4.6)$$

where $\mathbf{x}$ is the 3-D coordinates in the image space. Here, the specimen coordinates were normalized by the magnification to make the model Linear shift invariant (LSI) [32]. Making it "shift invariant" implies that the PSF is constant over the field of view. However, this assumption holds true only large NA objective and when imaging the central part of the field. Equation (4.6) can be simply written as

$$i(\mathbf{x}) = (h * o)(\mathbf{x}), \forall \mathbf{x} \in \Omega_s, \quad (4.7)$$

where the interaction between the functions h and o is a '3-D convolution'. As i is the known entity and o the unknown, from the computational viewpoint, this equation could be inverted with the knowledge of the scanning system properties (here the PSF $h(\mathbf{x})$).

When the imaging system has been a priori calibrated, there is almost negligible offset in the detector and the illumination is uniform; i.e., no misalignment in the laser. For an aberration-free model, the convolution is LSI, and can be written in a matrix notation,

$$(h * o)(\mathbf{x}) = \mathbf{Ho}, \tag{4.8}$$

where the matrix $\mathbf{H}$ is *block-circulant* (if periodic boundary conditions are used) [11]. It can be shown that if the matrix is block-circulant and when $\mathbf{o}$ is a vector containing all the pixel in the lexicographical order, it is diagonalized by the Discrete Fourier transform (DFT). In addition, we will assume that the encircled or volumetric energy is constant with respect to the blurring process. This assumption is valid because blurring is a passive process and hence

$$\sum_{\mathbf{x}\in\Omega_s} o(\mathbf{x}) = \sum_{\mathbf{x}\in\Omega_s} (h * o)(\mathbf{x}). \tag{4.9}$$

Imaging Statistics. The CLSM is a low-photon imaging technique and it's PMT works as a photon counter. That is, it collects the photons, converts the electrical impulse to a discrete intensity value, after analog to digital conversion [58]. There are several parameters that influence the final intensity. These variables include the quantum yield,[6] bleaching [25], saturation of the fluorescent molecules, the limited transmittance of signal throughout the entire optical path, and the QE of the detector.

In confocal microscopy, the detected signal is "quasi-random", as statistics determine both how many photons are emitted on any particular trial but also what fraction of these are detected. If we denote by N, the total number of photons collected, then the counting process follows a Poisson distribution so that $[N \sim (\mathcal{P}(q_s) + \mathcal{P}(q_d))]$, where q_s is the total flux due to the specimen fluorescence, while q_d is the dark current flux [52]. Due to the pixel-wise scanning of the specimen using a laser, the image acquisition process is slow compared to wide-field based methods in which the entire image is acquired at once. This can be compensated by increasing the scan speed, but will inevitably lead to the reduction in the photons per pixel producing noisy images. The pinhole aperture is a key cause why finally very few photons reach the PMT from the specimen. For example, in Fig. 4.3b, the maximum intensity drops to about 3,000 IU while in Fig. 4.3a the maximum intensity is about 20,000 IU. As the intensity is linearly proportional to the number of photons collected, low intensity means lower photon counts. Although the flux q_s can be increased by averaging, for many applications this is not recommended as multiple scans of the same region are required, resulting in increased phytotoxicity (the degree of toxic effect produced by chemical compounds) and photo-bleaching.

[6] The quantum yield of a radiation-induced process is the number of times that a photon is emitted per photon absorbed by the system. This is essentially the emission efficiency of a given fluorophore.

Also, under non-ideal conditions, dark current flux is generated due to thermionic emissions in the dynodes, leakage currents, field emissions, electronic emissions by cosmic rays and sometimes stray indoor illuminations. In modern day detectors, the dark current noise from the detector can be minimized to less than one photon per pixel, while the normal signal levels in CLSM are 10–20 photons/pixel even for the brighter intensities. Theoretically, a confocal image taken with 1 AU pinhole setting has 40 % higher resolution than the image taken with WFM, but in practice, it loses a third of its in-focus photons. The statistical noise becomes an important limitation on the contrast and the spatial resolution [74]. Due to these reasons, in practice, a confocal's resolution (sans deconvolution) is at best comparable, but usually lower than the WFM! The SNR can be improved by computationally denoising the images.

If $\{i(\mathbf{x}) : \mathbf{x} \in \Omega_s\}$ denote the observed intensity (bounded and positive) of a volume, for the Gaussian noise assumption, the observation model can be written as

$$\gamma i(\mathbf{x}) = \gamma (h * o)(\mathbf{x}) + w(\mathbf{x}), \forall \mathbf{x} \in \Omega_s, \tag{4.10}$$

where $w(\mathbf{x}) \sim \mathcal{N}(0, \sigma_g^2)$ is an Additive white Gaussian noise (AWGN) with zero mean and variance σ_g^2, $1/\gamma$ is the photon conversion factor, so that $\gamma i(\mathbf{x})$ is the photon count at the detector. If we were to approximate a Poisson process by a Gaussian noise, the variance of the noise will depend on the mean intensity, $\gamma(h * o)(\mathbf{x})$. As mentioned earlier in Sect. 4.2.1.2, the high SNR case can be addressed by employing the Central limit theorem (CLT) for large photon numbers, where the AWGN model fits well. However, under low SNR, the AWGN model provides a poorer description of the fluorescence microscopy imaging. In such a case, the following Poisson model needs to be adopted:

$$\gamma i(\mathbf{x}) = \mathcal{P}\left(\gamma\left([h * o](\mathbf{x}) + b(\mathbf{x})\right)\right), \forall \mathbf{x} \in \Omega_s, \tag{4.11}$$

where $\mathcal{P}(\cdot)$ denotes a voxel-wise noise function modeled as an i.i.d. Poisson process. $b : \Omega_s \mapsto \mathbb{R}$ is a uniformly distributed intensity that approximates the low-frequency background signal caused by scattered photons and autofluorescence from the sample. The models in Eqs. (4.10) and (4.11) represents the forward problem of observing the specimen, given the object, the imaging process, and a model of the instrument.

4.2.2 *Resolution and Contrast Improvement by Deconvolution*

Deconvolution algorithms were initially used to increase the quality of microscopy images by post-acquisition processing, to remove the blur and increase the image resolution and contrast. This approach involves the mathematical inversion of the PSF of the microscope imaging system. Although originally developed to improve images acquired with a WFM, it was eventually realized that confocal images could also as well benefit from deconvolution. We call the process of resolution improvement by deconvolution as the addition of an auxiliary computational lens because the system works almost like a virtual lens. The aim of this section is to firstly describe the challenges, and then the methodological developments to highlight the importance of post-processing of images from confocal microscopes.

4.2.2.1 Challenges

If the process of imaging (the forward problem) is expressed by convolving a given specimen with a PSF (Eq. 4.7), deconvolution (the inverse problem) aims to obtain the specimen from the image. This assumes that the PSFs used in the forward and in the inverse problems are the same, and also that both the signal and the PSF data are free from noise.

As mentioned in the previous section, the PSF is not known a priori and needs to calculated or measured. In addition, as the convolution is a linear operation (Eqs. 4.7 and 4.8), deconvolution involves multiplying by the inverse of the matrix $\mathbf{H}$ (if it exists). The difficulty lies in the fact that even if $\mathbf{H}^{-1}$ exists, applying it to the noisy observations will only amplify the noise as the eigenvalues in $\mathbf{H}$ are very small [11]. For example, if the observation in matrix notation is written as $\mathbf{i} = \mathbf{H}\mathbf{o} + \mathbf{w}$, where $\mathbf{w}$ is the AWGN term, then, the specimen function can be estimated as $\hat{\mathbf{o}} \approx \mathbf{H}^{-1}\mathbf{i} - \mathbf{H}^{-1}\mathbf{w}$. As $\mathbf{H}$ is ill-conditioned,[7] the term $\mathbf{H}^{-1}\mathbf{w}$ will drastically damage the unknown image $\mathbf{o}$ we are looking for by amplifying the noise.

The ill conditioning of the matrix $\mathbf{H}$ is due to very small eigenvalues of $\mathbf{H}$ that implies very high eigenvalues of $\mathbf{H}^{-1}$. The fact that $\mathbf{H}$ has always very small eigenvalues (when they are non-zero) comes from the fact that $\mathbf{H}$ models a low pass filtering. As $\mathbf{H}$ is block circulant, it is diagonalized by the 3-D Fourier transform, and the eigenvalues of $\mathbf{H}$ matrix are exactly given by the 3-D Fourier transform of the PSF. As the PSF is a low pass filter, the eigenvalues corresponding to its high frequency coefficients are quite null or exactly null. So, there are two cases that are probable:

- The PSF cuts off certain spatial frequencies, and $\mathbf{H}$ is not invertible. In this case, the lost frequency information cannot be regained by inversion.
- There are only non-null PSF spatial frequencies, but these are very small at the higher frequencies. In this case, the inversion is dominated by the noise amplification.

In the last case, the noise amplification can be prevented and the lost high frequencies can be restored, by introducing a priori constraints (for example by appropriate penalty functions or information on the search specimen $\mathbf{o}$ as in Sect. 4.2.3), during the inversion process.

---

[7] A given problem is said to be ill-conditioned when it has a high condition number or the solution changes by a very significant amount in proportion to very small changes in the input data.

4.2.2.2 Methodological Developments

The earliest application of a deconvolution algorithm in microscopy was the nearest-neighbor algorithm developed by Agard et al. [2]. It is based on the assumption that the most of the blur within the focal plane is due to the light emitted from its two neighboring planes. The focal plane intensities are restored by comparing the observed intensity with these two planes and using a parameter to control the amount that each contributes. The advantage is that computation is fast (only $2N_z$ number of 2-D forward and inverse Fourier transforms). The disadvantage is that, in the presence of an analytical PSF/OTF model, it only uses partial information, and hence the deconvolution could be said to be axially incomplete for the volume. Since 1989, there were several developments in this field and most of them could be overall classified under two large categories: direct inversion approaches and iterative approaches. A survey on deconvolution algorithms was carried out earlier by Cannell et al. [14], Meinel [47], Sarder and Nehorai [64] and more recently in [53]. For the convenience of the reader, we have listed in Table 4.1 the most significant methods that have been applied to fluorescence microscopy. In this chapter, we will present some of these approaches, but by adopting a general probabilistic framework.

If the number of photons are large, we can assume that observation model follows a Gaussian distribution hypothesis as in Eq. (4.10), and the likelihood of the observation $i(\mathbf{x})$ given the specimen $o(\mathbf{x})$ is:

$$\Pr(i|o) = \left(\frac{1}{2\pi\sigma_n^2}\right)^{\left(\frac{N_x N_y N_z}{2}\right)} \prod_{\mathrm{x}\in\Omega_s} \exp\left(-\frac{\|i(\mathbf{x}) - (h*o)(\mathbf{x})\|_2^2}{2\sigma_n^2}\right), \qquad (4.12)$$

where σ_n^2 is the variance of the Gaussian noise. The negative logarithm of the likelihood function in Eq. (4.12) is:

$$\mathcal{J}_{\text{obs}}(o(\mathbf{x})) = \|i(\mathbf{x}) - (h(\mathbf{x}) * o(\mathbf{x}))\|_2^2,\ \mathbf{x} \in \Omega_s, \qquad (4.13)$$

where the terms independent of i and o were dropped from Eq. (4.12). Here, $\mathcal{J}_{\text{obs}} : \Omega_s \mapsto \mathbb{R}$ is a measure of fidelity to the data and it has the role of pulling the solution towards the observation data. It specifies as well the penalty paid by the system for producing an incorrect estimate of the scene. The specimen function can be estimated by maximizing the likelihood in Eq. (4.12) or equivalently minimizing the function $\mathcal{J}_{\text{obs}}(o(\mathbf{x}))$ in Eq. (4.13). We remark that the minimization of this cost function is equivalent to the minimization of the Mean-squared error (MSE). It is straightforward to show that the gradient of this cost function can be written as:

$$\nabla_o \mathcal{J}_{\text{obs}} = h^*(\mathbf{x}) * h(\mathbf{x}) * o(\mathbf{x}) - h^*(\mathbf{x}) * i(\mathbf{x}), \qquad (4.14)$$

where $h^*(\mathbf{x}) = h(-\mathbf{x})$ is the Hermitian adjoint of $h(\mathbf{x})$. As the functional $\mathcal{J}_{\text{obs}}(o(\mathbf{x}))$ is convex w.r.t $o(\mathbf{x})$, a minimum of the function $\mathcal{J}_{\text{obs}}(o(\mathbf{x}))$ is calculated at the point

Table 4.1 Examples of deconvolution algorithms from the literature of microscopy, classified by the type of noise model and methodology

No noise	Gaussian noise	Poisson noise
– Nearest neighbors [1] – No neighbors [50] – Inverse filter [29]	*Closed-form solutions* – Tikhonov [79] – Regularized linear least squares [60] – Wiener filter [80] *Iterative solutions* – Jansson van Cittert (JVC) [1] – Nonlinear least squares (NLS) [15]	– Maximum likelihood (ML) [36] – Maximum a posteriori (MAP) [23, 31, 82, 83]

where the gradient is null. Accordingly, the estimate of the function o is given as $\hat{o}(\mathbf{x}) = (h^* * h)^{-1}(\mathbf{x}) * (h^* * i)(\mathbf{x})$. This estimation method and other inversion techniques such as the inverse/pseudo-inverse filters are fast because they have a closed form solution for the estimate $o(\mathbf{x})$. Though, they intrinsically assume that the noise is AWGN and are thus valid only for images with large photon counts. As an exact PSF inverse does not exist for CLSM, and the OTF has only a finite region of support, these algorithms have difficulty in restoring the information beyond a certain cut off frequency and cannot be used to extend the reconstruction to non-measured frequencies. On the other hand, gradient-based iterative algorithms can produce negative intensities during successive iterations even if the initial estimate is positive. For gradient-based algorithms, we have noticed a relative lowering of the contrast in the estimates, in addition to noise amplification.

4.2.2.3 Multiplicative Richardson-Lucy Algorithm

The principal idea behind the multiplicative Richardson-Lucy (RL) algorithm [44, 62] is to maximize the likelihood $\Pr(i|o)$, given that the statistics follows a Poisson distribution. As the intensities of the individual voxels $i(\mathbf{x})$ are conditionally independent, we can say that the overall likelihood is a multiplication of the individual likelihoods at each voxel. That is

$$\Pr(i|o) = \prod_{\mathbf{x}\in\Omega_s} \frac{((h * o)(\mathbf{x}) + b(\mathbf{x}))^{i(\mathbf{x})} \exp\left(-(h * o)(\mathbf{x}) + b(\mathbf{x})\right)}{i(\mathbf{x})!}. \quad (4.15)$$

The mean of the above Poisson process is $(h * o)(\mathbf{x}) + b(\mathbf{x})$. Using the idea of the negative logarithm as in the previous case, the data energy function to be minimized is

$$\mathcal{J}_{\text{obs}}(o(\mathbf{x})) = \sum_{\mathbf{x}\in\Omega_s} ((h * o)(\mathbf{x}) + b(\mathbf{x})) - i(\mathbf{x}) \log\left((h * o)(\mathbf{x}) + b(\mathbf{x})\right). \quad (4.16)$$

As this estimator is convex in $o(\mathbf{x})$, the global minimum for $o(\mathbf{x})$ can be obtained at the point where the gradient $\nabla \mathcal{J}_{\text{obs}}(o(\mathbf{x}))$ w.r.t o is null. This leads to solving the following Euler-Lagrange expression

$$1 - \left(\frac{i(\mathbf{x})}{(h * o)(\mathbf{x}) + b(\mathbf{x})}\right) * h^*(\mathbf{x}) = 0. \tag{4.17}$$

In an iterative form, the estimation can be written as:

$$\hat{o}^{(n+1)}(\mathbf{x}) = \left\{\left(\frac{i(\mathbf{x})}{(h * \hat{o}^{(n)})(\mathbf{x}) + b(\mathbf{x})}\right) * h^*(\mathbf{x})\right\} \cdot \hat{o}^{(n)}(\mathbf{x}). \tag{4.18}$$

This RL algorithm given in Algorithm 5, for the Poisson statistics, is a form of the Expectation-maximization (EM) algorithm for Maximum likelihood (ML). The first estimate in this iterative procedure is usually set to either the observed image/mean of the observation or a smoothed version of the observation. We also notice in Eq. (4.18) that if the initial estimate $\hat{o}^{(0)}(\mathbf{x})$ is positive, the successive estimates remain positive as well.

The earliest application of this algorithm was in astronomy and later it was used in biological image processing as discussed by Holmes in [36]. As the iterative algorithms are time consuming, Biggs et al. showed in [7] a simple way to accelerate this. Like the previous direct inversion, this iterative algorithm does not allow reconstruction of information outside the spatial frequency bandwidth. However as it includes a positivity constraint, it is less ill-posed than the previous inversion (in the Gaussian noise case). Moreover, as it is an iterative process, it can be stopped before convergence which prevents noise amplification. Undeniably, the RL algorithm is the most popular deconvolution algorithm for microscopy, but as the inversion process is ill-posed, when $n \to \infty$, the solution diverges due to noise amplification (ringing artefact). An *ad hoc* approach to avoid the divergence of the solution is to manually terminate the algorithm at a certain number of iterations.

Algorithm 5 The RL deconvolution algorithm

Input: Observation $i(\mathbf{x}) \forall \mathbf{x} \in \Omega_s$, background $b(\mathbf{x}) \geq 0$, criterion $\varepsilon \in \mathbb{R}^+$.
Output: Restored specimen $\hat{o}(\mathbf{x})$.
1: Calculate PSF $h(\mathbf{x}) \in \mathcal{O}$ (Eq. 4.5),
2: Initialize: $n \leftarrow 0$, $\hat{o}^{(n)}(\mathbf{x}) \leftarrow \text{Mean}(i(\mathbf{x}))$.
3: **while** $|\hat{o}^{(n)} - \hat{o}^{(n-1)}|/\hat{o}^{(n)} \geq \varepsilon$ **do**
4: {Deconvolve: $\hat{o}^{(n+1)}(\mathbf{x})$ by Eq. (4.18).}
5: {Sub-space projection (scale): $\hat{o}^{(n+1)}(\mathbf{x})$ for flux preservation Eq. (4.9) }
6: {Set: $\hat{o}^{(n)}(\mathbf{x}) \leftarrow \hat{o}^{(n+1)}(\mathbf{x})$ and $n \leftarrow (n+1)$.}
7: **end while**

4.2.3 Implicit Denoising by Regularization

To address the problem of noise amplification (Sect. 4.2.2.1), most deconvolution approaches adopt a strategy to reduce the noise. For example, in the iterative algorithms mentioned in the previous section, as the number of iterations increases, ideally, the images should appear sharper until the final solution is reached. However, in practical situations, the algorithm is terminated before divergence of the solution or amplification of the noise. There is thus a compromise to be made between the desired sharpness of the image (or high frequencies to be restored) and the amount of noise amplified. This happens because the algorithm is unable to find a stable noise-free solution. This can be overcome by introducing a smoothness constraint into the problem, as a priori, that causes the algorithm to stabilize.

Another approach is to adopt the Tikhononv methodology. In the 1960s, Tikhonov laid down the theoretical basis of modern inversion methods by introducing the concept of regularized solutions. That is, we only search a solution near the observations, as $||i(\mathbf{x}) - h(\mathbf{x}) * o(\mathbf{x})||^2 < \sigma^2$, and within this set of solutions we search for a smooth solution, for example by jointly minimizing $||\nabla(o)||^2$. Tikhonov formalizes this trade-off between fidelity to the data and regularity by defining regularized solutions as those that minimize a joint criterion. He showed that the problem becomes well-posed, if it is reformulated using this joint criterion. The encoding of uncertain or partial prior information can be envisaged within the variational framework (see for example [3] for a review), or within the following Bayesian probabilistic framework as we do below. Accordingly, the posterior probability is

$$\Pr(o|i) = \frac{\Pr(i|o)\Pr(o)}{\Pr(i)}, \tag{4.19}$$

where $\Pr(o)$ is a p.d.f (the prior) from which o is assumed to be obtained. By using the Bayesian formula in Eq. (4.19), a rigorous statistical interpretation of regularization immediately follows. $\hat{o}$ is obtained by using the Maximum *a posteriori* (MAP) estimate or by minimizing the negative logarithm of the a posteriori as

$$\hat{o}(\mathbf{x}) = \underset{o \geq 0}{\operatorname{argmax}} \Pr(o|i) = \underset{o \geq 0}{\operatorname{argmin}}(-\log[\Pr(o|i)]). \tag{4.20}$$

As $\Pr(i)$ does not depend on o or h, it can be considered as a normalizing constant, and it will hereafter be excluded from the estimation procedure. The minimization of the negative logarithm of $\Pr(o|i)$ in Eq. (4.20) can be rewritten as the minimization of the following joint energy functional:

$$\mathcal{J}(o(\mathbf{x})) = \mathcal{J}_{\text{obs}}(o(\mathbf{x})) + \mathcal{J}_{\text{reg}}(o(\mathbf{x})), \tag{4.21}$$

where:

- $\mathcal{J}_{\text{obs}}(o(\mathbf{x}))$ is the data fidelity term. It corresponds to the term $\Pr(i|o)$ from the noise distribution (discussed in previous section), and
- $\mathcal{J}_{\text{reg}}(o(\mathbf{x}))$ is the prior object energy. It corresponds to the penalty term $\Pr(o)$ on the object that ensures smoothness of the solution.

In Bayesian statistics, a MAP estimate is a mode of the posterior distribution. It is closely related to Fisher's method of ML that was mentioned earlier, but employs an augmented optimization objective which incorporates the prior distribution over the quantity one wants to estimate. MAP estimation can therefore be seen as a regularization of an ML estimation. In MAP, there is at least one variable parameter, $\lambda \in \mathbb{R}^+$, which usually controls the algorithm output by regulating the noise and the high frequencies. This controls the trade-off between the data fidelity term and the regularization term in Eq. (4.21).

If the Power spectral density (PSD) of the noise (denoted by $P_w(\mathbf{f})$) and the object (denoted by $P_o(\mathbf{f})$) respectively are known, then with the AWGN likelihood in Eq. (4.12) the Wiener deconvolution is [11, 80]:

$$\hat{o}(\mathbf{x}) = \mathcal{F}^{-1}\left(\frac{\text{OTF}^*(\mathbf{f})}{\left(|\text{OTF}(\mathbf{f})|^2 + \lambda\left(\frac{P_w(\mathbf{f})}{P_o(\mathbf{f})}\right)\right)}\mathcal{F}(i(\mathbf{x}))\right), \tag{4.22}$$

where $\mathcal{F}^{-1}$ is the 3-D inverse Fourier transform, $\text{OTF}(\mathbf{f})$ is the incoherent optical transfer function, and $\text{OTF}^*(\mathbf{f})$ is its complex conjugate. We can see that when $\lambda = 0$ in Eq. (4.22) or in Algorithm 6, the deconvolution is by a simple inverse filter.

The RLS filter [60, 79] in Table 4.1 is another approach that uses a pseudo-inverse and an a priori on o to find a smooth solution. The choice of regularization functionals $\mathcal{J}_{\text{reg}}(o(\mathbf{x}))$ in Eq. (4.21) is important because it integrates the a priori knowledge about the expected solution into the reconstruction process. In this respect, the quadratic regularization terms with ℓ_2 norms (as in Tikhonov-Miller [48, 79]) attracted most attention. This is primarily because they were the first to be introduced and also due to their computational simplicity. Nevertheless, such regularization techniques cannot generate reconstructions with sharp edges, and as a result singular regularization energies, especially those of ℓ_1 norm type (described in next section), have become quite popular.

Algorithm 6 The Wiener deconvolution algorithm

Input: Observation $i(\mathbf{x})\ \forall \mathbf{x} \in \Omega_s$, PSDs $P_w(\mathbf{f})\ P_o(\mathbf{f})$, parameter $\lambda \in \mathbb{R}^+$.
Output: Restored specimen $\hat{o}(\mathbf{x})$.
1: Calculate PSF $h(\mathbf{x})$ (Eq. 4.5) and the $\text{OTF}(\mathbf{f})$ by FFT,
2: Deconvolve: $\hat{o}(\mathbf{x})$ by Eq. (4.22).
3: Sub-space projection (real): $\hat{o}(\mathbf{x}) \leftarrow \text{Real}(\hat{o}(\mathbf{x}))$.

4.2.3.1 Total Variation Regularization

We observed a fixed region of the plant *Arabidopsis thaliana*, with a Zeiss LSM 510 confocal, for three different pinhole settings. The gradients of these observed image volumes were numerically calculated along the x-direction as $\nabla_x i(\mathbf{x})$ and their histograms are plotted in Fig. 4.9. We mentioned that as the size of the pinhole is increased, the observed images have lesser detail due to lowering of contrast and lesser resolution. From the histograms of the gradients of the observed images, we notice that the histogram tends to have longer tails when the pinhole sizes are smaller and smaller tails when the pinhole sizes are larger. We also notice that there is a large difference between the 1 AU distribution and the 2 AU, but a negligible difference between the 2 AU and the 5 AU distribution. This is another reason to confine the working pinhole size in the CLSM to 2–3 AU. Also, we can say that the observed volumes that have sharper edges tend to have longer tails in the gradients rather than volumes that are out-of-focus. With this as the basis, we introduce the following regularization model that can produce restored images with longer gradient distribution tails.

The object prior distribution in Eq. (4.19) is:

$$\Pr(o) = Z_\lambda^{-1} \exp\left(-\lambda \, \mathrm{TV}(o(\mathbf{x}))\right), \text{ with } \lambda \in \mathbb{R}^+, \tag{4.23}$$

where Z_λ is the *partition function*, $\mathrm{TV}(o)(\mathbf{x})$ is the Total variation (TV) operation on $o(\mathbf{x})$, and λ is the regularization parameter described earlier. The TV function was first described in [63] as an iterative denoising method. It was introduced as a spatial sparsity measure of images by calculating the total amplitude intensity variations or oscillations in the image. The discrete semi-norm definition of TV that we will use in this chapter, reads as follows:

$$\mathrm{TV}(o) = \sum_{\mathbf{x}} |\nabla o(\mathbf{x})|. \tag{4.24}$$

Although a large spectrum of regularization functionals exist in literature, in this chapter, we use an ℓ_1 norm based regularization because it is known for its edge preserving quality and convexity. It also does a non-isotropic smoothing that acts along the edge direction but not in the direction orthogonal to the edge. The TV can be used as a constraint as it provides an intuitive and precise mathematical framework to characterize the piecewise regularity of objects in an image. Minimizing the TV-norm as introduced by Charbonnier et al. [17], Rudin et al. [63], corresponds to constraining the number of discontinuities in an image. The TV regularization is also well suited for biological images, where the structures and the background provide very low gradient values, while a finite set of edges provide high gradient values. Other motivations for using TV are the reduction of noise and the realization of nearly homogeneous regions with sharp edges.

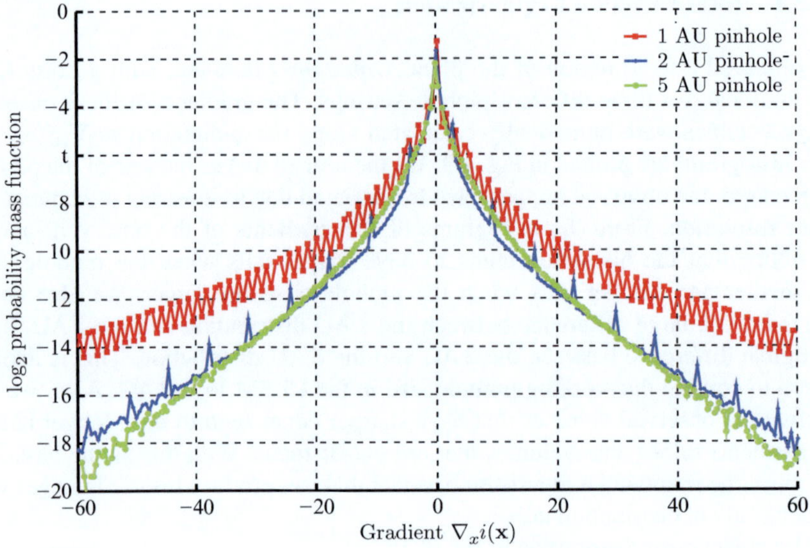

Fig. 4.9 Histograms of the gradient along the x-direction of the specimen under three different confocal pinhole settings. There is large difference between the histograms for 1 and 2 AU but very less difference between 2 and 5 AU

From a numerical perspective, $|\nabla o(\mathbf{x})|$ is not differentiable in zero. An approach to circumvent this problem is to instead consider the smoothed isotropic, discrete definition (by approximating the signal derivative in terms of finite differences over the sampling distance) as:

$$|\nabla o(x, y, z)|_{\varepsilon} = \sqrt{|\nabla o(\mathbf{x})|^2 + \varepsilon}, \tag{4.25}$$

where ε is an arbitrarily small value ($<10^{-3}$). When this is used as an a priori model for the object, we have the following smoothed TV regularization functional [63]:

$$\mathcal{J}_{\text{reg}}(o(\mathbf{x})) = \lambda \sum_{\mathbf{x}} |\nabla o(\mathbf{x})|_{\varepsilon}. \tag{4.26}$$

For the *partition function*, $Z_{\lambda} = \sum_{o \in \mathcal{O}} exp(-\lambda \sum_{\mathbf{x}} |\nabla o(\mathbf{x})|_{\varepsilon})$, to be finite, we restrict the possible values of $o(\mathbf{x})$ so that the numerical gradient $|\nabla o(\mathbf{x})|$ is also bounded. The resulting functional $\mathcal{J}(o(\mathbf{x}))$ (Eq. 4.21) is the sum of Eqs. (4.16) and (4.26)

$$\mathcal{J}(o(\mathbf{x})) = \sum_{\mathbf{x} \in \Omega_s} ((h * o)(\mathbf{x}) + b(\mathbf{x})) - \sum_{\mathrm{X} \in \Omega_s} i(\mathbf{x}) \log ((h * o)(\mathbf{x}) + b(\mathbf{x}))$$

$$+ \lambda \sum_{\mathbf{x} \in \Omega_s} |\nabla o(\mathbf{x})|_{\varepsilon}. \tag{4.27}$$

In Eq. (4.27), if we choose $h(\mathbf{x})$ to be a Dirac function, then the minimization of the above energy function gives a denoised image in the successive iterations, while a non-Dirac PSF leads to simultaneous deconvolution and denoising. The main challenge in minimizing this functional is the strong nonlinearity of the Poisson based negative log-likelihood functional in Eq. (4.16). We use the 3-D numerical implementation of the regularization algorithm for CLSM as described by Dey et al. [24], Pankajakshan [52], Pankajakshan et al. [54]. There are many approaches proposed in literature to minimize the above cost function. In the following paragraph, we will discuss one such method based on the RL algorithm with TV regularization

As in the case of the RL algorithm, the Euler-Lagrange equation for minimizing $\mathcal{J}(o|i)$, given by Eq. (4.27), with respect to $o(\mathbf{x})$ is

$$1 - \left(\frac{i(\mathbf{x})}{(h * o)(\mathbf{x}) + b(\mathbf{x})} \right) * h^*(\mathbf{x}) - \lambda \operatorname{div} \left(\frac{\nabla o(\mathbf{x})}{|\nabla o(\mathbf{x})|_\varepsilon} \right) = 0, \tag{4.28}$$

where $h^*(\mathbf{x})$ is the Hermitian adjoint operation on $h(\mathbf{x})$ and div stands for the divergence operator. Equation (4.28) can be solved for the object o by adopting the RL algorithm with TV regularization:

$$\hat{o}^{(n+1)}(\mathbf{x}) = \left(\frac{i(\mathbf{x})}{(h * \hat{o}^{(n)})(\mathbf{x}) + b(\mathbf{x})} * h^*(\mathbf{x}) \right) \cdot \frac{\hat{o}^{(n)}(\mathbf{x})}{1 - \lambda \operatorname{div} \left(\frac{\nabla \hat{o}^{(n)}(\mathbf{x})}{|\nabla \hat{o}^{(n)}(\mathbf{x})|_\varepsilon} \right)}, \tag{4.29}$$

where $(\cdot)$ denotes the Hadamard multiplication (component-wise) and n, the iteration number for the deconvolution algorithm. We stop the deconvolution algorithm as soon as the difference measure between two successive iterations is smaller than a specified threshold. The term $\operatorname{div}\left(\nabla \hat{o}^{(n)}(\mathbf{x}) / |\nabla \hat{o}^{(n)}(\mathbf{x})|_\varepsilon\right)$ can be numerically discretized with the central finite differences and the *minmod* scheme [23, 52]. Unlike in the RL algorithm, the introduction of regularization does not always ensure that the successive estimates are positive even if the initial estimate is positive. But, a fluorescence distribution is known to be greater than or equal to zero at all points in space. In order to constrain the reconstructed sample to positive values, we force values below a minimum intensity to that minimum intensity after each iteration step. Another alternative suggested in [83] is to operate on the real-valued square-root of the image rather than the intensity value. These steps are given in the Algorithm 7.

In order to test this algorithm, we simulated a fluorescent bead of size 100 nm with a radial sampling of 47 nm and axial sampling of 166 nm to respect the Nyquist criterion. This bead was assumed to be observed using a 1.3/40X oil objective lens and then corrupted by the photon statistics with a $\gamma = 5$. The three line profile intensities of the original object, the blurred object, and the observed object are plotted in the Fig. 4.10 along with the intensity profile of the restored bead after using the RL-TV algorithm. We noticed that the FWHM of the observed bead is much larger than that of the original intensity, while the restoration has almost the same intensity and FWHM as that of the original.

Algorithm 7 The RL-TV deconvolution algorithm

Input: Observation $i(\mathbf{x})\forall\mathbf{x} \in \Omega_s$, background $b(\mathbf{x}) \geq 0$, regularization parameter $\lambda \in \mathbb{R}^+$, criterion $\varepsilon \in \mathbb{R}^+$.
Output: Restored specimen $\hat{o}(\mathbf{x})$.
Calculate PSF $h(\mathbf{x}) \in \mathcal{O}$ (Eq. 4.5),
Initialize: $n \leftarrow 0$, $\hat{o}^{(n)}(\mathbf{x}) \leftarrow \mathrm{Mean}(i(\mathbf{x}))$.
while $|\hat{o}^{(n)} - \hat{o}^{(n-1)}|/\hat{o}^{(n)} \geq \varepsilon$ **do**
  {Calculate $\mathrm{div}(\nabla\hat{o}^n(\mathbf{x})/|\nabla\hat{o}^n(\mathbf{x})|_\varepsilon)$ by the *minmod* scheme.}
  {Deconvolve: $\hat{o}^{(n+1)}(\mathbf{x})$ by Eq. (4.29).}
  {Sub-space projection (scale): $\hat{o}^{(n+1)}(\mathbf{x})$ for total number of photons preservation $\sum_{\mathbf{x}} o^{(n)}(\mathbf{x}) = \sum_{\mathbf{x}} i(\mathbf{x})$ }
  {Sub-space projection (positivity): Set $\hat{o}^{(n+1)}(\mathbf{x}) < 0$ to zero.}
  {Set: $\hat{o}^{(n)}(\mathbf{x}) \leftarrow \hat{o}^{(n+1)}(\mathbf{x})$ and $n \leftarrow (n+1)$.}
end while

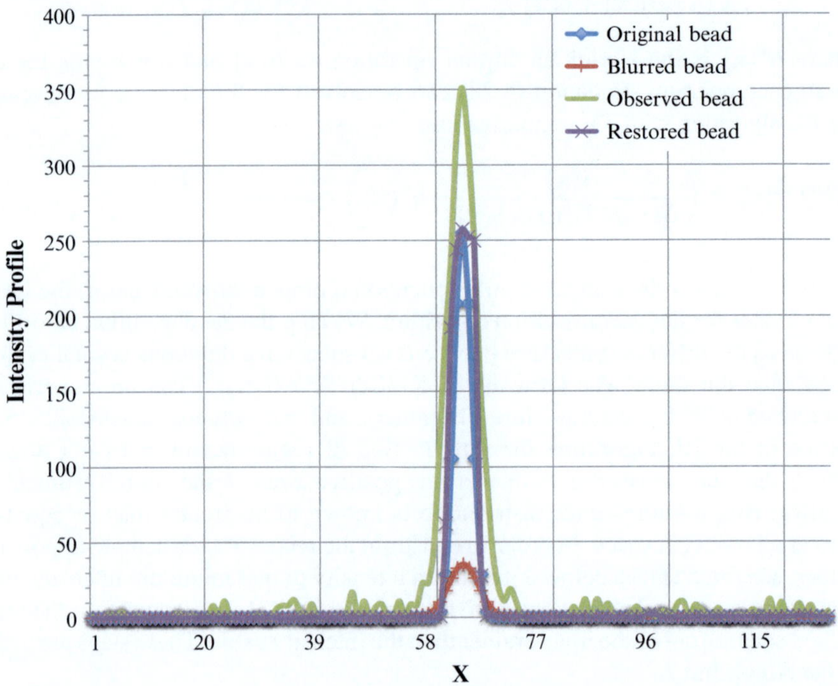

Fig. 4.10 Comparison of the radial intensity profiles for an original 100 nm fluorescent simulated bead sample, the blurred bead without any noise, the observed bead (after removing the background), and the restored bead using deconvolution. The result was obtained with a regularization parameter $\lambda = 0.002$, and the iterations were terminated at the 100th iteration while the convergence was after about 50 iterations

In addition to simulated data, we tested the algorithm on the *Convallaria* specimen of Fig. 4.3b in Sect. 4.2.1.1. This thin sample was imaged using a

Zeiss LSM 510 fit with a 1.3/40X oil immersion lens. The voxel sizes were (53.6, 53.6, 129.4)nm and the pinhole was set to about 33μm (backprojected radius 0.55 AU). In comparing the result of the deconvolution algorithm in Fig. 4.11b with the original image in Fig. 4.11a, we can see that a significant amount of background fluorescent was rejected after the deconvolution, even though the sample was imaged on a confocal. There are some cytoplasmic components that are better resolved in the images Fig. 4.11b, d than in Fig. 4.11a, c. From the Fourier transforms in Fig. 4.11e, f, we also see that there is a significant increase in the band width (or resolution) after deconvolution.

4.2.3.2 Comments on Other Priors

In the literature, apart from TV regularization, many other methods were studied and compared, especially when applied to fluorescence microscopy. For example, in [82, 83], both Gaussian and Poisson models were considered for the noise in combination with Tikhonov, Entropy, Good's roughness regularization and also with no regularization (ML with uniform prior). The Gaussian prior is widely used in fluorescence imaging because it prevents noise amplifications and has excellent convergence properties.

In [48, 79], the authors introduced a regularization based on the ℓ_2 norm of the gradient of the image. However, it has been observed that natural image statistics rarely follow a ℓ_2 or Gaussian priors. In addition, ℓ_2 norms are isotropic as they smooth the image along both the tangent and the normal to the contour surface. In addition, from the example given in Fig. 4.9, it was shown that the gradients of biological specimens have heavy tailed distributions rather than Gaussian.

The images reconstructed with TV-based methods suffer from a loss in contrast. A way to overcome is by combining the TV with an additional constraint using for example Bregman [12] iteration. Wavelet decomposition can perform quite well by analyzing the different scales separately in the deconvolution process. In literature, there are many papers that are based on deconvolution of 2-D or 3-D images by using wavelets. The wavelet denoising can be substituted as a form of regularization before or after 3-D deconvolution [18, 73], or performed in an optimizing context [31, 84]. Denoising, as a pre-processing step, before deconvolution is not recommended as the relationship between the denoised image and the specimen is no longer related by a linear convolution operation.

4.2.4 Success Stories

In the recent edition of the *Handbook of Biological Confocal Microscopy* [57], the editor Dr. James B. Pawley summarizes the main message of the handbook with the following words: "As ever more studies are done in the field of 3-D fluorescence microscopy, deconvolution becomes even more important for reducing the apparent

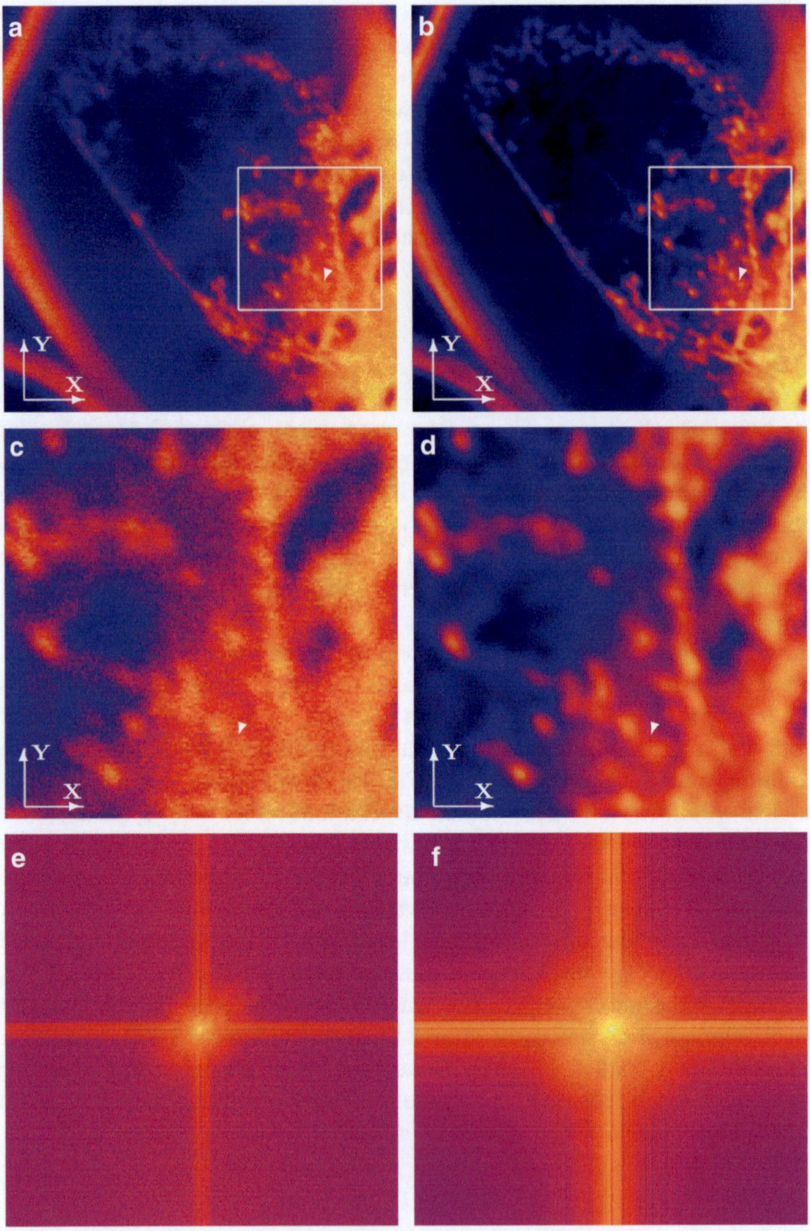

Fig. 4.11 A *Convallaria* cell from the *parenchyma* tissue shown in Fig. 4.3, but focussed on the cytoplasmic threads of a cell with small globular extensions. The fluorescence seen here is from the cell's cytoplasm, and is compared in (**a**) original confocal image, with (**b**) the restored image. (**c**) A zoom of the region in the original image and in (**d**) the restored image. The contrast is much improved and the *white arrow points* to one example of a cytoplasmic component that was resolved in the restored image. The 2-D Fourier transform of (**a**) is shown in (**e**), and the 2-D Fourier transform of (**b**) is shown in (**f**). We can note the increase in the band-limit in the frequency domain of the restored image

effect of Poisson noise, reducing the excitation light, and as a consequence reducing the effect of phototoxicity in living cells. The deconvolution should be considered as an essential tool in this regard."

In the following few paragraphs, we list some of the recent success stories from the literature which show the usefulness of deconvolution to biology.

4.2.4.1 Contrast Improvement for Imaging Plant Cells

Moreno et al. [51] described in an elegant review on "Imaging Plant Cells" that with particular biological samples, only deconvolution could reveal structures not observable using CLSM and multi-photon microscopy. The authors made use of the auto-fluorescence of a unicellular green algae to create an image of the ultrastructure of this organism. Both, confocal and multi-photon microscopy failed to produce sharp images taken over the entire cell depth. The main problem was that the high intensity illumination needed for confocal imaging caused signal bleaching as well as an emission shift from red to green [20]. Since wide-field imaging demands much less excitation power, deconvolution of the obtained wide-field data stack generated the contrast improvement needed to faithfully observe cell features.

4.2.4.2 Quantitative Analysis

In [78], the authors described that they could not resolve microtubular cytoskeleton components labeled with yellow fluorescent proteins fused to *tubulins* in the parasite. By contrast, when they deconvolved the image obtained from the wide-field microscope, they were able to visualize weak fluorescent signals from individual *microtubules* of the cytoskeleton. This allows the quantitative studies of very dim fluorescent structures in these thin living cells.

4.2.4.3 Improvement of Signal-to-Noise Ratio

The use of iterative deconvolution algorithms in live-cell imaging was also demonstrated by Platani et al. [59] for exploring the dynamics of the *Cajal body* within the nucleus of living human cells. They used a HeLa derived cell line stably expressing a fluorescent coilin protein to specifically label Cajal bodies. This line showed to be highly photosensitive, which required the use of short exposure times and imaging conditions with very low light levels. Once the SNR was improved, the detection of these tiny Cajal bodies became easy, because it permitted an automated segmentation and tracking algorithm to characterize their mobility in living cells. They also demonstrated that iterative deconvolution algorithms when applied in life cell microscopy proved to be essential to better visualize and explore the dynamics of these bodies within the nuclei of living human cells.

4.2.4.4 Localizing Macromolecules

As mentioned in Sect. 4.1.2, cell biologists are highly interested to precisely localize macromolecules in their sub-cellular compartments. To find out if two or more proteins possibly act in the same biological pathway, it is important to determine if they are co-localizing. Two proteins can be said to colocalize if one protein is present in the same voxel as the second protein. Since co-localization is of course resolution dependent, deconvolution may be able to increase the precision of an actual protein-protein relationship. In [67], the authors elegantly illustrates this and provides experimental data showing that deconvolution also decreases the possibility of false co-localization events. The authors used confocal microscopy and immunofluorescence labeling to determine the degree of co-localization of two receptor proteins Dihydropyridine (DHPR) and Ryanodine (RyR) as well as to follow the changes in their co-localization distribution during Cardiomyocytes development. This work clearly illustrated that deconvolution of confocal images has a positive impact on the reliability of quantitative co-localization measurements by increasing resolution (also verified in [41]) and decreasing false co-localization events.

4.2.4.5 Improving the Resolution of STED

In [85], the authors showed that linear deconvolution [79] could be used even in the STED microscopy technique, and revealed for the first time that nano scale arrangement of the protein SNAP-25 is not randomly spread out on the plasma membrane of a fixed neuronal cell, but organized in clusters of less than 60 nm in size. This important finding could have important implications for synaptic transmission in neurons. SNAP-25 is a part of the SNARE protein family, which co-determines the membrane site at which a vesicle may dock and fuse. In order to visualize the spatial distribution of this protein, it was labeled with primary and Atto532-stained secondary antibodies. Thus, STED in combination with deconvolution techniques not only provides sub-diffraction images but also gives new insights into the biological data.

4.2.4.6 Improvement in Volume Estimation

In [26], the authors compare the effective improvement in the quantitative image before and after image restoration. They measured the volume of the reference objects using the isotropic Fakir method. They found that in each experiment the estimates of the raw data volume were too large while measurements made on deconvolved images matched the actual object well. For example, when 2 μm fluorescent beads where imaged in oil immersion-medium, the radius of the specimen in the acquired image was calculated to be $2.28 \pm 0.66\,\mu\text{m}$ and after deconvolution as $1.99 \pm 0.49\,\mu\text{m}$.

4.3 Outlook

In the previous few sections, we have provided examples to show the usefulness of the deconvolution algorithm in improving the contrast, SNR and quantitative accuracy of chemically-fixed and living specimens. In addition, these algorithms complement recent microscopic techniques like STED and SIM. In fact, we recommend that deconvolution should be applied to all 3-D fluorescence microscope images irrespective of the modality used. In spite of the enormous research base that we have summarized above, we now highlight a few of the challenges that remain.

4.3.1 Algorithmic Developments

Since 2003, there has been many papers describing fast algorithms for minimizing the criterion such as the one in Eq. (4.27) or those with a general norm, such as ℓ_1 norm on wavelet basis coefficients, frame coefficient, or dictionary coefficients. ℓ_1 norm promotes sparse solution as an approximation of the ℓ_0 norm which simply counts the number of non zero coefficients. The difficulty of such a minimization is that the first term (data term) is convex but has no Lipschitz gradient, and that the second term (regularization) is not differentiable. In the context of convex analysis, several fast algorithms were developed, sometimes using proximal operators, duality, and augmented Lagrangian techniques as in [6, 19, 27, 30].

4.3.2 Search for an Ideal Prior

In Sect. 4.2.3, while we established the primordial role that prior knowledge plays in restoring the specimen intensities, an ideal representation of the specimen does not currently exist. In spite of the many existing priors for the specimen, there are none that can absolutely and completely define one biological specimen and yet be applicable to another sample. While the search for an ideal representation of the specimen is still an open problem, in computer vision, the estimation of the regularization parameter λ in Eq. (4.29) for non-Gaussian energy functions is another work in progress [5, 86].

4.3.3 Blind Shift-Varying Deconvolution

Most deconvolution software uses either a theoretical PSF (based on physical parameters) or an empirical PSF (deduced from the specimen). There are a few deconvolution approaches [13, 42], known as blind deconvolution, that estimate the PSF in addition to the object intensities from the observation. As in the case of the non-blind deconvolution algorithm, the blind approaches were developed primarily in the field of astronomy where the blurring kernel is often unknown and cannot be easily measured. This method was also adapted to other applications facing similar situations. We see from Eq. (4.16) that the object and the PSF are symmetric in the equation. So, intuitively any algorithm that estimates the object can also be used to estimate the PSF although with different solution constraints. The blind and non-blind algorithms assume that the PSF is shift invariant, i.e., it is the same for every point in the specimen. In reality, as the PSF will change depending on the specimen's optical properties, this assumption is not justified. At the very least, it varies in depth for microscopes with large NA. In [28], the authors showed that refractive index mismatch can result in intensity and phase variations in fluorescence microscopy. This means that the PSF at each pixel is a little different than that of its neighbor and an assumption of finding a universal PSF is an over-simplification. The current deconvolution methods [54] work very well for fixed specimens but not for live cells, because in the former case the refractive index had to be fixed to be constant and therefore the PSF is nearly space-invariant throughout the entire sample. More generally, the PSF can vary drastically within a living biological sample. As a consequence, there can be a serious deterioration of image quality particularly when deep 3-D imaging of live specimens.

To refine such data, computational methods involving space-varying deconvolution algorithms were developed. To a certain extent, these can take into account refraction index differences within the specimen. Currently, there are no known commercial or open-source software products that can correct the influence of the specimen-induced aberrations on the observed image. For example, accurately imaging the cell-components lying on the far side of a nucleus is virtually impossible without physically sectioning the sample. Nowadays, the development of more sensitive confocal microscopes and the advent of novel deep imaging devices such as Light Sheet based Fluorescence Microscope (LSFM), demonstrates that there is an increasing demand, in biomedicine, for high resolution deep imaging of living samples. Recently adaptive optics have been applied to pre-adjust the pupil phase and thereby correct for the aberrations induced by the specimen [8, 9, 38, 39, 71]. However, to achieve this computationally, using shift-variant blind deconvolution in 3-D is quite expensive in terms of Central processing unit (CPU) memory usage and time.

There are currently only a few libraries and toolboxes that utilize the power of the Graphical processing unit (GPU) to ease any of the calculations. These are listed in the next section. Apart from reducing the computational time, the iterative blind shift-varying algorithms need to be accelerated without affecting the solution of the optimization problem. Continued research to further refine deconvolution methods will be key to optimize the quality of images obtained by these methods.

4.4 Online Resources

The internet has a large volume of literature on fluorescence microscopy, deconvolution and softwares for analysis and visualization. While it is difficult to list them all, we have selected those that we consider relevant to this chapter.

4.4.1 Microscopy Primers

There are many online tutorials and demonstrations from the microscopy community which gives an idea of the working of the WFM and the CLSM.

Virtual Java Environment on WFM and CLSM
http://www.microscopyu.com/tutorials/java/virtual/confocal/index.html
http://www.olympusmicro.com/primer/virtual/confocal/index.html
From the given list, the user can select the specimen to image and the applet gives an impression of working with real microscope settings. The parameters that one can play with are the focusing into the samples, adjusting the other microscope parameters like the brightness, gain, and the scan speed.

4.4.2 Databases

There are also a few online sites where confocal images could be downloaded for visualization and computational analysis. The users may browse the data online or download it for reuse within the guidelines of each of the following site usage agreement. Here is a list of few of them.

Images used in this chapter
http://www-syscom.univ-mlv.fr/ANRDIAMOND/images.html
The images used in this chapter can be downloaded from the ANR DIAMOND project. These are primarily from a Zeiss LSM 510 confocal microscope and can be visualized using the softwares listed in the next section.

Simulating fluorescence microscopy images (CBIA, Masaryk University)
http://cbia.fi.muni.cz/simulator
The centre for biomedical image analysis (CBIA) also provides an on-line framework for simulating fluorescence microscopy images.

Images from BIG, EPFL
http://bigwww.epfl.ch/deconvolution/?p=bio
The EPFL Biomedical imaging group (BIG) have also a few trial images that are made available for validating the algorithms.

Image Database from La Jolia, California

http://ccdb.ucsd.edu/data/index.shtm

The Cell centered database (CCDB): The CCDB (La Jolla, California, USA) is a web accessible database for high resolution 2-D, 3-D and 4-D data from light and electron microscopy, including correlated imaging. Techniques range from wide field mosaics taken with multi-photon microscopy to 3-D reconstructions of cellular ultrastructure using electron tomography. While many of the data sets are derived from the nervous system, the CCDB contains data from many systems. For each dataset, the CCDB makes available the raw data, reconstructions derived from the raw data and any segmentations and analysis performed on the data. They also accept contributions of data from the scientific community.

4.4.3 Software

A large variety of software packages are available for deconvolution and visualization. The softwares can be either commercial or open-source packages, and each have their pros and cons. Roughly, deconvolution algorithms currently available in the market can be classified into two types. The first category increases the net contrast but might not remove the out-of-focus light. These are fast, involve moderate number of calculations, but the obtained images are non-quantitative. The second category represents more advanced image restoration algorithms that removes the out-of-focus light and conserve the quantitative fluorescence intensities.

4.4.3.1 Open-Source Packages

ImageJ developed at National institutes of health (NIH)

http://rsb.info.nih.gov/ij/

The single most widely used open-source package for microscopy is the Java-based engine ImageJ developed at the NIH. The ImageJ, when introduced, was truly revolutionary because it is light, easily portable and works on multi-OS without installation. The only requirement is that Java needs to be pre-installed. The software that is available online is the basic version and does not support all the microscope image formats. However, any additional feature can be easily added with the help of some 'plugins' and individuals can write their own 'macros' to automate the tasks.

McMaster Biophotonics plugin database

http://www.macbiophotonics.ca/downloads.htm

The ImageJ users community contributed a lot to the development of many plugins. A compiled list of some of the essential ones can be found at the McMaster Biophotonics facility site along with the instructions for the installation.

FIJI

http://fiji.sc/

As with any community maintained software, ImageJ quickly mushroomed with enormous plugins with very little organization and documentation. A group of researchers brought in a streamlined version of the ImageJ, the FIJI, with automatic updates, tested plugins and comprehensive documentation. It is important to mention that FIJI architecturally is the same as ImageJ. So all the plugins written on ImageJ work with FIJI as well. It can be described as a distribution of ImageJ together with Java, Java 3-D and many plugins organized into a coherent menu structure.

Iterative deconvolution plugin

http://www.optinav.com/Iterative-Deconvolution.htm

This is an iterative deconvolution plugin for ImageJ that was developed by Bob Dougherty.

Parallel iterative deconvolution plugin

http://sites.google.com/site/piotrwendykier/software/deconvolution/paralleliterativedeconvolution

The parallel iterative deconvolution by Piotr Wendykier, is essentially a deblurring plugin. It uses several iterative algorithms for deconvolution and it can also be customized to work on multiple computational threads.

DeconvolutionLab plugin

http://bigwww.epfl.ch/algorithms/deconvolutionlab

The DeconvolutionLab from the biomedical imaging group (BIG), EPFL, Lausanne is an ImageJ plugin that includes implementation of some of the state of the art algorithms (including the RL-TV algorithm mentioned in this chapter). However, the source of the plugin is not available in the public domain.

DeconvolutionJ plugin

http://rsbweb.nih.gov/ij/plugins/deconvolutionj.html

DeconvolutionJ is an ImageJ plugin that is based on the regularized Wiener Filter. It can handle 2-D and 3-D data sets.

Java-based GPU computing

http://www.jcuda.de/

To run some of these plugins on the NVIDIA® Compute unified device architecture (CUDA), the Java bindings may be required.

ICY

http://icy.bioimageanalysis.org/

ICY is an open source community software for bio-imaging. It is an integrated platform that aims to bridge the gap between developers and users. It is also based on the principle of developing plugins for various biological imaging applications. It offers a common set of tools to view and analyze data, an environment to efficiently develop algorithms, and a community-based website centralizing all the plugins and resources. Although, the plugin database is still in its nascent stages of development, it's platform is designed to natively handle large image data (2-D, 3-D, time-lapse sequences and unlimited channels, as memory might allow) and a wide variety of data formats.

Open microscopy environment (OME)

http://www.openmicroscopy.org

The main drawback of ImageJ or FIJI plugins are that the meta-data for each microscope are differently formatted and all the information cannot be made available in one universal format. The OME aims to bridge this gap and strives for a universal imaging format.

XCOSM

http://www.essrl.wustl.edu/~preza/xcosm/

http://www.ucalgary.ca/styslab/imagetrak

Technically speaking, the XCOSM was the earliest known deconvolution software available freely on the public domain. Although this software is not updated anymore and the source is unavailable in the public domain, it served as the predecessor for some of the other commercial and open-source softwares (like the ImageTrak) that were developed later.

MATLAB® codes

http://bioimageanlysis.org/~praveen/rltv/

There are other stand-alone open-source packages that are command-line based. One of it is the RL-TV algorithm implemented in this chapter. Although this is implemented in the commercial software MATLAB®, the codes are available as open-source.

DECONV

https://www.lsdcas.engineering.uiowa.edu/

http://code.google.com/p/iocbio/wiki/IOCBioMicroscope

Other examples of deconvolution algorithms run on command line are the DECONV [76] written in C++ accessible from the LSDCAS website, and the IOCBioMicroscope written in Python.

BioImage XD

http://www.bioimagexd.net/

BioImage XD is a collaborative open-source free software project, designed and developed by microscopists, cell biologists and programmers. It can be used for analysis and visualization. While writing this chapter, a deconvolution algorithm was unavailable but it is a feature that might be included in the next release.

Code acceleration by using GPUs

http://gp-you.org/

As mentioned earlier, some of these codes could be accelerated on some graphics cards. A starting place to look for them are the free GPU toolbox for the NVIDIA graphics card.

4.4.3.2 Commercial Packages

While the above packages are open-source, commercial microscope manufacturing companies have their own versions of visualization softwares that are free to download and use. Here are a few of them:

Zeiss LSM Browser and ZEN

http://www.zeiss.de/ZEN

The Zeiss LSM image browser as the name implies is the Zeiss' older version of the image browser for it's proprietary 'LSM' file format. The old version has been around for quite some time and was recently replaced by a new version, called ZEN. Both these versions can be downloaded from the Zeiss web site. However, we found the older version more intuitive to use than the newer version!

Leica LAS AF Lite

ftp://ftp.llt.de/softlib/LCSLite

Leica LAS AF Lite is the equivalent software for browsing the Leica SP2 and SP5 confocal images. The software can be downloaded from the FTP site of Leica.

Scientific volume imaging (SVI)

http://www.svi.nl/Download

http://www.svi.nl/FrontPage

http://www.svi.nl/BackprojectedPinholeCalculator

http://www.svi.nl/NyquistCalculator

Scientific volume imaging (SVI) are the foremost leaders in high-quality restoration and analysis programs for microscopy. Huygens software refers to the different image processing packages that are intended for restoration, interactive analysis and volume visualization of 2-D, 3-D, multi-channel and time series images from fluorescence microscopes. The deconvolution software is based on either the classical ML or the Tikhonov-Miller algorithm, and they need a commercial licence. However, they also have the freeware versions-the HuygensEssential and FreeSFP, which can be used for image reading/writing in many file formats, volume visualization, and a few basic image processing operations. This light version can be downloaded from their site. Apart from their softwares, they also have a public wiki site that provides a detailed explanation to the most frequently asked questions along with support tools such as Backprojected pinhole radius and Nyquist rate/PSF calculator.

MediaCybernetics AutoQuant

http://www.mediacy.com/index.aspx?page=AutoQuant

Other than SVI, the other commercial and popular vendor is MediaCybernetics that markets the AutoQuant deconvolution software. One of the essential features in this package, which Huygens does not have is AutoDeblur-the adaptive blind deconvolution routine for restoration. AutoQuant also added many other extensions for acceleration, constraining the estimation, and noise reduction.

Amira for scientific visualization and processing

http://www.amira.com/

Another commercial software for scientific visualization and processing is the Amira software. This boasts of a list of features such as slice alignment, image registration, Z-drop correction, etc that are not present in the above softwares. However, we found the 3-D deconvolution in Amira to be not very user friendly.

Volocity

http://cellularimaging.perkinelmer.com/downloads/

Volocity is designed to provide rapid, interactive, high resolution volume rendering of 3-D and 4-D data sets. It offers a choice of rendering methods so that the best results can be achieved from the given data set. A trial version can be downloaded from their online site.

Code acceleration by using GPUs

http://www.accelereyes.com/

http://www.mathworks.com/products/parallel-computing/

The final in the list of commercial software is code acceleration by using GPUs. Apart from the free version described earlier, the commercial ones are Accelereyes and Parallel computing toolbox.

Acknowledgements Part of the results presented in this chapter was funded by the P2R Franco-Israeli collaborative research program and the ANR DIAMOND project. The authors gratefully acknowledge Prof. Zvi Kam (Weizmann Institute of Sciences, Israel) and Dr. Jean-Christophe Olivo-Marin (Institut Pasteur, Paris, France) for several interesting discussions. Additionally, our sincere gratitude goes to the editors, and Prof. James B. Pawley (Department of Zoology, University of Wisconsin, Madison, USA) Dr. Francois Orieux (Paris Institute of Astrophysics, France) for their detailed reviews and suggestions.

List of Acronyms

ADMM	Alternating direction method of Multipliers
AU	Airy units
AWGN	Additive white Gaussian noise
BV	Bounded variations
CCDB	Cell centered database
CCD	Charge coupled device
CLSM	Confocal laser scanning microscope
CLT	Central limit theorem
CPU	Central processing unit
CUDA	Compute unified device architecture
DAPI	4',6-diamidino-2-phenylindole
DHPR	Dihydropyridine
DFT	Discrete Fourier transform
DOF	Depth-of-field
EM	Expectation-maximization
FFT	Fast Fourier transform
FT	Fourier transform
FWHM	Full-width at half maximum
GN	Gaussian noise
GFP	Green Fluorescent Protein

GPU	Graphical processing unit
IFFT	Inverse Fourier transform
i.i.d.	Independent and identically distributed
IU	Intensity units
JVC	Jansson van Cittert
LSFM	Light Sheet based Fluorescence Microscope
LSI	Linear shift invariant
MAP	Maximum *a posteriori*
MSE	Mean-squared error
MIP	Maximum Intensity Projection
ML	Maximum likelihood
NA	Numerical aperture
NIH	National institutes of health
NLS	Nonlinear least squares
OME	Open microscopy environment
OTF	Optical transfer function
PALM	Photon activated light microscope
PMT	Photomultiplier tube
PN	Poisson noise
PSF	Point-spread function
PSD	Power spectral density
QE	Quantum efficiency
RL	Richardson-Lucy
RLS	Regularized linear least squares
RL-TV	Richardson-Lucy Total Variation
RyR	Ryanodine
SIM	Structured-illumination microscope
SNR	Signal-to-noise ratio
STED	Stimulated emission depletion
SVI	Scientific volume imaging
TM	Tikhonov-Miller
TV	Total variation
WFM	Wide-field microscope

References

1. D.A. Agard. Optical sectioning microscopy: cellular architecture in three dimensions. *Ann. Rev. Biophys. Bioeng.*, 13:191–219, 1984.
2. D.A. Agard, Y. Hiraoka, P. Shaw, and J.W. Sedat. Fluorescence microscopy in three dimensions. *Methods Cell Biol.*, 30:353–377, 1989.
3. G. Aubert and P. Kornprobst. *Mathematical Problems in Image Processing: Partial Differential Equations and the Calculus of Variations*, volume 147 of *Applied Mathematical Sciences*. Springer Verlag, 2006.

4. M.R. Banham and A.K. Katsaggelos. Digital image restoration. *IEEE Sig. Proc. Mag.*, 14(2):24–41, March 1997.
5. J.M. Bardsley and J.J. Goldes. Regularization parameter selection methods for ill-posed poisson maximum likelihood estimation. *Inverse Problems*, 25(9):095005, 2009.
6. A. Beck and M. Teboulle. A fast iterative shrinkage thresholding algorithm for linear inverse problems. *SIAM J. Imaging Sciences*, 2(1):183–202, 2009.
7. D.S.C. Biggs and M. Andrews. Acceleration of iterative image restoration algorithms. *Appl. Opt.*, 36(8):1766–1775, 1997.
8. M.J. Booth. Adaptive optics in microscopy. *Philos. Transact. A Math. Phys. Eng. Sci.*, 365(1861):2829–2843, December 2007.
9. M.J. Booth, M.A. Neil, R. Juskaitis, and T. Wilson. Adaptive aberration correction in a confocal microscope. *Proc. Natl. Acad. Sci.*, 99(9):5788–5792, 2002.
10. M. Born and E. Wolf. *Principles of Optics*. Cambridge U. Press, 1999.
11. A.C. Bovik, editor. *Handbook of image and video processing*. Elsevier Academic Press, Amsterdam [u.a.], 2005.
12. L.M. Bregman. The method of successive projection for finding a common point of convex sets (Theorems for determining common point of convex sets by method of successive projection). *Soviet Mathematics*, 6:688–692, 1965.
13. P. Campisi and K. Egiazarian, editors. *Blind Image Deconvolution: Theory and Applications*. CRC Press, 2007.
14. M.B. Cannell, A. McMorland, and C. Soeller. Image enhancement by deconvolution. In J. B. Pawley, editor, *Handbook of Biological Confocal Microscopy*, chapter 25, pages 488–500. Springer, 3rd edition, 2006.
15. W.A. Carrington, K.E. Fogarty, and F.S. Fay. 3D Fluorescence Imaging of Single Cells Using Image Restoration. In J. K. Foskett and S. Grinstein, editors, *Noninvasive techniques in cell biology*, pages 53–72. Wiley-Liss, 1990.
16. T.F. Chan and J. Shen. *Image Processing and Analysis: Variational, PDE, Wavelet, and Stochastic Methods*. SIAM Publisher, 2005.
17. P. Charbonnier, L. Blanc-Féraud, and M. Barlaud. An adaptive reconstruction method involving discontinuities. In *IEEE Int. Conf. Acoust. Speech Signal Process.*, volume 5, pages 491–494, Minneapolis, MN, USA, April 1993.
18. C. Chaux, L. Blanc-Féraud, and J. Zerubia. Wavelet-based restoration methods: application to 3D confocal microscopy images. In *Proc. SPIE*, volume 6701, San Diego, USA, August 2007.
19. C. Chaux, J.-C. Pesquet, and N. Pustelnik. Nested iterative algorithms for convex constrained image recovery problems. *SIAM Journal on Imaging Sciences*, 2(2):730–762, 2009.
20. P.-C. Cheng, B.-L. Lin, F.-J. Kao, M. Gu, M.-G. Xu, X. Gan, M.-K. Huang, and Y.-S. Wang. Multi-photon fluorescence microscopy - The response of plant cells to high intensity illumination. *Micron*, 32:661–670, 2001.
21. J.-A. Conchello and J.W. Lichtman. Optical Sectioning Microscopy. *Nature Methods*, 2(12):920–931, 2005.
22. J. Boutet de Monvel, S. Le Calvez, and M. Ulfendahl. Image restoration for confocal microscopy: improving the limits of deconvolution, with application to the visualization of the mammalian hearing organ. *Biophys. J.*, 80(5):2455–2470, 2001.
23. N. Dey, L. Blanc-Féraud, C. Zimmer, Z. Kam, P. Roux, J.-C. Olivo-Marin, and J. Zerubia. Richardson-Lucy algorithm with total variation regularization for 3D confocal microscope deconvolution. *Microsc. Res. Tech.*, 69:260–266, 2006.
24. N. Dey, L. Blanc-Féraud, C. Zimmer, P. Roux, Z. Kam, J.-C. Olivo-Marin, and J. Zerubia. 3D microscopy deconvolution using richardson-lucy algorithm with total variation regularization. Research Report 5272, Inria, France, July 2004.
25. A. Diaspro, G. Chirico, C. Usai, P. Romoino, and J. Dobrucki. Photobleaching. In J. B. Pawley, editor, *Handbook of Biological Confocal Microscopy*, chapter 39, pages 690–702. Springer, 3rd edition, 2006.

26. F. Difato, F. Mazzone, S. Scaglione, M. Fato, F. Beltrame, L. Kubínová, J. Janácek, P. Ramoino, G. Vicidomini, and A. Diaspro. Improvement in volume estimation from confocal sections after image deconvolution. *Microscopy Research and Technique*, 64(2):151–155, 2004.
27. F.-X. Dupé, M.J. Fadili, and J.-L. Starck. A proximal iteration for deconvolving Poisson noisy images using sparse representations. *IEEE Trans. Image Proc.*, 18(2):310–321, 2009.
28. A. Egner, M. Schrader, and S.W. Hell. Refractive index mismatch induced intensity and phase variations in fluorescence confocal, multiphoton and 4Pi-microscopy. *Opt. Comm.*, 153:211–217, August 1998.
29. A. Erhardt, G. Zinser, D. Komitowski, and J. Bille. Reconstructing 3-D light-microscopic images by digital image processing. *Appl. Opt.*, 24:194–200, 1985.
30. M. Figueiredo and J. Bioucas-Dias. Restoration of Poissonian images using alternating direction optimization. *IEEE Transactions on Image Processing*, 19(12):3133–3145, 2010.
31. M.A.T. Figueiredo and R.D. Nowak. An EM algorithm for wavelet-based image restoration. *IEEE Trans. Image Process.*, 12(8):906–916, August 2003.
32. J.W. Goodman. *Introduction to Fourier Optics*. Roberts & Company Publishers, 2004.
33. B.M. Hanser, M.G. Gustafsson, D.A. Agard, and J.W. Sedat. Phase retrieval for high-numerical-aperture optical systems. *Opt. Lett.*, 28(10):801–803, May 2003.
34. S.W. Hell. Far-field optical nanoscopy. *Single Molecule Spectroscopy in Chemistry, Physics and Biology*, 96(7):365–398, 2009.
35. Y. Hiraoka, J.W. Sedat, and D.A. Agard. Determination of three-dimensional imaging properties of a light microscope system. *Biophys. J.*, 57:325–333, February 1990.
36. T.J. Holmes. Maximum-likelihood image restoration adapted for noncoherent optical imaging. *J. Opt. Soc. Am. A*, 5:666–673, May 1988.
37. S. Inoué. Foundations of Confocal Scanned Imaging in Light Microscopy. In J. B. Pawley, editor, *Handbook of Biological Confocal Microscopy*, chapter 1, pages 1–19. Springer, 3rd edition, 2006.
38. Z. Kam, B. Hanser, M.G.L. Gustafsson, D.A. Agard, and J.W. Sedat. Computational adaptive optics for live three-dimensional biological imaging. *Proc. Natl. Acad. Sci.*, 98(7):3790–3795, 2001.
39. Z. Kam, P. Kner, D. Agard, and J.W. Sedat. Modelling the application of adaptive optics to wide-field microscope live imaging. *J. Microsc.*, 226:33–42, 2007.
40. X. Lai, Z. Lin, E.S. Ward, and R.J. Ober. Noise suppression of point spread functions and its influence on deconvolution of three-dimensional fluorescence microscopy image sets. *J. Microsc.*, 217(1):93–108, 2005.
41. L. Landmann and P. Marbet. Colocalization analysis yields superior results after image restoration. *Microscopy Research and Technique*, 64(2):103–112, 2004.
42. A. Levin, Y. Weiss, F. Durand, and W.T. Freeman. Understanding and evaluating blind deconvolution algorithms. In *Proc. IEEE Comp. Vis. and Pat. Recog.*, Miami, FL, USA, June 2009. to appear.
43. J.W. Lichtman and J.-A. Conchello. Fluorescence Microscopy. *Nature Methods*, 2(12):910–919, 2005.
44. L.B. Lucy. An iterative technique for the rectification of observed distributions. *Astron. J.*, 79:745–754, 1974.
45. J.G. McNally, T. Karpova, J. Cooper, and J.A. Conchello. Three-Dimensional Imaging by Deconvolution Microscopy. *Methods*, 19:373–385, 1999.
46. J.G. McNally, C. Preza, J.Á. Conchello, and L.J. Thomas Jr. Artifacts in computational optical-sectioning microscopy. *J. Opt. Soc. Am. A*, 11:1056–1067, March 1994.
47. E.S. Meinel. Origins of linear and nonlinear recursive restoration algorithms. *J. Opt. Soc. Am. A*, 3(6):787–799, 1986.
48. K. Miller. Least squares methods for ill-posed problems with a prescribed bound. *SIAM J. Math. Anal.*, 1(1):52–74, 1970.
49. M. Minsky. Memoir on inventing the confocal scanning microscope. *Scanning*, 10:128–138, 1988.

50. J.R. Monck, A.F. Oberhauser, T.J. Keating, and J.M. Fernandez. Thin-section ratiometric Ca2+ images obtained by optical sectioning of fura-2 loaded mast cells. *J. Cell Biol.*, 116(3):745–759, 1992.
51. N. Moreno, S. Bougourd, J. Haseloff, and J. A. Feijò. Imaging Plant Cells. In J. B. Pawley, editor, *Handbook of Biological Confocal Microscopy*, chapter 44, pages 769–787. Springer, 3rd edition, 2006.
52. P. Pankajakshan. *Blind Deconvolution for Confocal Laser Scanning Microscopy*. PhD thesis, Université de Nice-Sophia Antipolis, December 2009.
53. P. Pankajakshan, L. Blanc-Féraud, Z. Kam, and J. Zerubia. Point-spread function retrieval in fluorescence microscopy. In *Proc. IEEE International Symposium on Biomedical Imaging*, pages 1095–1098, Boston, USA, July 2009.
54. P. Pankajakshan, B. Zhang, L. Blanc-Féraud, Z. Kam, J.-C. Olivo-Marin, and J. Zerubia. Blind deconvolution for thin-layered confocal imaging. *Appl. Opt.*, 48(22):4437–4448, 2009.
55. G.H. Patterson. Fluorescence microscopy below the diffraction limit. *Seminars in cell & developmental biology*, 20(8):886–893, 2009. Imaging in Cell and Developmental Biology; Planar Cell Polarity.
56. J.B. Pawley. Fundamental limits in confocal microscopy. In J. B. Pawley, editor, *Handbook of Biological Confocal Microscopy*, chapter 2, pages 20–42. Springer, 3rd edition, 2006.
57. J.B. Pawley, editor. *Handbook of Biological Confocal Microscopy*. Springer, 3rd edition, 2006.
58. J.B. Pawley. Points, pixels, and gray levels: digitizing image data. In J. B. Pawley, editor, *Handbook of Biological Confocal Microscopy*, chapter 4, pages 59–79. Springer, 3rd edition, 2006.
59. M. Platani, I. Goldberg, J.R. Swedlow, and A.I. Lamond. In Vivo analysis of Cajal body movement, separation and joining in live human cells. *J.Cell.Biol.*, 151:1561–1574, 2000.
60. C. Preza, M.I. Miller, L.J. Thomas Jr., and J.G. McNally. Regularized linear method for reconstruction of three-dimensional microscopic objects from optical sections. *J. Opt. Soc. Am. A*, 9(2):219–228, 1992.
61. E.H. Ratzlaff and A. Grinvald. A tandem-lens epifluorescence macroscope: hundred-fold brightness advantage for wide-field imaging. *J Neurosci. Methods.*, 36:127–137, 1991.
62. W.H. Richardson. Bayesian-based iterative method of image restoration. *J. Opt. Soc. Am. A*, 62(1):55–59, January 1972.
63. L.I. Rudin, S. Osher, and E. Fatemi. Nonlinear total variation based noise removal algorithms. *Phys. D.*, 60:259–268, 1992.
64. P. Sarder and A. Nehorai. Deconvolution methods for 3-D fluorescence microscopy images. *IEEE Signal Process. Mag.*, 23(3):32–45, May 2006.
65. B.A. Scalettar, J.R. Sweldow, J.W. Sedat, and D.A. Agard. Dispersion, aberration and deconvolution in multi-wavelength fluorescence images. *J. Microsc.*, 182:50–60, 1996.
66. L. Schermelleh, R. Heintzmann, and H. Leonhardt. A guide to super-resolution fluorescence microscopy. *J. Cell Biol.*, 190:165–175, 2010.
67. F. Sedarat, E. Lin, E.D.W. Moore, and G. F.Tibbits. Deconvolution of confocal images of dihydropyridine and ryanodine receptors in developing cardiomyocytes. *J. Appl. Physiol.*, 97:1098–1103, 2004.
68. J.W. Shaevitz and D.A. Fletcher. Enhanced three-dimensional deconvolution microscopy using a measured depth-varying point-spread function. *J. Opt. Soc. Am. A*, 24(9):2622–2627, 2007.
69. S.L. Shaw. Imaging the live plant cell. *The Plant Journal*, 45(4):573–598, 2006.
70. C.J.R. Sheppard. Depth of field in optical microscopy. *J. Microsc.*, 149:73–75, 1988.
71. L. Sherman, J.Y. Ye, O. Albert, and T.B. Norris. Adaptive correction of depth-induced aberrations in multiphoton scanning microscopy using a deformable mirror. *J. Microsc.*, 206(1):65–71, 2002.
72. J.B. Sibarita. Deconvolution microscopy. *Advances in Biochemical Engineering and Biotechnology*, 92:201–243, 2005.
73. J.-L. Starck and A. Bijaoui. Filtering and deconvolution by the wavelet transform. *Signal Process.*, 35(3):195–211, 1994.

74. E.H.K. Stelzer. Contrast, resolution, pixelation, dynamic range and signal-to-noise ratio: fundamental limits to resolution in fluorescence light microscopy. *J. Microsc.*, 189:15–24, January 1998.
75. P.A. Stokseth. Properties of a defocused optical system. *J. Opt. Soc. Am. A*, 59:1314–1321, october 1969.
76. Y. Sun, P. Davis, E.A. Kosmacek, F. Ianzini, and M.A. Mackey. An open-source deconvolution software package for 3-D quantitative fluorescence microscopy imaging. *J. Microsc.*, 236(3):180–193, 2009.
77. T. Suzuki, T. Matsuzaki, H. Hagiwara, T. Aoki, and K. Takata. Recent Advances in Fluorescent Labeling Techniques for Fluorescence Microscopy. *Acta Histochem. Cytochem.*, 40:131–137, 2007.
78. J.R. Swedlow, K. Hu, P.D. Andrews, D.S. Roos, and J.M. Murray. Measuring tubulin content in Toxoplasma gondii: a comparison of laser-scanning confocal and wide-field fluorescence microscopy. *Proc. Soc. Natl. Acad. Sci.*, 99(4):2014–2019, February 2002.
79. A.N. Tikhonov and V.A. Arsenin. *Solution of Ill-posed Problems*. Winston and Sons, 1977.
80. T. Tommasi, A. Diaspro, and B. Bianco. 3-D reconstruction in optical microscopy by a frequency-domain approach. *Signal Process.*, 32(3):357–366, 1993.
81. R.Y. Tsien, L. Ernst, and A. Waggonnr. Fluorophores for confocal microscopy: photophysics and photochemistry. In J. B. Pawley, editor, *Handbook of Biological Confocal Microscopy*, chapter 39, pages 690–702. Springer, 3rd edition, 2006.
82. G.M.P. van Kempen, L.J. van Vliet, P.J. Verveer, and H.T.M. van Der Voort. A quantitative comparison of image restoration methods for confocal microscopy. *J. Microsc.*, 12:354–365, March 1997.
83. P.J. Verveer, M.J. Gemkow, and T.M. Jovin. A comparison of image restoration approaches applied to three-dimensional confocal and wide-field fluorescence microscopy. *J. Microsc.*, 193:50–61, 1999.
84. R.M. Willett, I. Jermyn, R.D. Nowak, and J. Zerubia. Wavelet-based superresolution in astronomy. In F. Ochsenbein, M.G. Allen, and D. Egret, editors, *Proc. Astronomical Data Analysis Software and Systems*, volume 314 of *Astronomical Society of the Pacific*, pages 107–116, Strasbourg, France, July 2003.
85. K.I. Willig, J. Keller, M. Bossi, and S.W. Hell. STED microscopy resolves nanoparticle assemblies. *New J. Phys.*, 8:106–113, 2006.
86. R. Zanella, P. Boccacci, L. Zanni, and M. Bertero. Efficient gradient projection methods for edge-preserving removal of poisson noise. *Inverse Problems*, 25(4):045010, 2009.
87. B. Zhang, J. Zerubia, and J.C. Olivo-Marin. Gaussian approximations of fluorescence microscope point-spread function models. *Appl. Opt.*, 46(10):1819–1829, 2007.

Chapter 5
Statistical Shape Analysis of Surfaces in Medical Images Applied to the Tetralogy of Fallot Heart

Kristin McLeod, Tommaso Mansi, Maxime Sermesant, Giacomo Pongiglione, and Xavier Pennec

5.1 Introduction

During the past 10 years, biophysical modeling of the human body has been a topic of increasing interest in the field of biomedical image analysis. The aim of such modeling is to formulate personalized medicine where a digital model of an organ can be adjusted to a patient from clinical data. This virtual organ would enable to estimate parameters which are difficult to quantify in clinical routine and to test therapies in-silico as well as to predict the evolution of the organ over time and with therapy.

While biophysical modeling has been successfully developed and validated at the cell level, the integration at the organ level is still challenging because the interaction between the physiology and the organ shape is complex. There is thus a strong need to understand the organ shape characteristics and their evolution at the population level, which is the topic of Computational Anatomy. By shape, we intend here, and in the sequel of this paper, geometric data such as curves and surfaces (interfaces of organs) that describe the anatomy. On top of this, Computational Physiology is aiming at describing and modeling the function of these organs and how they interact together. Anatomy can be observed in most medical image modalities, such as Computational Tomography (CT) or Magnetic Resonance Imaging (MRI), while measuring the physiology requires more specific imaging modalities that are often more

K. McLeod (✉) · X. Pennec · M. Sermesant
Inria Sophia Antipolis Méditerranée, Asclepios project-team, 2004 Route des Lucioles, 06902 Sophia Antipolis, France
e-mail: Kristin.Mcleod@inria.fr; Xavier.Pennec@inria.fr; Maxime.Sermesant@inria.fr

T. Mansi
Siemens Corporate Research, Image Analytics and Informatics, Princeton, NJ, USA
e-mail: Tommaso.Mansi@tmansi.net

G. Pongiglione
Ospedale Pediatrico Bambine Gesù, Rome, Italy

F. Cazals and P. Kornprobst (eds.), *Modeling in Computational Biology and Biomedicine*, 165
DOI 10.1007/978-3-642-31208-3_5,

difficult to acquire in a clinical context. The expectation is that anatomical shapes can then be used to better understand physiology, or at least some of its pathologies.

The biological mechanisms involved in the formation and evolution of the shape of organs are often too complex to model. Thus, one has to rely on robust statistical approaches to produce generative models from data. Given the shapes of a number of subjects, we can create a generalized model of an organ, called an atlas, which can be used to distinguish normal from pathological organs or can be deformed to give patient-specific models driven by the clinically available parameters of interest. Such statistical analyses can both provide a predictive model and guide the biophysical approach. However, computing statistics on dynamic 3D shapes is very challenging. Traditional methods rely on point based parameterizations of the shapes where the point-to-point correspondences can be an important limiting factor for the usability of the method since the correspondences must be correct and consistent over all the shapes. New approaches were recently developed to compute such statistics without this limitation [6, 10, 17, 26–28, 34, 35]. Here, the statistical shape analysis tools are based on currents, a non-parametric representation of shapes. These tools have a wide range of applications, and provide a well-posed framework for statistical shape analysis of groups. Due to the fact that the methods do not assume point correspondences between structures (and in fact assigning landmarks to structures such as the heart are arbitrary), a wider range of data can be used. For example, one can use surfaces to model organs such as the heart, brain, and lungs, curves to model sulcal lines on the brain cortex, and sets of curves to represent fibre bundles from diffusion MRI in the brain. The goal of this chapter is to give an overview of this methodology in the context of a specific clinical problem: the prediction of the cardiac shape remodeling in repaired tetralogy of Fallot patients due to chronic regurgitation. The underlying assumption is that changes in heart morphology may reveal structural abnormalities and dysfunctions due to the pathology. In particular, we aim to establish which shape patterns are related to the pathology in order to give further insights into the condition.

5.1.1 Repaired Tetralogy of Fallot

Tetralogy of Fallot (ToF) is a congenital heart defect that affects approximately 4 out of every 10,000 babies [9]. The primary defect associated with this condition is a ventricular-septal opening which allows blood to flow freely between the ventricles. Secondary is stenosis in the pulmonary artery which restricts the blood flow from the right ventricle to the lungs. Due to a misalignment of the aorta over the ventricular-septal defect, the aorta is fed by both the left and right ventricle rather than just the left ventricle. Patients may also have hypertrophy in the right ventricle that causes a boot-like shape of the ventricle which is characteristic of this condition.

These abnormalities require open heart surgical repair early in infancy. As part of this surgery, the stenosis in the pulmonary artery is cleared to allow blood to flow more freely through the artery. As a consequence, the pulmonary valves

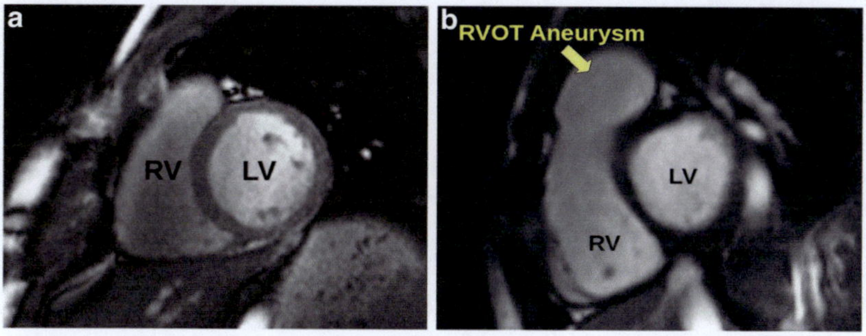

Fig. 5.1 Comparison between (**a**) a normal heart and (**b**) a Tetralogy of Fallot heart

which control one way blood flow from the right ventricle to the pulmonary artery can be destroyed completely or damaged, causing blood to leak back to the right ventricle. The long-term outcome of this condition is right ventricular re-modeling caused by the regurgitated blood (Fig. 5.1). Nowadays, patients undergo follow up operations to reduce this post-operative sequelae. In particular, the valves are replaced with an artificial device to reduce the regurgitation. A clinical challenge for these patients is determining the optimal time for intervention. On the one hand, it is preferable to wait as long as possible before performing such operations since the artificial devices have a limited life-span resulting in the need for repeat follow-up operations. Either open heart surgery or percutaneous valve replacement may be required, depending on the diameter of the pulmonary annulus. These therapies carry some danger, so minimizing the frequency and total number of surgeries is crucial for these patients. On the other hand, the heart undergoes re-modeling due to the regurgitated blood producing volume overload in the right ventricle and if left long enough this damage can be irreversible. Determining the optimal time for intervention is a trade-off between maximizing the time between operations and minimizing damage to the heart.

Therefore, the clinical application for this condition is to better understand how the heart evolves over a large period of time, often decades, with and without surgery, with the main goal to be able to predict the optimal time for intervention. However, due to the large variability of shape, the optimal time, placement and size of the artificial device can be difficult to determine. Furthermore, the complexity of the biological mechanisms involved in heart growth hinders the development of a direct model of cardiac re-modeling. Statistical shape analysis can therefore be employed to aid in further understanding the pathology to assist cardiologists with diagnosis, therapy planning, and long term prognosis. More particularly, we are interested in determining quantitative measures of the shape that correlate with the cardiac function and the severity of the disease in these patients (bio-markers of the disease). Such bio-markers could be used to determine the severity of the disease when regurgitation cannot be assessed or to provide structural information that may appear before deterioration of the blood dynamics.

5.1.2 Chapter Overview

Public

This chapter is targeting people interested in statistics of surfaces, the surface being considered as the random variable itself. The reader is assumed to have good notions of multivariate statistics (expectation, moments, PCA, regression, hypothesis tests) and some notions of analysis in infinite dimensional vector spaces, in particular with reproducing kernel Hilbert spaces (RKHS). This last notion is borrowed from machine learning theory and is applied here to the characterization of geometric shapes through geometric integration theory. A few notions from Riemannian geometry are used but no specific knowledge is required.

Outline

In the following sections, we will first review in Sect. 5.2 statistical shape analysis methods with a particular focus on the statistical analysis of surfaces. We will introduce the formalism of deformations which is at the center of most of the current works in computational anatomy [21]. Then, we will detail the framework of currents to represent surfaces, and show how this can be turned into an effective shape analysis technique. Section 5.3 will apply this methodology to the shape of the heart of 13 patients with repaired tetralogy of Fallot. Correlating shape with clinical variables will illustrate how we can extract some insight about the relationship between morphology and physiology. Last but not least, we will exemplify how the lack of longitudinal measurements can be bypassed by building a statistical generative growth model from cross-sectional data which summarizes the heart shape remodeling at the population level.

5.2 Statistical Shape Analysis

5.2.1 Shapes, Forms and Deformations

There is generally no physical model that can faithfully relate the shape of organs in different patients. Thus, to analyze their variability in a population, one usually extracts some anatomically representative landmarks (or more generally geometric features), and models their statistical distribution across the population, via a mean shape and covariance structure analysis after a group-wise matching for instance. One of the earliest methods [2, 3] consists in studying the variability of anatomical landmark positions among a population: after a global pose (position

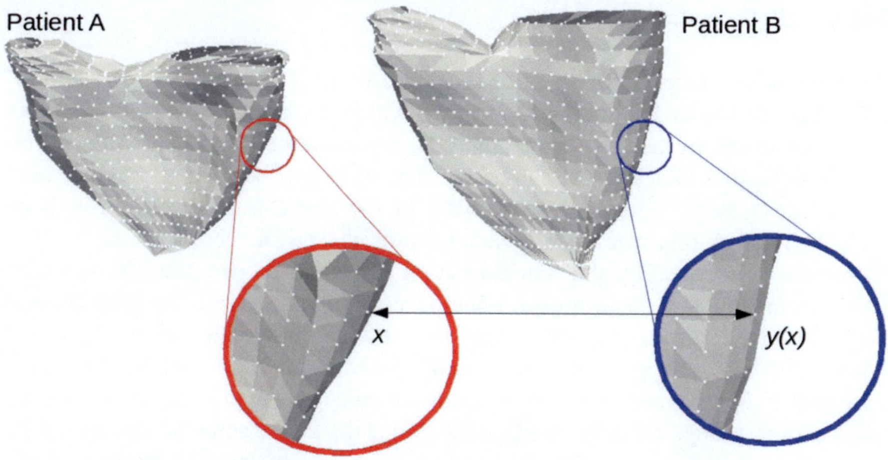

Fig. 5.2 Surface correspondences. Using landmark-based methods to register one surface mesh to another surface mesh, for each point x in patient A, we require the corresponding point $y(x)$ in patient B. This requires a full parameterization of the patient surface meshes. This is illustrated on the right ventricle of two patients with rToF

and orientation) normalization and the consistent identification of landmarks within all patients (Fig. 5.2), a principle component analysis (PCA) is performed to extract the main modes of variation of the shape.

5.2.1.1 Shapes

In such a process, the global pose of our objects is often considered as a nuisance factor related to the arbitrary coordinate system of the acquisition device: generally speaking, the shape of an object is understood as the geometric information that is invariant under translation, rotations and rescaling. Thus, one is often interested only in the *shape*, i.e. what remains if we quotient out the object space by a group action, usually rigid body, similarity or affine transformations. For instance, when we consider the equivalent classes of sets of N (labeled or unlabeled) points under the action of a global similarity (resp. rigid body) transformation, we obtain the celebrated shape (resp. size and shape) spaces of Kendall et al. [12, 13]. One can define similarly the shape spaces of curves by removing the effect of re-parameterizations and the global pose of the curve [29–31, 40]. However, in medical image analysis, the natural coordinate system of an organ is its position and orientation with respect to surrounding tissues and organs in the image. Thus, once images are normalized by aligning them to a reference anatomy, the pose of organs is a variable of interest and should be kept for analysis.

5.2.1.2 Deformations

An alternative modeling of shapes was proposed by D'Arcy Thompson in 1917 [37]. The idea is to assume that there is a template object, which represents the reference shape. In medical image analysis, the template is most often built from real observations and is called an atlas.[1] The atlas can be seen as an equivalent of the mean shape. Then, the variability of the shape is analyzed through the deformations of this reference object towards the actual observations: a shape difference is encoded by the transformation that deforms one onto the other. As the deformation of a smooth object should be a smooth object, we have to work with diffeomorphisms (invertible, one-to-one mappings with smooth inverses). This formalism was promoted to a generic shape analysis tool by Grenander and Miller [8, 20] based on advanced mathematical tools to compute on infinite dimensional groups of diffeomorphisms [38]. One key feature of this lift of the shape characteristics from the object space to the transformation space is that it allows to apply the typical deformations to other objects than the ones analyzed in the first place (provided that they also live in the coordinate system of the atlas). For instance, the surface of the cortex in the brain has deep foldings called *sulci* which are important anatomical features. Trouvé et al. [28] analyzed the variability of the trace of these sulci on the surface of the brain cortex to extract the main 3D deformation modes. These modes could then be used to deform accordingly the surface of the cortex or the full 3D volume of the brain.

However, the problem is even more complex than previously as we want here to perform statistics on large deformations, which are known to belong to a non-linear and infinite dimensional manifold. The Riemannian setting is one of the most powerful structures to generalize simple statistics to non-linear spaces: it provides a definition of the distance between points of our manifold and a notion of shortest path/straight lines using geodesics [22, 23]. The main difficulty is that the mean value cannot be defined through an integral or a sum as in Euclidean spaces. Instead, one must look for points in the manifold that minimize the dispersion of the other points around it, which is commonly measured by the variance (the mean squared distance). This is what is called the Fréchet or Karcher mean (Fig. 5.3).

Then, one can compute the covariance matrix (the directional dispersion around the mean) by developing the manifold onto its tangent space at the mean point. Basically, this amounts to representing each data point by the momentum (initial speed vector) needed to shoot[2] a geodesic to it from the mean.

This generalization of statistics has to be slightly modified to fit the atlas deformation model: the 'distance' between the atlas and a shape is given here by the length of the shortest path in the space of deformations. The atlas is called unbiased

---

[1] In this paper, atlas is always taken in the sense of template and not in the sense of the atlas of differential geometry.

[2] We use the term "geodesic shooting" to define the integration of the Euler-Lagrange equations, which plays the role of the exponential map in Riemannian geometry.

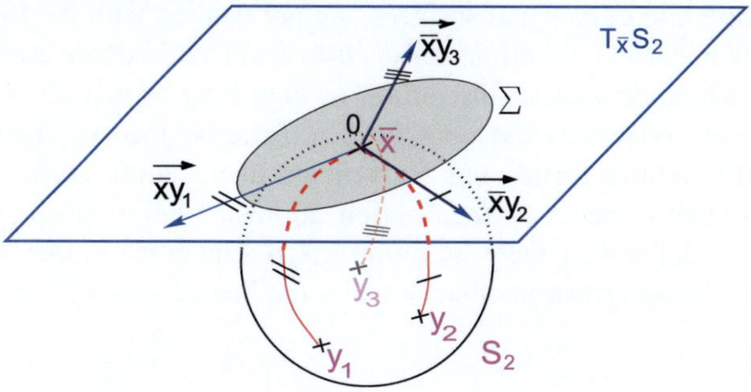

Fig. 5.3 Riemannian statistics on the sphere: the Fréchet mean is the point minimizing the squared Riemannian distance. It corresponds to the point for which the development of the geodesics to the data points on the tangent space is optimally centered (the mean in that tangent space is zero). The covariance matrix is then defined in that tangent space (Figure adapted from [22])

if it is centered, i.e. if it minimizes (among all possible atlases) the sum of squared distances to the shapes [32]. In this sense, this is the Fréchet mean of the shapes. Then, each shape is represented by the momentum of the deformation that allows to regenerate the shape by deforming the atlas. These momentum all belong to the tangent space (more precisely to the cotangent space) of the diffeomorphisms at the identity, which is a vector space in which we can perform a PCA [15]: shooting in the space of deformations along the first eigenvectors of the covariance matrix of the momentum gives deformation modes that represent the main shape variability when applied to the atlas.

5.2.2 From Points to Surfaces: The Formalism of Currents

Landmarks can be encoded by the probability of their location in space (typically a Gaussian around their expected value). When the noise is going to zero, then this probability density function (pdf) becomes singular. However, we can continue to deal with these types of singularities by considering distributions (generalized functions) instead of functions, which include Diracs. Mathematically, a Dirac is not a function, but an object which can be characterized by the result of its integration against any function of a sufficiently smooth functional space: $\forall f \subset W, \int \delta_x(y).f(y).dy = f(x)$. In a finite dimensional Euclidean vector space, any linear form ϕ (function from the vector space to $\mathbb{R}$) can be characterized using the scalar product: Thanks to the Rietz representation theorem, there exists a vector x such that $\phi(w) = \langle x \mid w \rangle$. The equivalent in infinite dimensional functional spaces is the Dirac distribution, which is actually an element of the linear functionals over the space W. In that framework, a set of N points x_k can be represented by the 'pdf' $p(x) = 1/N \sum_i \delta_{x_i}(x)$, and its evaluation on a function $f(x)$ results in the mean of the values of f at the points x_k.

When we move to curves and surfaces, we are dealing with the locally singular locus of points in certain directions only: curves and surfaces are continuous along their tangent. The extension of the notion of distributions that allows to take that information into account is the geometric integration theory. The basic idea is to integrate differential forms, but for objects like smooth surfaces in 3D, we shall simply define currents by their action on vector fields, similarly to the way distributions are defined by their action on scalar functions. In fact, one could see currents as the characterization of surfaces by the flux of vector fields through it.

5.2.2.1 Surfaces Seen as Currents

Let W be a Hilbert space of test vector fields. This is a possibly infinite complete vector space provided with a scalar product. For a given surface S, we can measure the flux of any vector field $\omega \in W$ through this surface:

$$\mathcal{S}(\omega) = \int_{x \in S} \langle\, \omega(x) \mid n(x) \,\rangle \, d\sigma(x),$$

where $n(x)$ is the normal to the smooth surface at point $x \in S$ and $d\sigma(x)$ the surface element around point x. The shape of the surface S is characterized by the variation of the flux as the test vector field ω varies in W. Thus, the surface actually defines a continuous linear form on W which can be identified as an element of the dual space W^*, which is the space of linear functionals from W to $\mathbb{R}$ (currents). The nice property of currents is that it is a vector space: we can add or subtract currents from each other, or multiply them by a scalar. However, we should keep in mind that the space of currents is larger than the space of smooth surfaces: one can for instance add many pieces of surfaces together in a non-continuous way to create a non-continuous object.

5.2.2.2 The Kernel Trick for Currents

Now that we have identified surfaces to currents, we need to define more carefully what is the space W that we consider. The core element proposed by Glaunès [7] is to consider a kernel metric (typically the Gaussian kernel $K_W(x,z) = \exp(-\|x - z\|^2/\lambda_W^2)$) in order to turn W into a Reproducible Kernel Hilbert Space (RKHS). The reproducibility property implies that W is the dense span of basis vector fields of the form $\omega_x^\alpha(.) = K_W(x,.)\alpha$. This means that any vector field of W can be written as an infinite linear combination of such basis vectors. The kernel induces a scalar product which is easily computed on two basis vectors: $\langle\, \omega_x^\alpha \mid \omega_y^\beta \,\rangle_W = \alpha^{\mathrm{T}} K_W(x,y)\beta$.

Since W is a Hilbert space, by the Rietz representation theorem, there exists an isometric linear mapping $\mathcal{L}_W$ from W to W^* which maps to each $\omega \in W$ the linear form that reproduces the scalar product: $(\mathcal{L}_W(\omega))(\omega') = \langle \omega \mid \omega' \rangle_W$. This canonical isomorphism allows to define the dual of an element, or conversely to map to each surface the vector field that optimizes the flux. For instance, the dual of the above basis vectors are the Dirac delta currents $\delta_x^\alpha = \mathcal{L}_W(\omega_x^\alpha)$ as we have $\langle \omega_x^\alpha \mid \omega \rangle_W = \alpha^{\mathrm{T}} \omega(x)$. These can be seen as vector fields whose spatial support is concentrated at one point only. The space of currents W^* is the dense span of these basis elements. For instance, the surface S is represented by the current $\mathcal{S} = \int_{x \in S} \delta_x^{n(x)} d\sigma(x)$. The usual way to define a norm on the dual space is to take the operator norm: $\|\mathcal{S}\|_{W^*} = \sup_{\omega \in W, \|\omega\|=1} |\mathcal{S}(\omega)|$. The distance induced by this norm might seem difficult to use, but thanks to the RKHS properties, we have a closed form for basis vectors:

$$\langle \delta_x^\alpha \mid \delta_y^\beta \rangle_{W^*} = \alpha^{\mathrm{T}} K_W(x,y) \beta \quad \text{and} \quad \|\delta_x^\alpha - \delta_y^\beta\|_{W^*}^2 = \|\delta_x^\alpha\|_W^2 + \|\delta_y^\beta\|_W^2 - 2 \langle \delta_x^\alpha \mid \delta_y^\beta \rangle_W.$$

5.2.2.3 Numerical Issues

It is interesting to notice that this distance can be approximated by

$$\|\delta_x^\alpha - \delta_y^\beta\|_{W^*}^2 = \|\alpha - \beta\|^2 + 2\|x-y\|^2/\lambda_W^2 \alpha^{\mathrm{T}} \beta + O\left(\|x-y\|^4/{\lambda_W}^4\right),$$

when the points x and y are within a fraction of λ_W, while the distance is essentially constant (and equal to $\|\alpha\|_{L_2}^2 + \|\beta\|_{L_2}^2$) when the points x and y are more than a few λ_W apart. This behavior is typical of a robust distance in statistics, meaning that outliers (over a few λ_W in distance) will have (almost) no effect on the optimization of the distance between the surfaces as it is an almost constant penalty.

In practice, surfaces are often represented by discrete triangulated meshes. Assuming that each face has a support which is smaller than ε times the scale λ_W, we can approximate the surface by the current $\mathcal{S} = \sum_k \delta_{x_k}^{\alpha_k}$, where x_k is the barycenter and α_k the normal weighted by the area of the face. From the above Taylor expansion of the distance, we can see that the approximation error is less than ε^2 for each face in the W^* norm.

The scalar product between two discrete currents is obtained by linearity:

$$\left\langle \sum_k \delta_{x_k}^{\alpha_k} \,\middle|\, \sum_j \delta_{y_j}^{\beta_j} \right\rangle_{W^*} = \sum_{k,j} \alpha_k^{\mathrm{T}} K_W(x_k, y_j) \beta_j,$$

and the distance between two discrete surfaces is simply $\mathrm{dist}^2(S, S') = \|\mathcal{S} - \mathcal{S}'\|_{W^*}^2$.

5.2.2.4 Effective Approximation of Currents Using Matching Pursuit Algorithm

Representing shapes using currents gives a nice theoretical framework that allows us to compute simple statistics such as mean and principal modes. However, the complexity of the distance computation is quadratic in the number of Dirac delta currents used to represent the shape. This number can be quite high if we take highly detailed surface meshes, even though it adds nothing for the comparison of surfaces at a given scale λ_W. Therefore, in order to remain computationally as efficient as possible, we require a low number of degrees of freedom representation of currents that retains the information needed for the scale of the analysis. This is the essence of the matching pursuit algorithm as described in [26].

In brief, the matching pursuit algorithm is a greedy approach to find the best projection of multi-dimensional data on an over-complete family of basis functions [16]. In our case, we want to find an approximation of the current T that solves $\mathcal{L}_W^{-1}(T) = \gamma$, for a given vector field $\gamma \in W$. This amounts to finding N points (x_k) and vectors (α_k) such that the current $\Pi(T) = \sum_{k=1}^{N} \delta_{x_k}^{\alpha_k}$ is as close as possible to T. If the points are known, then $\Pi(T)$ is the orthogonal projection of T onto $Span(\delta_{x_k}^{\epsilon_q}; q = 1, 2, 3, k = 1 \dots N)$, where ϵ_q is the canonical basis of $\mathbb{R}^3$. The orthogonality condition: $\langle T, \delta_{x_k}^{\epsilon_q} \rangle_{W^*} = \langle \Pi(T), \delta_{x_k}^{\epsilon_q} \rangle_{W^*}$ leads to a linear set of $3N$ equations

$$\sum_{p=1}^{N} (K_W(x_i, x_p)\alpha_p)_k = \gamma(x_i)_k, \tag{5.1}$$

which can be solved iteratively over the continuous space of Dirac delta currents to find the point positions (x_k) as well as their associated momenta and the residual vector field. Additionally, we can also sample the vector field on a grid Λ. Applying the matching pursuit in the discrete case constrains the estimated momenta to lie on the nodes of the grid which forces the estimated current to belong to a discrete set of currents W_Λ^*. This simplifies the problem in both complexity as well as computation time. Figure 5.4 shows a compressed representation of the right ventricle.

In this approximation algorithm, the interest is not to fix the number of points N and to look for the optimal representation (this cannot be achieved properly with a greedy algorithm), but rather to fix an error threshold (typically 1–5 % of the norm of the original surface) and to stop the approximation whenever the number of points is sufficient to guaranty an error below that threshold. This typically results in compressions of 80 % with less than a few percent of error for one surface, but compression by order of magnitudes for the mean of several currents.

5.2.3 *An Algorithm for Surface Registration Using Currents*

Having defined a non-parametric representation of surfaces, we now require an algorithm to register one surface to another. Here, the term registration is meant

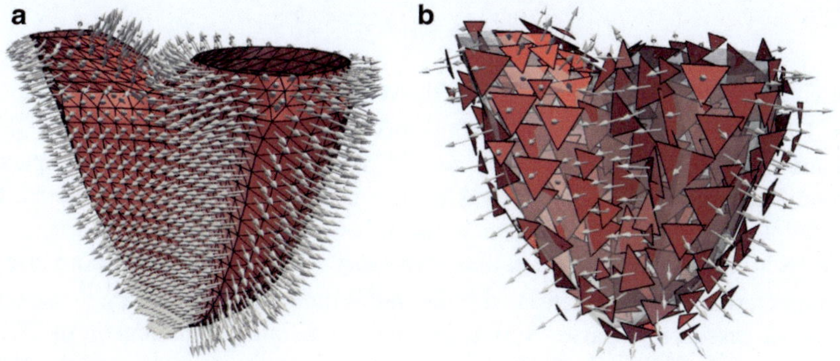

Fig. 5.4 Surfaces seen as integrators of vector fields. In a triangulated mesh, the atomic vector field integrators, the Dirac delta currents, are the normal vectors of every face, centered at the face barycenters and weighted by the face surface. Subfigure (**a**) displays the original mesh of the right ventricle with 1,476 Diracs. In subfigure (**b**), a greedy algorithm reduces the amount of delta currents needed to represent the shape while preserving the accuracy of the representation to 281 Diracs, i.e., a compression of 80 % with less than 5 % error

to define the action of transforming two objects into the same co-ordinate system. The basic idea is to rephrase registration into a geodesic problem where the initial conditions are related to the formalism of currents.

Since we are comparing topologically similar shapes, the transformations are restricted to those which preserve the topology of the object and give a smooth one-to-one (invertible) transformation (i.e. a diffeomorphism). The space of diffeomorphisms give non-linear deformations that allow local smooth variations to be captured in the registration. However, diffeomorphic transformations have an infinite number of degrees of freedom, therefore optimizing over the whole group of diffeomorphisms may not be possible. So, we use a smaller infinite group of diffeomorphisms to allow computations with discrete parameterizations using the Large Deformation Diffeomorphic Metric Mapping (LDDMM) method described in [18, 19] for images.

The LDDMM framework uses a group of diffeomorphisms constructed through integration of time-varying vector fields that belong to a RKHS. This gives a geodesic flow of diffeomorphisms ϕ_t for a continuous parameter t within the interval $[0, 1]$. At time $t = 0$, we have the identity mapping ϕ_0. The mapping at time 1 gives the desired transformation ϕ_1 which is required for mapping one image to the other. The path of any point x is defined by $\phi_t(x)$ and leads to the final position $\phi(x) = \phi_1(x)$. By following the reverse path, we can compute the inverse deformation.

The basic idea of the LDDMM framework is to minimize the squared distance between objects after transformation (this is the similarity energy or data attachment term) with a penalization for the energy of the deformation trajectory (regularization). A time t, the speed at point $y = \phi_t(x)$ is $v_t(y) = d\phi_t(y)/dt$. This suggests to define the energy of a velocity field at deformation ϕ_t using a right-invariant metric:

$$\|v_t\|_{\phi_t}^2 = \|v_t \circ \phi_t^{(-1)}\|_V^2,$$

where $\|.\|_V$ is the norm of the RKHS of the velocity fields at the identity transformation. The energy of the deformation trajectory is thus $E(\phi_t) = \int_0^1 \|v_t \circ \phi_t^{(-1)}\|_V^2 dt$, and the optimal curves joining $\phi_0 = Id$ to ϕ_1 are geodesics in the space of diffeomorphisms. Since optimal transformations are geodesics, we know that they are completely determined by their initial value, here the velocity field $v_0(x)$. As the velocity fields are most often non-zero everywhere, it is usually more efficient to parameterize the deformation with the initial momentum $\mu_0 = K_V^{(-1)} v_0$, which is in most cases very sparse. Notice that the momentum is an element of V^*, the dual space of V. Thus, it is a current, just as the ones we used to represent surfaces. However, even if the mathematical structures are identical, the objects "velocity fields" and "surfaces" are quite different and the kernels K_V and K_W are usually very different. Denoting ϕS the action of the transformation ϕ on the source object S, the registration criterion to the target object T is thus

$$C(\phi) = \text{dist}(\phi_1 S, T)^2 + \int_0^1 \|v_t \circ \phi_t^{(-1)}\|_V^2 dt.$$

For our surface registration problem, the objects are surfaces represented by their associated currents $\mathcal{S} = \sum_i \delta_{x_k}^{\alpha_k}$ and $\mathcal{T}$ and the distance is taken in the space of currents W^*. In this setting, one can show that the optimal initial momentum of the deformation has the same point-wise support as $\mathcal{S}$: $\mu_0(x) = \sum_i \delta_{x_k}^{\beta_k}$ (i.e. $v_0(x) = \sum_i K_V(x, x_k)\beta_k$) for some set of vectors β_k. This means that we are left with a finite dimensional optimization problem [7, 39]. Notice that the number of unknowns is directly linked to the number of Diracs in the source surface $\mathcal{S}$, which is usually the atlas. This emphasizes the interest of compressing the atlas before registering it to target surfaces.

5.2.4 *Building an Unbiased Atlas*

When we have more than two surfaces, one has to choose a reference (called the atlas) to compare all the others consistently. Choosing this atlas randomly generally leads to biases in the forthcoming statistical analysis because the atlas shape might be far from the mean shape. The atlas is called unbiased if it is centered with respect to all the surfaces, i.e. if the mean deformation from the atlas to the observed surfaced is close to zero.

For building the optimally centered atlas, we took the 'forward' strategy that models the set of surfaces as the deformation of an unknown 'ideal' atlas plus some residuals [24–26]. For the patient number i, this can be expressed as

$$T_i = \phi_i \hat{T} + \varepsilon_i, \tag{5.2}$$

where $\hat{T}$ is the atlas we are estimating, ϕ_i is the deformation that maps the atlas to the surfaces T_i, and ε_i represents the residuals (shape features not captured by the atlas such as changes in topology etc.). The mean shape information is described in the atlas $\hat{T}$ while the shape variability is encoded in the transformation ϕ_i.

The atlas is first initialized by taking the mean of the patient meshes. This initial atlas is then registered to each of the patients individually. A new atlas $\hat{T}$ that minimizes the error $\Lambda(\hat{T})$ is computed, with

$$\Lambda(\hat{T}) = \sum_{i=1}^{M} \|T_i - \phi_i(\hat{T})\|^2_{W^*}. \tag{5.3}$$

We then register the updated atlas to the individuals, recompute the atlas and loop until convergence (Algorithm 8).

Algorithm 8 Atlas Estimation

Input: M surface meshes T_i extracted from patient images.
1: Rigidly align meshes to a reference patient using rotations and translations.
2: Create initial atlas $\hat{T}_0$ as the mean of the patient meshes(this is a current and no longer a mesh)
3: **loop** $\{$*over* $k\}$
4: Estimate the M transformations ϕ_i and their initial momenta μ_i that register the atlas $\hat{T}_{k-1}$ to the individual T_i.
5: Update the atlas by minimizing the error in Eq. 5.3 using the estimated transformations ϕ_i and the atlas $\hat{T}_{k-1}$
6: Compute bias $\bar{\mu} = \frac{1}{M}\sum_{i=1}^{M}\mu_i$ and variance $\sigma_\phi^2 = \frac{1}{M}\sum_{i=1}^{M}\|\mu_i\|^2_{V^*}$
7: **end loop**[*Stop whenever* $\|\bar{\mu}\|_{V^*} < \eta\sigma_\phi$ *where* $\eta = 5\,\%$ *typically*]
8: **return** Final atlas $\hat{T}_k$ and the related transformations ϕ_i parameterized by their momenta μ_i

---

This algorithm produces an atlas which is a current and not any more a surface mesh. To constrain the atlas to remain a mesh, one usually restrains the search of the atlas to the subset of currents generated by all deformations of the data meshes. This means that the atlas can be written $\hat{T} = \phi T_l$ for some diffeomorphism ϕ and some index l. In practice, the atlas is initialized as the data mesh that has the smallest variance (sum of square distance to all other data meshes without deformation), and the update of the atlas becomes $\hat{T}_k = \bar{\mu}\hat{T}_{k-1}$.

5.3 Shape Analysis of ToF Data

To demonstrate the usefulness of our statistical shape analysis in a clinical context, we consider a population of patients with repaired tetralogy of Fallot. Recall that shape here means an atlas, which is a surface represented by a current, along with a deformation of that atlas. One of the key ideas of the "atlas framework" is to transfer the analysis from the space of currents (shapes) to the space of deformations.

As mentioned in the introduction, the clinical problem in rToF patients is in better understanding the shape re-modeling over time to aid in determining the optimal time for surgical intervention. In view of this problem, we are first interested in identifying the clinically relevant shape patterns by investigating the relationship between shape and given clinical indices. The assumption here is that changes in heart morphology may reveal structural abnormalities and dysfunctions due to the chronic regurgitation. In particular, we aim to establish which shape patterns are related to the pathology in order to give further insights into the condition. We can then go beyond the identification of pathological shape patterns, to estimate a predictive growth model of the heart to give an indication of how the heart will grow and re-model in time for this pathology.

Making use of currents to represent shapes, and an atlas to give an average representation of the population, we can apply classical statistical methods on the initial deformation momenta to address key questions for diagnosis, prognosis and prediction.

5.3.1 *The Analysis Pipeline*

Patient data: We consider a data-set of 13 patients (10 males, mean age $\pm$ SD $=$ 19 ± 9) with repaired tetralogy of Fallot. Steady-state Free Precision cine MRI (Fig. 5.5a) were acquired using a 1.5T scanner (Avanto, Siemens AG, Erlangen) in the short axis view covering entirely both ventricles (10–15 slices; isotropic in-plane resolution: 1.1×1.1 mm to 1.7×1.7 mm; slice thickness: 5–10 mm; 25–40 phases).

Surface mesh delineation: In the case of ToF patients, we are interested in studying the shape of the left and right ventricles of the heart. Using the statistical shape analysis methods described in the previous section, we can consider the ventricles as surfaces represented by triangulated meshes. The surface meshes of the right and left ventricle endocardium (inner layer of heart tissue), and left ventricle epicardium (outer layer of heart tissue) are defined using image segmentation by delineating the boundaries of each ventricle at end-diastole using the methods proposed by Zheng et al. [41] (Fig. 5.5). The right ventricle epicardium is not segmented because the thickness of the right ventricle is often not sufficient (with respect to the image resolution) to clearly distinguish the inner from the outer wall. This method defines the delineated boundaries using an anatomical model and thus establishes a point correspondence between meshes. Therefore, in this instance we used a standard least-squares method to first rigidly align the surface meshes to reduce the effect of patient positioning.

Mean atlas construction: Using the algorithm for computing a mean shape with the currents method described in Sect. 5.2, we estimated an atlas for the ToF data, see Fig. 5.6. Eleven iterations of the alternate minimization for the shape atlas

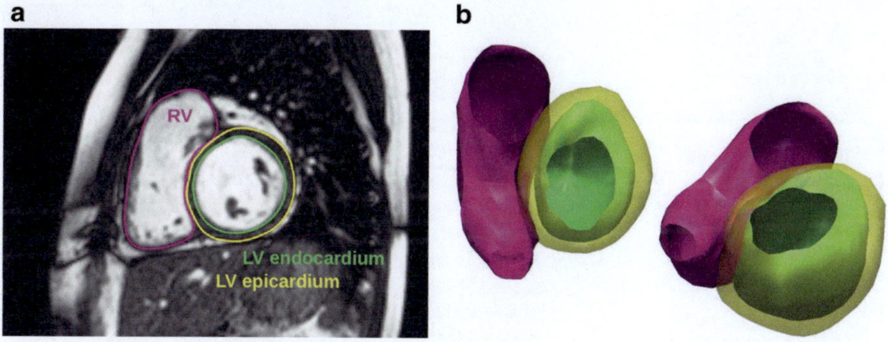

Fig. 5.5 Delineated boundaries of the right ventricle (*pink*), left ventricle endocardium (*green*) and left ventricle epicardium (*yellow*) shown on: (**a**) One image slice; (**b**) Two views of a 3D reconstruction

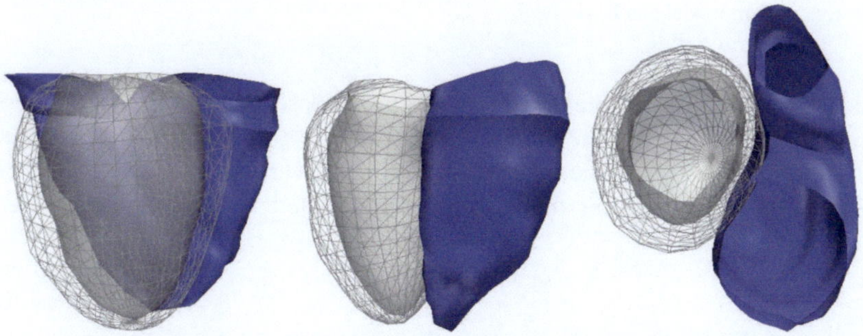

Fig. 5.6 Three views of the mean atlas estimated from 13 patients with repaired ToF with the right ventricle endocardium in *blue*, left ventricle endocardium in white and left ventricle epicardium in wire-frame

(Algorithm 1) were needed to reach convergence. The resulting atlas $\hat{T}$ was well centered (mean over standard deviation of the deformations was *0.36946*). Atlas-to-patient registration is implemented in parallel on a cluster of computers, which means that the atlas creation time is minimally dependent of the number of patients used to create it.

5.3.2 Diagnosis Parameters

A key topic in computational shape analysis is identifying pathologically specific shape features in populations of diseased patients compared to controls, see [11] and references therein for instance. Beyond simply identifying the pathological shape features, we would like to also quantify the degree to which the shape is altered due to the pathology. One way in which this can be done is by correlating the shapes with

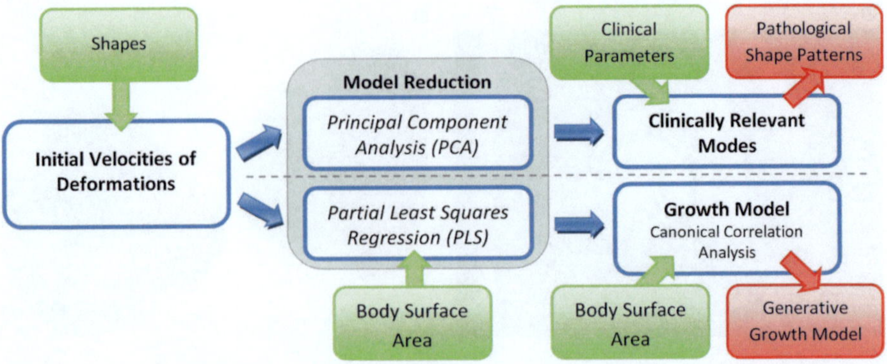

Fig. 5.7 Shape variation analysis pipeline with both clinically relevant model reduction and growth model generation. Using PCA modes and standard correlation analysis, the pathological shape patterns can be identified. Using a combination of PLS regression and CCA a statistical generative growth model can be derived

clinical features to determine the severity of the disease, which requires a consistent representation of the patient shapes. For this we use firstly principal component analysis (PCA) to extract the main modes of shape variation followed by standard statistical design to exhibit those that are correlated to the pathology (Fig. 5.7 top row).

5.3.2.1 Model Reduction Using Principal Component Analysis

Since statistical shape analysis is a high dimensional problem with a large number of parameters and variables to solve for (despite the matching pursuit reduction, a shape can still be represented by hundreds of moments), we first reduce the dimension of the problem by applying principal component analysis (PCA). This gives the modes of deformation that describe the amount of variation of shape observed in the population.

PCA is applied on the initial velocity fields $v_0^{(i)}$ to extract the main deformation modes observed in the population. PCA finds basis vectors, the modes, of the space of variables (here the initial velocities) that best explain their variance. The modes p are calculated by solving the eigenvalue problem $\Sigma p = \mu p$, where the elements σ_{ij} of the covariance matrix Σ are calculated in the kernel space W. Assuming that the deformation from the atlas to patient i is parameterized by the initial vector field $v_0^{(i)}(x) = \sum_k K_W(x, x_k)\beta_k^{(i)}$, where the x_k are the point positions of the delta Dirac currents of the atlas, and $\beta_k^{(i)}$ the moment vector at x_k, then the mean initial vector field is $\bar{v}_0(x) = \sum_k K_W(x, x_k)\bar{\beta}_k$ and the covariance is

$$\sigma_{ij} =< v_0^i - \bar{v}_0, v_0^j - \bar{v}_0 >_V = \sum_{k,l} (\beta_k^{(i)} - \bar{\beta}_k) K_W(x_k, x_l)(\beta_l^{(j)} - \bar{\beta}_l). \quad (5.4)$$

The principal components are obtained by computing the spectral decomposition $\Sigma = PMP^T$. M is the diagonal matrix of the eigenvalues σ_m, or variances, sorted in decreasing order and P is the orthonormal matrix (in the L_2 -norm sense) of the eigenvectors p^m. The mth loading l^m of the PCA decomposition is given by the formula:

$$l^m = \sum_i p^m[i]\beta^{(i)}. \tag{5.5}$$

In this equation, $p^m[i]$ is the ith element of the mth eigenvector of Σ, $\beta^{(i)} = (\beta_1^{(i)} - \bar{\beta}_1, \ldots \beta_n^{(i)} - \bar{\beta}_n)^T$ is the $n \times 3$ matrix that gathers the n centered moment of patient i. As a result, the initial velocity field of the mth mode is $v_{0_{l^m}}(x) = \sum_k K_W(x, x_k)l^m$. The variability captured by this m^{th} deformation mode between $[-\eta\sigma^m; +\eta\sigma^m]$ is visualized by deforming the atlas T with the deformations ϕ^{-m} and ϕ^{+m} parameterized by the moments $\bar{\beta} - \eta l^m$ and $\bar{\beta} + \eta l^m$ respectively. Selecting the first p modes only among the $N-1$ possible modes (where N is the number of patients) allows to explain a percentage $\sum_{m=1}^{p} \sigma^m/\text{trace}(\Sigma)$ of the total variance.

The orthogonal projection of each patient's initial velocity field onto the selected PCA subspace gives a unique shape vector. This simply corresponds to the coordinates of the projection in the basis constituted by the chosen eigenmodes:

$$s^{i,m} =< v_0^{(i)} - \bar{v}_0, v_{0_{l^m}} >_W = \sum_{j,k} [\beta_j^{(i)} - \bar{\beta}] K_W(x_j, x_k) l_k^m. \tag{5.6}$$

Using PCA we have reduced the amount of data needed to represent the shape of a patient by two or more orders of magnitude. The precision of the representation is controlled by the number of components of the PCA subspace. However, it has to be observed that modes with low variances may still be relevant to external clinical parameters. For instance, a mode that captures a local bulging is probably more related to the pathology than a global scaling of the shape although this bulging is not very visible in the population and could be considered as noise in the model. Consequently, we are fairly conservative in the selection of the PCA subspace and select the modes based on their relationship with the clinical parameters of interest and not their variance, as described in the following sections.

5.3.2.2 Identifying Factors Between Shape and Clinical Features

The $s^{i,m}$'s quantify the amount of variability along the mth mode present in patient deformation. We can thus investigate the heart shape by relating these shape vectors to clinical parameters that quantify the pathology. Ordinal clinical parameters are investigated using non-parametric rank-based statistics. Kruskal-Wallis analysis of variance is applied to find effects between the investigated parameters and shape [36]. If an effect is found, post-hoc two-sample Wilcoxon test is used to determine which levels differ [36]. Continuous clinical parameters are investigated using linear regression and Akaike Information Criterion (AIC) model reduction [1] to detect relevant modes and the direction of correlation.

To illustrate this method, in [35] we showed that these methods enable one to identify shape features related to the severity of the regurgitation for a data-set of 49 repaired ToF patients. The relationship between RV shape and pulmonary regurgitation were investigated by relating the PCA shape vectors with tricuspid regurgitation, trans-pulmonary valve regurgitation and pulmonary regurgitation volume indices taken from color Doppler ultrasound and phase-contract magnetic resonance images (PC-MRI). 90 % of the spectral energy was explained by 18 PCA modes.

5.3.3 Building an Evolution Model

As explained beforehand, understanding and quantifying heart remodeling in these patients is crucial for planning pulmonary valve replacement. Given that there is a lack of longitudinal data available for these patients, we make use of the atlas as the mean of the population and cross-sectional statistics to formulate a generative growth model. Such a model could be used as reference, from which the pathology evolution of one patient could be quantitatively compared. In cross-sectional statistical design, one does not propagate the evolution over time for a single patient but rather considers the image of each patient acquired at "time" t as an instance in the growth evolution of the group. In this way we can model the growth of the population given these instances using regression analysis (Fig. 5.8). By making use of the initial velocities v_0^i that parametrize the deformations ϕ^i computed using the methods described in the previous sections, we can regress the velocities against an index of patient growth using standard statistical techniques.

In order to obtain statistically significant results, we first need to reduce the dimensionality of the problem to consider just the factors related to patient growth, while also removing any co-linearity between factors. In the previous section the model reduction was performed using PCA. In this case we chose instead to use partial least squares (PLS) regression since it has the added advantage of computing the components that are most related to a given external parameter (i.e. patient growth). Using PLS allows us to compute the components that best describe the variance of both the matrix of predictors (X) and the matrix of responses (Y), as well as the covariance between X and Y, in a manner such that the regression $Y = f(X)$ is optimal.

In the case of ToF patients, we would ideally like to model the atlas deformation of a patient as a function of growth (i.e. $deformation = f(growth)$, however solving this problem is not possible due to the large number of deformation parameters that would need to be predicted with a single, one-dimensional, parameter (growth). Rather, we revert the problem to be a function of the deformations: $growth = f(deformations)$ which has a much lower number of parameters to predict. The output values are then projected onto the reduced PLS subspace and

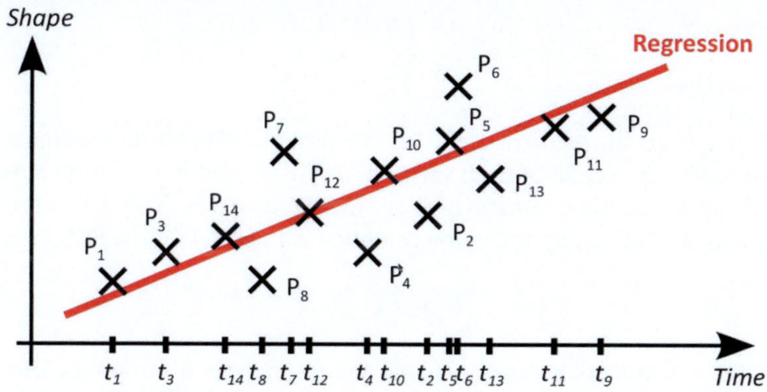

Fig. 5.8 Cross-sectional regression of shapes. Each patient is associated to a point in time (patient age for instance). A regression model is derived from the temporal data

from that we can revert the problem to the desired form as a function of the shape using canonical correlation analysis (CCA) to give a generative growth model of the heart.

For this example we use body surface area (BSA) as the index of growth rather than patient age to better represent the growth given the variable age at which children enter puberty. We use the Dubois formula [5] to compute the BSA for each patient:

$$BSA(m^2) = 0.007184 \times weight(kg)^{0.425} \times height(cm)^{0.725}. \qquad (5.7)$$

The pipeline we have just described for computing a generative growth model is shown in Fig. 5.7 (bottom row).

5.3.3.1 Model Reduction Using Partial Least Squares Regression

As introduced in the previous section, regression using the partial least squares regression (PLS) method is based on finding an optimal basis of the predictor variables X that maximizes the variances of X and Y as well as their covariances. The method can be considered as the optimal estimation of two weight vectors r and s that satisfy

$$\max_{|r|=|s|=1} cov(Xr, Ys) = \max_{|r|=|s|=1} var(Xr) corr(Xr, Ys)^2 var(Ys), \qquad (5.8)$$

under the constraint that the regression between X and Y is optimal. Mathematically, the generative model is

$$X = \bar{X} + TP^T + E,$$
$$Y = \bar{Y} + UQ^T + F,$$

where T and U are the matrices of the PLS modes, P and Q are the loading matrices which describe the weight of each variable in $X - \bar{X}$ and $Y - \bar{Y}$ respectively and E and F are the residual terms which are the same size as X and Y respectively. Additionally the following regression condition is imposed for the PLS modes:

$$U = TD + G, \tag{5.9}$$

where D is a diagonal matrix of weights and G a matrix of residuals. Due to this added condition, the PLS loadings P and Q are not necessarily orthogonal as is the case for PCA modes.

Algorithm 9 Partial least squares regression (PLS)

Input: Variables X and Y, number of components $p \leq N - 1$.
$X_1 = X - \bar{X}, Y_1 = Y - \bar{Y}$
for $n = 1$ *to p* **do**
  $r_n \leftarrow$ first eigenvector of $X_n{}^T Y_n Y_n{}^T X_n$
  $t_n \leftarrow X_n r_n / \|r_n\|$ n^{th} *PLS component of X*
  $s_n \leftarrow Y_n t_n / (t_n{}^T t_n)$
  $u_n \leftarrow Y_n s_n / \|s_n\|$ n^{th} *PLS component of Y*
  $p_n \leftarrow X_n t_n / (t_n{}^T t_n)$ n^{th} *loading of X*
  $q_n \leftarrow Y_n u_n / (u_n{}^T u_n)$ n^{th} *loading of Y*
  $X_{n+1} \leftarrow X_n - t_n p_n{}^T$ *deflation of X*
  $Y_{n+1} \leftarrow Y_n - t_n [t_n{}^T Y_n / t_n{}^T t_n]$ *deflation of X*
end for
return $T = (t_n)_{n=1\ldots p}$, $P = (p_n)_{n=1\ldots p}$, $U = (u_n)_{n=1\ldots p}$, $Q = (q_n)_{n=1\ldots p}$

Several algorithms have been proposed to compute the PLS modes. In this work, we use the PLS1 method, an efficient iterative algorithm that does not require matrix inversion as summarized in Algorithm 9. X is the matrix of the initial velocity field moments for all patients and Y is the vector of the BSA values for all patients. The first five PLS modes are shown in Fig. 5.9 and the explained variance and correlation of modes is shown in Fig. 5.10. These modes account for 99 % of the covariance between shape and BSA in the population and 61 % of the shape variability and are oriented along increasing BSA. Visually we can see that Modes 1 and 3 display an overall dilation in both the left and right ventricles. The second mode shows a narrowing in the right ventricular outflow tract with a noticeable dilation in the left ventricle which can be seen in the top view of the mode. Modes 4 and 5 show an elongation at the right ventricular outflow tract. Mode 5 also shows an aneurysm in the right ventricle outflow tract which is characteristic of rToF patients, although it is still not clear if this is due to pulmonary regurgitations (although both are correlated) or the initial surgical patch.

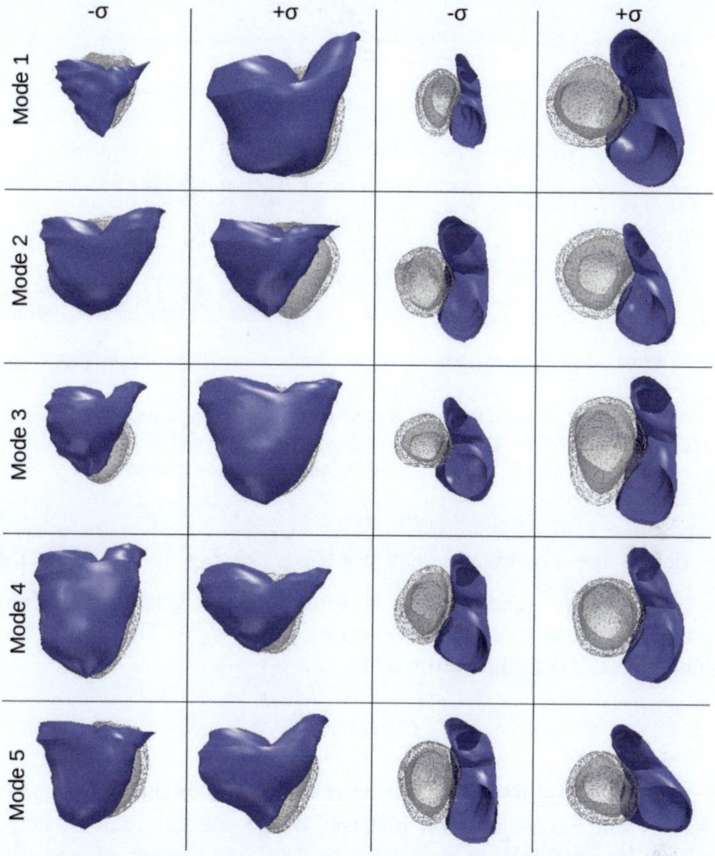

Fig. 5.9 The first five PLS modes of variation that describe 99 % of the observed BSA variability and 61 % of the observed shape variability in the population

5.3.3.2 Generating a Growth Model Using Canonical Correlation Analysis

Using PLS as described above allows us to predict BSA given the shape, however what we would like is to estimate the shape given BSA. To reverse the relationship we use canonical correlation analysis (CCA) on the PLS shape vectors (the $t_i's$ of Algorithm 9). CCA computes the vectors r and s that maximize the correlation between the two sets X and Y

$$\max_{|r|=|s|=1} corr(Xr, Ys)^2, \tag{5.10}$$

where Y is the vector of BSA values and $X = [t_1, \ldots t_N]^{\mathrm{T}}$ is the matrix of shape descriptors.

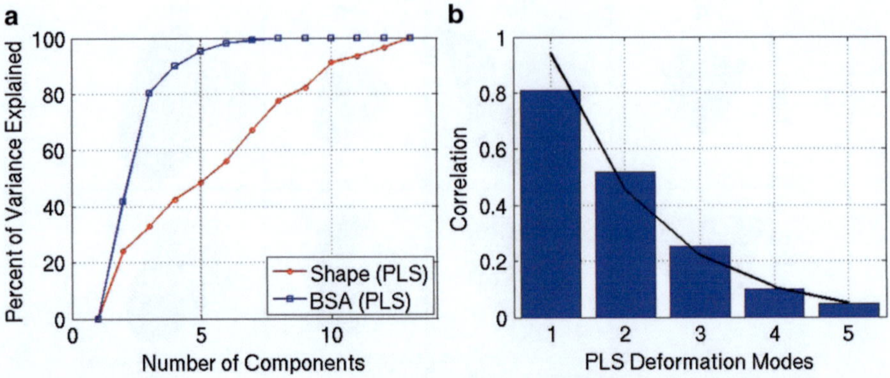

Fig. 5.10 Analysis of the variance and dependency of the shape and covariates using PLS. (**a**) Cumulative variance of PLS modes with respect to shape and BSA. (**b**) CCA correlation coefficients between BSA and PLS modes with exponential fitted curve in *black*

If we define the covariance matrices $V_{UZ} = \frac{1}{N-1}U^{\mathrm{T}}Z$, then the matrix $\Gamma = V_{XX}^{-1/2}V_{XY}V_{YY}^{-1/2}$ can be seen as a multi-variate generalization of the uni-dimensional correlation coefficient $r = \sigma_{XY}/\sqrt{\sigma_{XX}\sigma_{YY}}$. The sought correlations are obtained by SVD decomposition of Γ:

$$\Gamma = ASB^T, \tag{5.11}$$

where S is the diagonal matrix of the correlation coefficients between correlation vectors and A and B are rotation matrices of correlation vectors, i.e., $A^TA = B^TB = Id$. In our application, Y is a one-column matrix. Hence, S has only one non-null coefficient R, which is the overall correlation between the PLS shape vectors X and BSA. B is a scalar equal to ± 1 that determines the direction of BSA correlation. The elements of the first correlation vector of A, denoted by ρ, relate to the amplitude and direction of correlations of each predictor, namely each PLS mode, when Y varies along the direction defined by the sign of B. In other words, when BSA varies by 1, the kth predictor varies by $BR\rho[k]$. We can therefore compute a generative average model of heart growth by artificially increasing BSA and deforming the atlas T with the growth deformation Φ parametrized by the moments $\mu = B.R.\Sigma_k\rho[k]p^k$, where p^k is the kth PLS loading.

5.3.3.3 Interpretation

The growth model computed on the rToF data-set is shown in Fig. 5.11. This model shows an expected overall growth of both ventricles as body surface area increases. We can also see the caving of the septum into the right ventricle as time passes and the elongation of the right ventricular outflow tract which is observed in these patients over time.

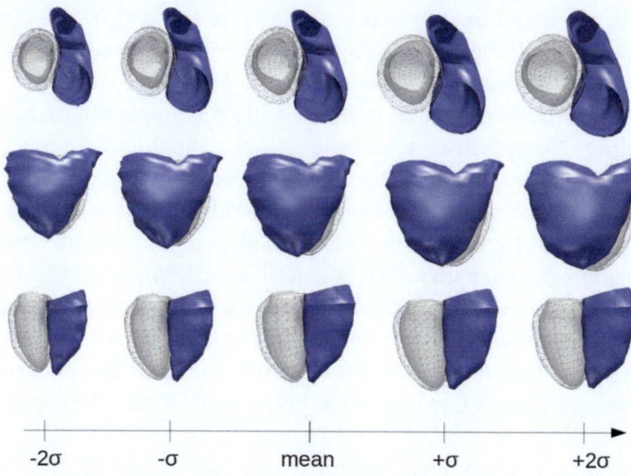

Fig. 5.11 Mean growth model computed from a population of 13 rToF patients. Both ventricles grow as BSA increases

Given a larger data-set this model can be improved further by increasing the certainty in the model and by capturing more variability in shape observed in a wider population. As well, given more data we can divide the patients according to an external parameter such as treatment group to formulate a growth model for a given course of treatment. In the case of ToF, this could allow clinicians to have a better idea of how the heart re-models after different types of valve replacement surgery and more importantly, the effect of the initial surgery on the long-term outcome. This is the key question.

5.4 Conclusion

In this chapter we presented methods for computing statistics on shapes. The proposed methods rely on currents to represent the anatomical shapes in a consistent way without the need for defining landmarks or point correspondences between shapes. This allows the computation of population statistics such as the mean and variance observed within a population and to correlate shape with clinical indices, for instance to quantify the severity of a disease. We also present a framework for formulating a generative statistical growth model to simulate the growth of an organ over time. This is especially important for our clinical application where the shape remodeling of the heart in congenital diseases like Tetralogy of Fallot is crucial to better understand the evolution of the disease and ultimately to aid for therapy planning.

From a clinical perspective, most of the shape changes that could be put into evidence in our longitudinal study were corresponding to some previously reported observations in the literature. However, we have for the first time a combined and quantitative model of all the changes together. Of course, the number of parameters that could influence the shape of the heart is so large that we would need much more subjects in order to obtain an accurate and reliable quantitative model of the heart remodeling. Nevertheless, the shape analysis pipeline we proposed is more than a black box that learns the statistical relationship between sets of variable: it gives a generative shape model that can be explored. For instance, the visualization of the remodeling (or the deformation correlated to other clinical variables) could allow to draw some intuition about the potential underlying links. Ideally this would lead to a simplified model that could be tested against new data, thus gradually improving the knowledge that we have about the disease. As the anatomy is the support of the physiology, one of the goals is now to couple the anatomical changes with the physiological evolution in order to better understand how one influences the other.

Other interesting clinical questions related to rToF could be investigated using our approach. For instance, there is nowadays a clinical consensus that a longer QRS complex duration measured on the electrocardiogram (see for instance Fig. 3.3 in Chap. 3) correlates with RV dilation. Correlating this parameter with the RV shape could reveal how the abnormal conduction impacts the RV anatomy and function. Similarly, studying body-mass index (BMI) jointly with BSA could provide a more comprehensive representation of the patient growth. One could apply the PLS method between two sets of multivariate variables (BMI, BSA and QRS on the one hand, the shape vectors on the other hand) and get a growth model that explains both features. Unfortunately, QRS durations and BMI were not available for all the patients as the population was retrospective, keeping these questions for future work. The effect of genetic factors that regulate myocardium stiffness on the long-term RV remodeling could also be analyzed. Patients with stiffer myocardium are known to be more protected against RV dilation. These patients may be the outliers of the model with "abnormally" smaller RV than the average. The decision for valve implant may be delayed and based on different features for these patients. Finally, tools are now been developed and integrated in several hospitals to keep track of the patient records during a long time-period, which opens the possibility to now correlate the shape features directly with the clinical outcome. Such a shape-based model of the outcome, when validated, would be immediately usable by the clinicians as a computer aided diagnosis tool.

5.5 Online Resources

Integrated Models of Paediatric Heart Diseases
http://www-sop.inria.fr/asclepios/projects/Health-e-Child/DiseaseModels/content/cardiac/
This site summarizes some modeling issues in pediatric cardiac diseases. It explains the approach that was adopted in the European FP6 Health-e-Child project.

Shape analysis software: Exoshape

http://www-sop.inria.fr/asclepios/projects/Health-e-Child/ShapeAnalysis/

This web site contains the source of the Matlab software developed for doing the analysis described in this chapter, and a description of its application to the analysis of the shape right ventricule of the heart in ToF.

The Virtual Physiological Human

http://ercim-news.ercim.eu/en69/special

April 2007 issue of the ERCIM News journal on "The Digital Patient" describing many different aspects of the Virtual Physiological Human initiative to develop *in silico* models of human physiology and pathology.

http://en.wikipedia.org/wiki/Virtual_Physiological_Human

The wikipedia description of the VPH with links to other resources

Acknowledgements The computational tools used in this chapter were originally developed within the context of the European FP6 project Health-e-Child (http://www.health-e-child.org/). The software was made available to the community in collaboration with the EU network of Excellence Virtual Physiological Human (http://www.vph-noe.eu/). The extension to the analysis of the bi-ventricular shape of the heart in rToF was performed in the context of the European ITEA2 Care4Me project (www.care4me.eu/).

List of Acronyms

BMI	Body Mass Index
BSA	Body Surface Area
CCA	Canonical Correlation Analysis
LDDMM	Large Deformations Diffeomorphic Metric Mapping
MRI	Magnetic Resonance Imaging
PCA	Principal Component Analysis
PLS	Partial Least Squares
RKHS	Reproducing Kernel Hilbert Space
rToF	repaired Tetralogy of Fallot
RV	Right Ventricle
ToF	Tetralogy of Fallot

References

1. H. Akaike. A new look at the statistical model identification. *IEEE transactions on automatic control*, 19(6):716–723, 1974.
2. F.L. Bookstein. *The Measurement of Biological Shape and Shape Change*, volume 24 of *Lecture Notes in Biomathematics*. Springer-Verlag, 1978.
3. F.L. Bookstein. Size and shape spaces for landmark data in two dimensions (with discussion). *Statistical Science*, 1:181–242, 1986.

4. I.L. Dryden and K.V. Mardia. Theoretical and distributional aspects of shape analysis. In *Probability Measures on Groups, X (Oberwolfach, 1990)*, pages 95–116, New York, 1991. Plenum.
5. D. DuBois and E. DuBois. The measurement of the surface area of man. *Archives of Internal Medicine*, 15(5), 1915.
6. S. Durrleman. *Statistical models of currents for measuring the variability of anatomical curves, surfaces and their evolution*. Thèse de sciences (phd thesis), Université de Nice-Sophia Antipolis, 2010.
7. J. Glaunès. *Transport par difféomorphismes de points, de mesures et de courants pour la comparaison de formes et lanatomie numérique*. Thèse de sciences, Université Paris 13, November 2005.
8. U. Grenander. *General Pattern Theory: A Mathematical Study of Regular Structures*. Oxford University Press Inc., New York, NY., 1993.
9. J. Hoffman and S. Kaplan. The incidence of congenital heart disease. *Journal of the American College of Cardiology*, 39(12):1890–1900, 2002.
10. H. Hufnagel. *A probabilistic framework for point-based shape modeling in medical image analysis*. Phd thesis, University of Lübeck, 2010.
11. P. Fletcher J. Cates and R. Whitaker. A hypothesis testing framework for high-dimensional shape models. In *In MICCAI Workshop on Mathematical Foundations of Computational Anatomy*, page 170, 2008.
12. D.G. Kendall. A survey of the statistical theory of shape (with discussion). *Statistical Science*, 4:87–120, 1989.
13. D.G. Kendall, Shape manifolds, Procrustean metrics, and complex projective spaces, *Bull. London Math. Soc*, 16 (1984), 81–121.
14. H. Le and D.G. Kendall. The Riemannian structure of Euclidean shape space: a novel environment for statistics. *Annals of Statistics*, 21:1225–1271, 1993.
15. L. Younes M. Vaillant, M.I. Miller and A. Trouvé. Statistics on diffeomorphisms via tangent space representations. *NeuroImage*, 23(Supp. 1):S161–S169, 2004.
16. S. G. Mallat and Z. Zhang. Matching pursuits with time-frequency dictionaries. *IEEE Transactions on Signal Processing*, 41(12):3397–3415, 1993.
17. T. Mansi. *Image-Based Physiological and Statistical Models of the Heart, Application to Tetralogy of Fallot*. Thèse de sciences (phd thesis), Ecole Nationale Supérieure des Mines de Paris, 2010.
18. A. Trouvé M.F. Beg, M.I. Miller and L. Younes. Computing large deformation metric mappings via geodesic flows of diffeomorphisms. *Int. Journal of Computer Vision*, 61(2):139–157, 2005.
19. A. Trouvé M.I. Miller and L. Younes. On the metrics and Euler-Lagrange equations of computational anatomy. *Annual Re-view of Biomedical Engineering*, pages 375–405, 2003.
20. M.I. Miller and L. Younes. Group actions, homeomorphisms, and matching: A general framework. *International Journal of Computer Vision*, 41(1/2):61–84, 2001.
21. J. T. Ratnanather R. A. Poldrack Th. E. Nichols J. E. Taylor P. M. Thompson, M. I. Miller and K. J. Worsley, editors. *Mathematics in Brain Imaging*, volume 45 of *NeuroImage, special issue*. Springer, March 2004.
22. X. Pennec. Intrinsic statistics on riemannian manifolds: Basic tools for geometric measurements. *Journal of Mathematical Imaging and Vision*, 25(1):127–154, July 2006. A preliminary appeared as Inria RR-5093, January 2004.
23. X. Pennec. Statistical computing on manifolds: from riemannian geometry to computational anatomy. In *Emerging Trends in Visual Computing*, volume 5416 of *LNCS*, pages 347–386. Springer, 2008.
24. Y. Amit S. Allassonnire and A. Trouv. Towards a coherent statistical framework for dense deformable template estimation. *Journal Of The Royal Statistical Society Series B*, 69(1): 3–29, 2007.
25. A. Trouvé S. Durrleman, X. Pennec and N. Ayache. A forward model to build unbiased atlases from curves and surfaces. In X. Pennec and S. Joshi, editors, *Proc. of the International Workshop on the Mathematical Foundations of Computational Anatomy (MFCA-2008)*, September 2008.

26. A. Trouvé S. Durrleman, X. Pennec and N. Ayache. Statistical models on sets of curves and surfaces based on currents. *Medical Image Analysis*, 13(5):793–808, 2009.
27. A. Trouvé G. Gerig S. Durrleman, X. Pennec and N. Ayache. Spatiotemporal atlas estimation for developmental delay detection in longitudinal datasets. In *Medical Image Computing and Computer-Assisted Intervention (MICCAI'09), Part I*, volume 5761 of *Lecture Notes in Computer Science*, pages 297–304, London, UK, 2009. Springer.
28. A. Trouvé P. Thompson S. Durrleman, X. Pennec and N. Ayache. Inferring brain variability from diffeomorphic deformations of currents: an integrative approach. *Medical Image Analysis*, 12(5):626–637, 2008.
29. A. Srivastava S. H. Joshi, D. Kaziska and W. Mio. Riemannian structures on shape spaces: A framework for statistical inferences. In Hamid Krim and Anthony Yezzi, editors, *Statistics and Analysis of Shapes*, Modeling and Simulation in Science, Engineering and Technology, pages 313–333. Birkhäuser Boston, 2006.
30. A. Srivastava S. H. Joshi, E. Klassen and I. Jermyn. A novel representation for riemannian analysis of elastic curves in rn. *Computer Vision and Pattern Recognition, IEEE Computer Society Conference on*, 0:1–7, 2007.
31. A. Srivastava S. H. Joshi, E. Klassen and I. Jermyn. Removing shape-preserving transformations in square-root elastic (sre) framework for shape analysis of curves. In *EMMCVPR'07*, pages 387–398, 2007.
32. M. Jomier S. Joshi, B. Davis and G. Gerig. Unbiased diffeomorphic atlas construction for computational anatomy. *NeuroImage*, 23(Supplement 1):S151–S160, 2004. Mathematics in Brain Imaging.
33. C.G. Small. *The Statistical Theory of Shapes*. Springer series in statistics. Springer, 1996.
34. B. Bernhardt M. Sermesant H. Delingette I. Voigt T. Mansi, S. Durrleman, J. Blanc Y. Boudjemline X. Pennec P. Lurz, A. M. Taylor, and N. Ayache. A statistical model of right ventricle in tetralogy of fallot for prediction of remodelling and therapy planning. In *Proc. Medical Image Computing and Computer Assisted Intervention (MICCAI'09)*, volume 5761 of *Lecture Notes in Computer Science*, pages 214–221. Springer, 2009.
35. B. Leonardi X. Pennec S. Durrleman M. Sermesant H. Delingette A. M. Taylor Y. Boudjemline G. Pongiglione T. Mansi, I. Voigt and N. Ayache. A statistical model for quantification and prediction of cardiac remodelling: Application to tetralogy of fallot. *IEEE Transactions on Medical Images*, 2011.
36. R Development Core Team. R: A language and environment for statistical computing. *R Foundation for Statistical Computing, Vienna, Austria*, Online. Available: http://www.R-project.org, 2009.
37. W. D'Arcy Thompson. *On Growth and Form*. Cambridge University Press, England., 1917.
38. A. Trouvé. Diffeomorphisms groups and pattern matching in image analysis. *International Journal of Computer Vision*, 28(3):213–221, 1998.
39. M. Vaillant and J. Glaunes. Surface matching via currents. In *Proc. of IPMI'05*, pages 381–392, 2005.
40. A. Srivastava W. Mio and S. H. Joshi. On shape of plane elastic curves. *International Journal of Computer Vision*, pages 307–324, 2007.
41. B. Georgescu M. Scheuering Y. Zheng, A. Barbu and D. Comaniciu. Four-chamber heart modeling and automatic segmentation for 3-d cardiac ct volumes using marginal space learning and steerable features. *IEEE Trans. on Medical Imaging*, 27(11):1668–1681, 2008.

Chapter 6
From Diffusion MRI to Brain Connectomics

Aurobrata Ghosh and Rachid Deriche

6.1 Introduction

The main objective of this chapter is to present some mathematical models and computational tools for analyzing and modeling the complex central nervous system's (brain and spinal cord) neural connectivity. These models and tools will help to better understand the white matter architecture of the human *central nervous system* (CNS) and in a long term, will also help in addressing important and challenging clinical and neuroscience questions. Indeed, due to our aging society, diseases like Alzheimer's, Parkinson's disease (PD) and depression will affect a large population. These examples of CNS diseases as well as others, like multiple sclerosis have characteristic abnormalities in the microstructure of brain's tissues such as its white matter, which are not apparent and cannot be revealed reliably by standard imaging techniques. Diffusion MRI (dMRI), a recent imaging modality based on the measurement of the random thermal movement (diffusion) of water molecules within samples, can make visible these co-lateral damages to the fibers of the CNS white matter that connect different brain regions. This is why in this chapter, dMRI is the major anatomical imaging modality that will be considered to recover the neural connectivity in the CNS.

We begin this chapter by presenting the CNS, in particular the brain, before delving into the mathematical framework for dMRI. Section 6.1 is dedicated to a perusal of its general structure and organization, the tissues constituting it, and in highlight, the brain's major neuronal pathways interconnecting its various regions. It aims to provide a context for understanding the general physical problem dMRI attempts to solve.

A. Ghosh (✉) · R. Deriche
Inria Sophia Antipolis Méditerranée, Athena project-team, 2004 Route des Lucioles, 06902, Sophia Antipolis, France
e-mail: aurobrata.ghosh@inria.fr; rachid.deriche@inria.fr

F. Cazals and P. Kornprobst (eds.), *Modeling in Computational Biology and Biomedicine*, DOI 10.1007/978-3-642-31208-3_6, © Springer-Verlag Berlin Heidelberg 2013

6.1.1 The Central Nervous System

The human nervous system can be divided into the CNS, which consists of the brain and spinal cord and the *peripheral nervous system* (PNS) which consists of the cranial and spinal nerves and their ganglia. The CNS is the largest part of the nervous system and is composed of the spinal cord and the brain which comprises a lower part, the brainstem, and an upper part, the prosencephalon or forebrain composed of two main units. One is known as the diencephalon. It is located in the midline of the brain and contains the thalamus and the hypothalamus. The other is called the telencephalon or cerebrum and holds the lateral ventricles, the basal ganglia and the cerebral cortex. The brainstem is also composed of two units: the mesencephalon or midbrain and the rhombencephalon or hindbrain which connects the forebrain and midbrain to the spinal cord.

The nervous system is made of about 100 billion nerve cells, or neurons, able to generate and propagate electrical signals to process and transmit neural information. Neurons can receive electrical stimulation from other neurons on their soma, through their multiple dentrites. They can integrate this information and propagate it to more or less distant locations of the cerebrum by an extension called an axon. Nerve signal communication is performed at specialized loci called synapses. Each neuron has on average 1,000 synaptic connections with other neurons. This yields about 100 trillion connections within a human brain. All these synapses result in an impressively dense and complex network between functional areas, which can be understood as aggregates of nerve cells' soma and dendrites. They are essentially located in the grey matter while the underlying wiring constitutes the white matter. According to the connectionist point of view, the human brain is organized into distinct processing regions interconnected by a network of anatomical relays. Processing units handle the execution of primary cognitive functions, and higher cognitive tasks arise from a global coordination between these processing units. Neural signal is processed in the cerebral cortex and transmitted to various regions of the brain through the white matter. So we distinguish:

- *Grey matter* essentially forms the outer part of the cerebrum, some nuclei within the brain, as well as the deeper part of the spinal cord. It is made of neurons and their unmyelinated fibers. The cerebral cortex is the most important structure of the grey matter and plays a major role in various functions such as memory, attention and language.
- *White matter* is composed of axonal nerve fibers, covered by a myelin sheath giving its distinctive colour in MRI. It is found in the inner layer of the cortex, the optic nerves, the central and lower areas of the brain and surrounding the central shaft of grey matter in the spinal cord.

The CNS white matter axons can be distributed diffusely or concentrated in bundles, also referred to as tracts or fiber pathways. The brain's white matter pathways are generally categorized into commissural, association and projection fibers, depending on the areas they connect (Fig. 6.1).

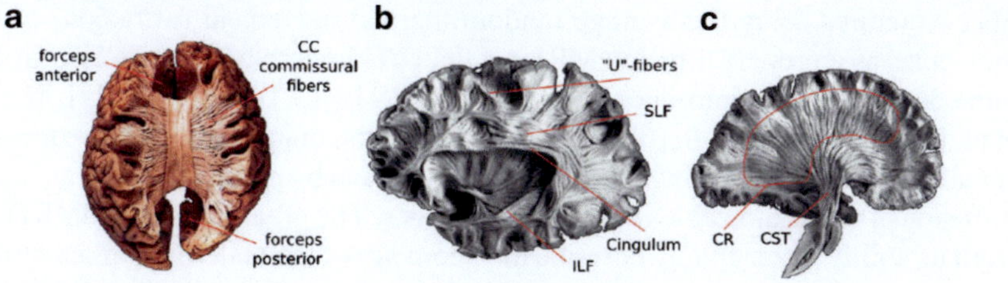

Fig. 6.1 Major white matter fiber pathways in the brain. (**a**) Commissural tracts—the Corpus Callosum (*CC*). (**b**) Association tracts—short "U"-fibers connecting neighbouring gyri, the Superior Longitudinal Fasciculus (*SLF*), the Cingulum and the Inferior Longitudinal Fasciculus (*ILF*). (**c**) Projection tracts—the Corona Radiata (*CR*) and the Corticospinal Tract (*CST*) (Adapted from [74])

- The *commissural tracts* connect a region in one hemisphere to another region of the opposite hemisphere.
- The *association tracts* connect various cortical areas within a given hemisphere.
- The *projection tracts* connect the cortex to deep brain regions such as the thalamus or the spinal cord.

In the spinal cord, it is the grey matter that is located in the center with a typical H-shaped appearance in transverse sections. It is surrounded by white matter which contains long ascending and descending pathways.

6.1.2 In Vivo CNS Connectivity

Compared to the understanding of neural circuitry in animals as cats or mice where the use of invasive tracers is possible [59], the knowledge of the human brain organization is relatively poor. Therefore, how to study the connectivity information about the CNS anatomy, in particular, about the cerebral and spinal cord white matter? Anatomical MRI allows us to distinguish and classify grey matter and white matter. However, with this contrast, white matter retains a homogeneous aspect, preventing any observation of neural fibers and thus of neuronal connectivity. Cerebral and spinal dissection used to be the only means of accessing the neural architecture [22, 30, 74]. Then, anatomists started using chemical markers to do neuronography [55, 59]. More recently, neural fiber tractography based on local injection of chemical markers and subsequent observation of the induced propagation yielded high-quality connectivity mapping in the cat and monkey cerebral cortices [59, 75].

The relatively recent development of dMRI, which uses the orientational preference of water molecules in diffusing along the directions of the fibers has brought in the last 20 years great hopes for the non-invasive exploration of the neural anatomy of the CNS. Indeed dMRI provides a non-invasive way of estimating in vivo CNS fiber structures using the average random thermal movement (diffusion) of water molecules as a probe. Diffusion MRI is a field of research with a history of roughly three decades. It was introduced in the mid 1980s by Le Bihan et al. [39], Merboldt et al. [43] and Taylor et al. [63]. As of today, it is the unique non-invasive technique capable of describing the neural connectivity in vivo by quantifying the anisotropic diffusion of water molecules in biological tissues. The great success of dMRI comes from its ability to accurately describe the geometry of the underlying microstructure and to probe the structure of the biological tissue at scales much smaller than the imaging resolution.

The diffusion of water molecules is Gaussian in an isotropic medium and under normal unhindered conditions, but in fibrous structure such as white matter, the diffusion is very often directionally biased or anisotropic and water molecules tend to diffuse along fibers. For example, a molecule inside the axon of a neuron has a low probability to cross a myelin membrane. Therefore the molecule will move principally along the axis of the neural fiber. Conversely if we know that molecules diffuse locally principally in one direction, we can infer that this corresponds to a set of fibers.

6.1.3 Chapter Overview

Public

This chapter is an essential reading for MSc students and scientists who want to learn about the key mathematical tools and methods underlying diffusion MRI from modelling the MRI signal and measuring diffusion properties to reconstructing the underlying architecture of anatomical connections in the human brain.

Outline

Section 6.2 briefly covers the historical development of Nuclear Magnetic Resonance (NMR) and MRI to set the stage. We explore the physics of the signal generation in NMR and how diffusion properties can be measured non-invasively from NMR in Sect. 6.3. Section 6.3 begins by presenting the fundamental ideas of NMR. Section 6.3.1 then presents the crucial spin echo experiment proposed by Hahn, which is a corner stone experiment that led to the development of diffusion NMR. Next in Sect. 6.3.2 we describe diffusion in considerable detail. Section 6.3.3 then presents the pulse-field-spin-echo (PGSE) experiment that was proposed by Stejskal and Tanner. This important experiment is the modern and practical approach for measuring diffusion from NMR. The Stejskal-Tanner model for the diffusion NMR signal is based on Fick's laws of diffusion. Section 6.3.4 presents next the q-space approach for modelling the diffusion NMR signal, which is based on Einstein's random walk approach to Brownian motion.

Section 6.4 is dedicated to dMRI reconstruction algorithms that allow to go beyond simply measuring intrinsic diffusion properties from NMR/MRI to inferring the microstructure of the underlying tissue non-invasively. Section 6.4.1 presents diffusion tensor imaging (DTI), the most widely used dMRI algorithm with the simplest experimental requirements, but with powerful applications. DTI is however limited under certain microstructure configurations. Section 6.4.2.1 presents diffusion spectrum imaging (DSI), which allows to overcome these limitations. Finally, Sect. 6.4.2.2 presents Q-Ball imaging, which can recover complex microstructures like DSI, but isn't limited by the latter's lengthy acquisition schemes.

Section 6.5 introduces briefly the computational framework consisting of sophisticated mathematical tools that have been developed for processing dMRI data and images, since these represent complex mathematical objects. In particular we present various metrics for a field of diffusion tensors in Sect. 6.5.1, an algorithm that ensures that the estimated diffusion tensors are in the proper space of symmetric positive definite matrices using a Riemannian metric in Sect. 6.5.2, and a segmentation algorithm for segmenting fields or images of diffusion tensors in Sect. 6.5.3.

Section 6.6 is dedicated to tractography—algorithms that integrate local microstructure information to reconstruct white matter fiber pathways. Section 6.6.1 presents deterministic approaches while Sect. 6.6.2 presents probabilistic approaches. The penultimate section—Sect. 6.7, is dedicated to applications. It presents the impacts of dMRI in both clinical and neuroscientific scenarios, and highlights its usefulness.

The chapter concludes with a summary in Sect. 6.8. In Sect. 6.9 we provide an incomplete list of important internet resources in terms of available softwares and database for processing dMRI data.

6.2 A Brief History of NMR and MRI

The scientific heritage of NMR and MRI is reflected in the list of Nobel laureates who contributed to their developments. The theoretical underpinnings that made NMR possible were proposed in 1924 by Wolfgang Pauli who suggested a new quantum degree of freedom that later came to be known as spin. He formulated the mathematical theory by 1927, and was awarded the Nobel prize in physics in 1945 for his contributions. The concept of spin implies that atomic nuclei bearing spins exhibit magnetic moments. The fact that protons exhibit magnetic moments had already been discovered in 1922 by Otto Stern prior to the concept of spin. Stern was awarded the Nobel prize in physics in 1943. Pauli's theory was verified in 1938 by Isidor Rabi in molecular beams. From his experiments Rabi was able to both detect the effects of spin and measure the gyromagnetic ratio that is the characteristic signature of an atomic nucleus due to its spin. His experiments also established the concept and the technique of NMR for manipulating spins. Rabi was awarded the Nobel prize in physics in 1944.

In 1946 Bloch [12] and Purcell [56] independently extended the techniques established by Rabi. They successfully demonstrated the magnetic resonance effect in liquids and solids. Bloch and Purcell shared the Nobel prize of 1952 in physics, and NMR was established. In his seminal paper of 1950 [31] Erwin Hahn proposed the spin echo experiment, which used a combination of 90° and 180° electromagnetic or radio frequency pulses to filter out effects of magnetic field inhomogeneities in the measurement of the transverse signal. Further works of Herman Carr and Purcell in 1954 [14] led to the full development of the radio frequency pulse technique introduced by Hahn. This formed the foundations of NMR.

It must be noted at this point that both the papers of Hahn [31] and Carr and Purcell [14] critically point out the observed effects of diffusion of the spin bearing nuclei in magnetic resonance experiments with a succession of radio frequency pulses. Although these papers generally perceive the diffusion effect as an unfortunate phenomenon resulting in a loss of signal, Carr and Purcell [14] in fact demonstrate that diffusion can be directly measured from NMR and go on to actually measure the diffusion constant of water at 25 °C. This forms the corner-stone of diffusion NMR.

Although NMR became a well established technique for studying various materials, it took almost three decades since the experiments of Bloch and Purcell in 1946, for MRI to be invented. NMR by itself is capable of examining a single spin ensemble or a tiny region of a sample, but it can't image the whole sample to recreate a 2D slice or a 3D volumetric image necessary to study entire biological samples like the human body. Paul Lauterbur in 1973 [37] proposed the use of magnetic gradient fields to spatially encode the positions or voxel regions of the spin ensembles. This was a remarkable invention, which made it possible to reconstruct entire slice or volumetric images from NMR data. Spatial encoding was improved in terms of frequency encoding by Richard Ernst in 1978, and phase encoding by Bill Edelstein in 1980 using pulsed gradients. In 1977 Peter Mansfield [42] developed the mathematical framework for rapidly switching gradients for spatial encoding, greatly speeding up the process of reconstructing images of an entire biological sample. This is known as *echo planar imaging* (EPI). Lauterbur and Mansfield were jointly awarded the Nobel prize in medicine in 2003 for making MRI possible. Thus modern MRI was developed from the phenomenon of NMR coupled with the method of spatial encoding.

6.3 Nuclear Magnetic Resonance and Diffusion

The principles of NMR are based on *spin*, a fundamental quantum characteristic possessed by electrons, protons, and neutrons, like electrical charge and mass. Spins come in multiples of $1/2$ and can be positive or negative. In grouped particles, e.g. atomic nuclei, opposite spin-signs can pair up to eliminate the total spin of the group. But the net spin of unpaired particles or atomic nuclei imparts a

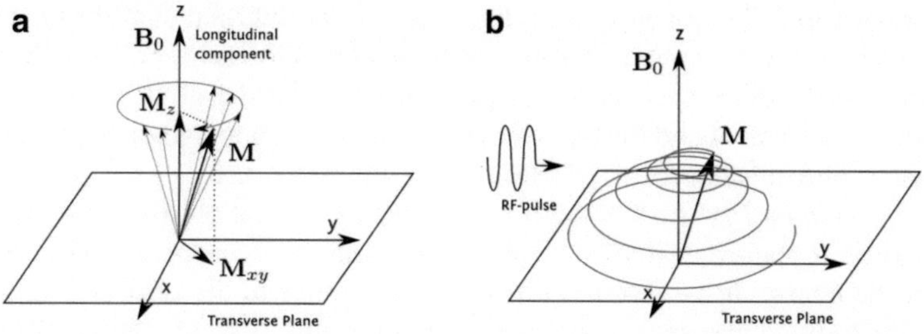

Fig. 6.2 Nuclear Magnetic Resonance. (**a**) A net magnetization dipole moment vector $\mathbf{M}$ forms when spins are exposed to an external magnetic field $\mathbf{B_0}$. (**b**) The NMR signal is generated by exciting the ensemble of spins precessing along $\mathbf{B_0}$ by exposing them to a radio-frequency pulse—the magnetization vector $\mathbf{M}$ spirals down to the transverse plane in the fixed frame of reference

magnetic dipole moment. In other words such particles or such atomic nuclei can be influenced by an external magnetic field. In the presence of a strong magnetic field $\mathbf{B_0}$ with magnitude B_0, the magnetic dipole moment vector or the spin vector of the particle or nucleus aligns itself with $\mathbf{B_0}$ and precesses around it with an angular frequency known as the *Larmor frequency* $\omega_0 = \gamma B_0$, where γ is the *gyromagnetic ratio*, characteristic of the particle or the nucleus. The effect is detectable when it becomes pronounced in the presence of an ample collection of spin bearing particles or nuclei with the same gyromagnetic ratio. From a macroscopic perspective, when such a collection is subjected to a magnetic field, the randomly oriented individual magnetic dipole moment vectors align themselves along $\mathbf{B_0}$. Laws of thermodynamics ensure that a greater number of spins point along the magnetic field (low energy configuration) than opposite to it (high energy configuration). This forms a resultant ensemble magnetic dipole moment vector $\mathbf{M}$. Conventionally the external magnetic field $\mathbf{B_0}$ is considered to be aligned with the Z-axis. The XY-plane is then known as the transverse plane, and the net magnetization vector $\mathbf{M}$ can be separated into the *longitudinal* component $\mathbf{M}_z$, along the Z-axis (or $\mathbf{B_0}$), and the *transverse* component $\mathbf{M}_{xy}$, in the transverse plane (Fig. 6.2).

Of particular interest is the hydrogen nucleus ^1H, which is found abundantly in nature, accounting for 99.98 % of all hydrogen atoms, and also constituting water. 60 % of the human body and 78 % of the brain is water. Therefore, ^1H is a natural spin bearing nucleus of choice for MRI. ^1H is an unpaired proton with a net spin of $1/2$, and has a gyromagnetic ratio of $\gamma = 42.58$ MHz/T.

The NMR signal is generated by exposing the ensemble of spins precessing along $\mathbf{B_0}$ to an oscillating magnetic field or an electro-magnetic (radio-frequency: RF) pulse. This is known as the excitation phase. The energy absorbed by the low energy configuration spins from this pulse tilts the magnetization vector $\mathbf{M}$ away from $\mathbf{B_0}$ towards the high energy configuration. The oscillation of the secondary magnetic field ensures that the spins (and hence $\mathbf{M}$) continue to precess around $\mathbf{B_0}$ even tilted away from it—along the surface of a cone (Fig. 6.2). Once the RF pulse is switched off, the spins begin to recover their alignment with the main magnetic field $\mathbf{B_0}$, and to return to their low energy configuration or the thermal equilibrium. This is known as the relaxation phase. The signal is created as the spins precess tilted away from $\mathbf{B_0}$, and it decays as the spins relax, dissipating the absorbed energy. The longitudinal relaxation and the transverse relaxation of $\mathbf{M}$ are governed by different phenomena and are characterized by different time signatures.

The *longitudinal relaxation* is known as the T1 relaxation since it is described using a time signature denoted T1. The T1 relaxation occurs as the spin ensemble radiates the energy it had absorbed from the RF pulse to the surrounding thermal reservoir or lattice and regains its thermal equilibrium with the lattice. Therefore, the T1 relaxation is also known as the *spin-lattice* relaxation. In this process the spins realign themselves with $\mathbf{B_0}$. In terms of the net magnetization vector $\mathbf{M}$, this implies that the longitudinal component $\mathbf{M}_z$ progressively regains its initial magnitude, while the transverse component $\mathbf{M}_{xy}$ progressively becomes null again.

The *transverse relaxation* involves the phenomenon of the spins regaining their thermal equilibrium amongst themselves, and is characterized by the time signature T2. Therefore it is also known as the *spin-spin* relaxation or the T2 relaxation. In the transverse plane this is interpreted by the spins losing their initial coherence. From an initial coherent transverse magnetization vector $\mathbf{M}_{xy}$, they progressively dephase as they radiate the energy they had absorbed to neighbouring spins. Transverse relaxation is, however, a complex phenomenon. Although theoretically $\mathbf{B_0}$ is supposed homogeneous, in reality minor inhomogeneities exist. These inhomogeneities are relevant enough to also contribute to spins dephasing in the transverse plane, though this is not a true relaxation. Transverse relaxation is therefore a combination of spin-spin relaxation and field inhomogeneity dephasing. The pure spin-spin relaxation time is known as T2. The combined transverse relaxation time is known as T2*.

The *Bloch equations* are a coupled set of three differential equations that combine the effects of NMR and describe the evolution of the net magnetization vector $\mathbf{M}$ over time. These are macroscopic and phenomenological equations that include the effects of Larmor precession and T1 and T2 relaxations. They are written in the fixed frame of reference in terms of the relaxation time constants as:

$$\frac{d\mathbf{M}(t)}{dt} = \gamma\mathbf{M}(t) \times \mathbf{B}(t) + \begin{pmatrix} -\frac{1}{T_2} & 0 & 0 \\ 0 & -\frac{1}{T_2} & 0 \\ 0 & 0 & -\frac{1}{T_1} \end{pmatrix} \mathbf{M}(t) + \begin{pmatrix} 0 \\ 0 \\ \frac{M_0}{T_1} \end{pmatrix}, \tag{6.1}$$

where $\mathbf{B}(t)$ is the total external magnetic field.

6.3.1 The Hahn Spin Echo Experiment

Erwin L Hahn was the first to notice the effects of diffusion when he conceived the *spin echo* experiment to remove the effects of field inhomogeneities or T2* from the signal [31]. Diffusion NMR is derived from Hahn's original spin echo experiment of 1950. Hahn put forth the idea that following a 90° RF pulse that tilts the net magnetization vector to the transverse plane, the dephasing that follows caused by the field inhomogeneities, could be refocused using a second RF pulse of 180°, thus removing the effects of the field inhomogeneities.

After the 90° RF pulse, the spins precessing in the transverse plane should appear static in a frame of reference rotating at the Larmor frequency. However, due to field inhomogeneities, as the spins begin to dephase, some would appear to speed up (or move ahead clockwise in the rotating frame of reference), while some would slow down (or fall back anti-clockwise in the rotating frame of reference). This transverse dephasing is known as the *free induction decay* (FID) and causes the signal to decay faster than pure T2 effects. However the application of a second RF pulse of 180° has the effect of flipping the individual spins in the transverse plane such that the "slow" spins that had fallen behind the rotating frame of reference are flipped ahead of it, while the "fast" spins that had moved ahead are flipped behind the rotating frame of reference. Indeed, the 180° RF pulse causes the spins to refocus after a certain length of time as the fast spins catch up with the slow spins, which regenerates the signal. This is known as the *echo* and it is free of the $T2^*$ effects due to field inhomogeneities (Fig. 6.3).

It must be noted, however, that the echo regenerates the signal completely only under the assumption that none of the spins in the ensemble have moved. If they move then the 180° RF pulse doesn't completely invert the spin and this results again in signal decay. However, this is not due to field inhomogeneities. As noted by Hahn [31] and Carr and Purcell [14], this is due to the translational motion of diffusion. This forms the basis of diffusion NMR.

6.3.2 Diffusion

Diffusion NMR (dNMR) is a modality of NMR that is sensitive to the Brownian motion of the particles in a sample. The dNMR experiment can therefore be used to measure the diffusion properties of the underlying sample. This makes dNMR central to diffusion MRI. At the heart of dNMR is the diffusion process, and understanding diffusion helps to understand how it can be measured from NMR. It leads to the critical improvements that were made by Stejskal and Tanner to the original spin echo experiments of Hahn and Carr and Purcell that opened up the domain of dNMR.

Diffusion is a process of mass transport that describes the random spreading of molecules or particles generally in the presence of a concentration gradient. The process of diffusion was observed, studied and mathematically described over the entire nineteenth century. It was initially observed in three different forms, namely heat diffusion in the presence of a temperature gradient, molecular diffusion in the presence of a concentration gradient, and Brownian motion, which occurs even in the absence of any gradients. These, apparently very different phenomena—the

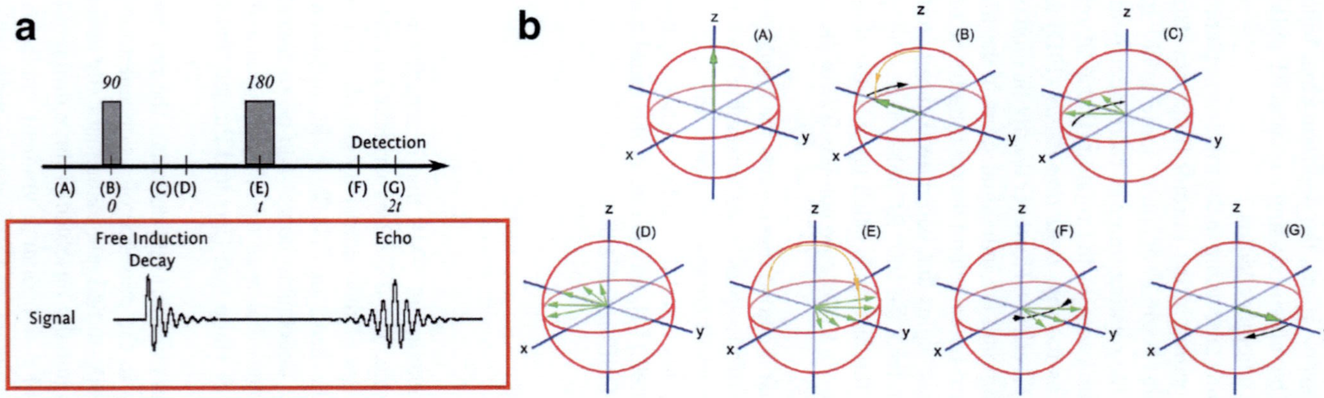

Fig. 6.3 The Hahn Spin Echo experiment. (**a**) Pulse sequence with Free Induction Decay and signal echo. (Adapted from Wikipedia.) (**b**) Concept of spin echo refocussing in the rotating frame of reference. (*A*) The net magnetization vector **M** (*green vector*), initially aligned with the Z-axis under the influence of $\mathbf{B_0}$. (*B*) 90° RF pulse tilts (*orange arrow*) **M** on to the transverse plane. (*C*) and (*D*) Spins dephasing due to local field inhomogeneities. Slow spins fall back, while the fast spins move ahead in the transverse plane. (*E*) 180° RF pulse (*orange arrow*) flips the spins around such that now the slow spins are ahead and the fast spins are behind. (*F*) Refocussing begins as the fast spins begin to catch up the slow spins from behind. The echo begins to form. (*G*) The spins are completely refocussed. This is the centre of the echo and the signal is free of the effects of field inhomogeneities (Drawing by A.G. Filler, image source Wikipedia)

first, concerning the spreading of energy in a solid medium, the second, concerning the spreading of molecules from a region of high concentration to a region of low concentration in fluids, and the third, concerning the random motion of molecules and particles in fluids due to the ambient temperature—can all be described by the same diffusion equation. However, while today their correspondence is widely accepted, establishing this connection wasn't always an easily demonstrable task.

Fick's Laws of Diffusion: The phenomenological equations of diffusion were proposed by Joseph Fourier in 1822 to describe the diffusion of heat in solids, and then adapted by Adolf Fick in 1855 to describe the diffusion of molecules in fluids in the presence of a concentration gradient [21]. Fick derived his "laws of diffusion" from Fourier's laws by analogy, while attempting to describe the experiments conducted by Thomas Graham in 1831 on the diffusion of gases. These laws describe the molecular transfer or diffusion that takes place in a system from regions of high concentration to regions of low concentration due to the concentration gradient.

Fick's first law relates the rate of transfer of the diffusing substance per unit area, or flux $\mathbf{J}$, to the concentration gradient $\mathbf{C}$ causing the diffusion:

$$\mathbf{J} = -D\nabla\mathbf{C}, \tag{6.2}$$

where D is the diffusion coefficient. Conservation of mass during the diffusion process implies $-\nabla\mathbf{J} = \partial\mathbf{C}/\partial t$. This leads to Fick's second law of diffusion:

$$\frac{\partial\mathbf{C}}{\partial t} = D\nabla^2\mathbf{C}. \tag{6.3}$$

Fick's second law describes the change of the concentration field over time due to the diffusion process. Equation (6.3), which relates the time derivative of the concentration to the second order spatial derivative of the concentration is known as the *diffusion equation*—it describes diffusion phenomenologically.

D being a scalar quantity in Eqs. (6.2) and (6.3) is an indication that diffusion is equal in all directions. This is known as *isotropic* diffusion. However, certain media such as crystals, textile fibers, etc. can be inherently anisotropic and can favour diffusion in a certain spatial direction while hindering it in others. This results in *anisotropic* diffusion, which is described by replacing the scalar diffusion coefficient D by a generalized *diffusion tensor* $\mathbf{D}$ (3×3 matrix) in Fick's laws [21]:

$$\mathbf{J} = -\mathbf{D}\nabla\mathbf{C}, \tag{6.4}$$

$$\frac{\partial\mathbf{C}}{\partial t} = \nabla\cdot(\mathbf{D}\nabla\mathbf{C}). \tag{6.5}$$

Diagonalizing the diffusion tensor $\mathbf{D}$ into its eigenvalues and eigenvectors provides a local orthogonal coordinate system that indicates the preferential diffusion direction favoured by the anisotropy of the underlying material. This is the budding idea that indicates that diffusion can be considered as a probe of the underlying medium's microstructure. Isotropic diffusion can be understood as a special case of anisotropic diffusion when $\mathbf{D} = D\mathbf{I}$, where $\mathbf{I}$ is the identity matrix. The idea of the diffusion tensor is central to dMRI, since the fibrous quality of the cerebral white matter also exhibits directional anisotropy.

Brownian Motion and Einstein's Random Walk Approach: Although Fick's laws are concerned with the diffusion of molecules from regions of high concentration to regions of low concentration, they essentially describe the evolution of the concentration gradient over time and space, and aren't concerned with the movements of the molecules themselves. The molecular description of diffusion emerged with Albert Einstein in 1905 when he related the molecular-kinetic theory of heat to the observations made by Robert Brown in 1828. Brown had noted the perpetual erratic motion of pollen grains suspended in water while observing them under a microscope. This erratic movement came to be known by his name as Brownian motion. When Einstein proposed [28] that due to the thermal kinetic energy of molecules, particles suspended in a liquid large enough to be observed under a microscope would exhibit random movements governed by the probabilistic law he derived, his idea was quickly recognized to be the theoretical description of Brownian motion. It turned out that the probabilistic law of Brownian motion derived by Einstein also satisfied the diffusion equation. This provided the final link and showed that diffusion was driven by the thermal kinetic energy of molecules due to the ambient temperature, implying that diffusion, in the form of Brownian motion also occurred in the absence of a temperature or a concentration gradient. The special case of diffusion when the suspended particles belong to the liquid is known as *self diffusion*.

To describe the erratic movement of a large number of particles undergoing Brownian motion, Einstein adopted the probabilistic approach of a random walk model [28]. He modelled diffusion using two Probability Density Functions (PDF)s—$f(\mathrm{x}, t)$, the probability of finding a particle at the position x at a time t, and $P(\Delta\mathrm{x}, \Delta t)$, the transition probability or the probability of finding a particle at a distance $\Delta\mathrm{x}$ from its initial position after a time Δt. Considering $P(\Delta\mathrm{x}, \Delta t)$ symmetric, such that $P(\Delta\mathrm{x}, \Delta t) = P(-\Delta\mathrm{x}, \Delta t)$, Einstein proposed the relation between $f(\mathrm{x}, t)$ and $P(\Delta\mathrm{x}, \Delta t)$:

$$f(\mathrm{x}, t + \Delta t) = \int_{-\infty}^{\infty} f(\mathrm{x} - \Delta\mathrm{x}, t) P(\Delta\mathrm{x}, \Delta t) d\Delta\mathrm{x}. \tag{6.6}$$

He then showed that $f(\mathrm{x}, t)$, which can also be considered as the local particle concentration, satisfies the diffusion equation:

$$\frac{\partial f(\mathrm{x}, t)}{\partial t} = D\nabla^2 f(\mathrm{x}, t), \tag{6.7}$$

which introduces the diffusion coefficient D, showing that the random walk approach can model diffusion. In the isotropic case discussed by Einstein, he further showed that the diffusion coefficient is directly proportional to the variance of the particle displacement $\langle(\Delta x)^2\rangle = 2\Delta t D$, where without loss of generality x is any chosen spatial direction. A similar development in the anisotropic case along the lines proposed by Einstein results in the anisotropic diffusion equation and connects the diffusion tensor to the covariance tensor:

$$\frac{\partial f(\mathrm{x},t)}{\partial t} = \nabla \cdot (\mathbf{D}\nabla f(\mathrm{x},t)), \tag{6.8}$$

$$\mathbf{D} := \frac{1}{2\Delta t}\begin{pmatrix} \langle(\Delta x)^2\rangle & \langle\Delta x\Delta y\rangle & \langle\Delta x\Delta z\rangle \\ \langle\Delta y\Delta x\rangle & \langle(\Delta y)^2\rangle & \langle\Delta y\Delta z\rangle \\ \langle\Delta z\Delta x\rangle & \langle\Delta z\Delta y\rangle & \langle(\Delta z)^2\rangle \end{pmatrix}. \tag{6.9}$$

Finally Einstein also derived that under the initial condition $f(\mathrm{x},0) = \delta(\mathrm{x})$, which corresponds to free diffusion, the local particle concentration $f(\mathrm{x},t)$ is a Gaussian function with the derived variance $2tD$. This, however, implies that the Green's function of the diffusion equation, or the transition probability is also a Gaussian in the case of free diffusion:

$$P(\Delta\mathrm{x},\Delta t) = \frac{1}{(4\pi\Delta t)^{3/2}|\mathbf{D}|^{1/2}}\exp\left(-\frac{\Delta\mathrm{x}^T\mathbf{D}^{-1}\Delta\mathrm{x}}{4\Delta t}\right). \tag{6.10}$$

In the anisotropic case it is an oriented Gaussian parameterized by the covariance tensor.

6.3.3 The Stejskal-Tanner PGSE Experiment

After Hahn who first noticed the effects of diffusion in NMR in his spin echo experiment [31], Carr and Purcell measured the diffusion coefficient for the first time from NMR. In their modification to Hahn's experiment they employed a temporally constant magnetic gradient field and modelled the diffusion of spin bearing particles with discrete jumps [14]. However, the continuous description was formulated by Torrey in 1956. He modified the phenomenological Bloch equations by adding to it Fick's diffusion equation (Eq. 6.3) [64]. This came to be known as the Bloch–Torrey equation for describing the net magnetization vector $\mathbf{M}$ (without flow):

$$\frac{\partial\mathbf{M}}{\partial t} = \gamma\mathbf{M}\times\mathbf{B} + \begin{pmatrix} -\frac{1}{T_2} & 0 & 0 \\ 0 & -\frac{1}{T_2} & 0 \\ 0 & 0 & -\frac{1}{T_1} \end{pmatrix}\mathbf{M} + \begin{pmatrix} 0 \\ 0 \\ \frac{M_0}{T_1} \end{pmatrix} + D\nabla^2\mathbf{M}, \tag{6.11}$$

where Fick's law is employed to describe the self diffusion of the net magnetization.

About a decade later, in 1965, Stejskal and Tanner designed the *pulsed gradient spin echo* (PGSE) experiment by modifying Hahn's spin echo experiment with two identical magnetic gradients around the 180° RF pulse to encode the transverse

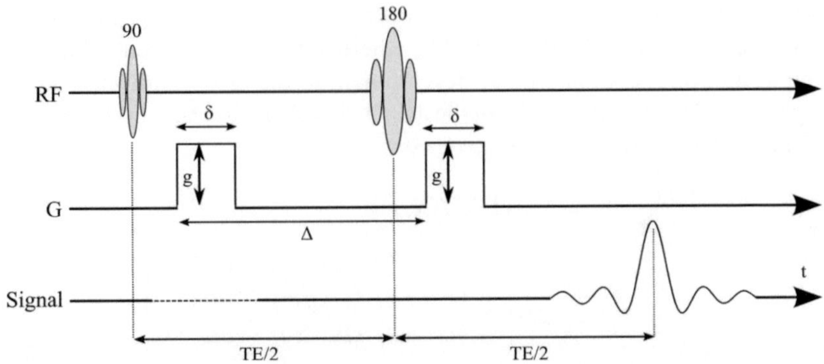

Fig. 6.4 The pulsed gradient spin echo (PGSE) sequence. Two identical gradients are applied around the 180° RF pulse of Hahn's spin echo experiment. This encodes the transverse phase of the diffusing spin bearing particles. It then becomes easier to measure the decay of the signal due to diffusion

phase of the diffusing spin bearing particles [61, 62] (Fig. 6.4). This made it easier to measure the decay in the transverse signal due to diffusion, and from there the diffusion coefficient. The PGSE experiment established the field of dNMR.

In the PGSE experiment the first gradient $\mathbf{G}$ of duration δ spatially encodes the phase of the individual spins (by dephasing them by an amount dependent on their position), and the effects of this gradient are undone by the second identical gradient after the 180° RF pulse which flips the spins around (implying an effect $-\mathbf{G}$ from the second gradient). This results in a complete recovery of the signal since the magnitude of transverse magnetization vector $\mathbf{M}_{xy}$ depends on the phase coherence of the individual spins. However, if the individual spins move due to diffusion during the period Δ, between the two pulsed gradients, then the effects of the second gradient isn't the exact opposite of the first gradient ($-\mathbf{G}$) that was used to encode their phases. This leads to a partial phase incoherence—resulting in a reduced transverse magnetization $\mathbf{M}_{xy}$, implying a loss in the spin echo signal. Since the signal decay is related to the rate of diffusion or the diffusion coefficient, measuring the signal decay makes it possible to measure the diffusion coefficient.

Stejskal and Tanner provided the mathematical solution to the Bloch-Torrey differential equation for their PGSE experiment, which became the corner stone equation for dNMR as the Stejskal-Tanner equation for the signal:

$$S = S_0 \exp\left(-\gamma^2\delta^2 g^2\left(\Delta - \frac{\delta}{3}\right)D\right) = S_0 \exp\left(-bD\right), \qquad (6.12)$$

where S is the magnitude of the signal decay due to diffusion, S_0 is the magnitude of the signal in the absence of a diffusion encoding gradient, Δ is the time between the two gradients, δ is the application time of each gradient, and $b = \gamma^2\delta^2 g^2\left(\Delta - \frac{\delta}{3}\right)$ is the b-value. The modifications introduced by Stejskal and Tanner in the PGSE experiment, therefore, makes it possible to measure the diffusion coefficient D a regular spin-echo experiment in NMR.

In the same year, Stejskal further considered diffusion in anisotropic media by employing the anisotropic Fick's law (Eq. 6.5) instead of the isotropic law (Eq. 6.3), in the Bloch-Torrey equation, which introduces the diffusion tensor $\mathbf{D}$ [61]. He was able to derive the modified Stejskal-Tanner equation incorporating the diffusion tensor in anisotropic media [61]:

$$S = S_0 \exp\left(-\gamma^2\delta^2 g^2 \left(\Delta - \frac{\delta}{3}\right) \mathbf{g}^T\mathbf{D}\mathbf{g}\right) = S_0 \exp\left(-b\mathbf{g}^T\mathbf{D}\mathbf{g}\right). \tag{6.13}$$

However, Stejskal in this seminal paper fell short of providing a method for measuring the diffusion tensor from NMR, which could have preempted diffusion tensor imaging by almost three decades. But he did lay the foundations of the q-space formalism with the "pulsed" gradient assumption.

6.3.4 *Narrow Gradient Pulse PGSE: q-Space Formalism*

Diffusion in the PGSE experiment can also be modelled from a probabilistic or random-walk model driven by the thermal kinetic energy of the spin bearing particles. The PGSE experiment (Fig. 6.4) spatially encodes or labels the transverse phase of the spins using the first gradient, which results in a deliberate dephasing of the transverse magnetization. The purpose of the second gradient after the 180° RF pulse is to undo the effects of the first gradient and rephase the transverse magnetization. However, if the spins diffuse away from their position between the two gradients, then the transverse magnetization isn't entirely rephased after the second gradient, resulting in a loss of the transverse signal. This can be described by using a random-walk approach for the spin bearing particles.

Under the assumption $\delta \ll \Delta$, which is known as the *narrow gradient pulse* (NGP) condition, which implies that the spins are static during the application of the diffusion encoding gradients $\mathbf{G}(t)$, the dephasing accrued by a spin in the initial position $\mathbf{r}_0$ during the first gradient is [44] $\phi_1 = \gamma \int_0^\delta \mathbf{G}(t) \cdot \mathbf{r}_0 dt = \gamma\delta\mathbf{G} \cdot \mathbf{r}_0$, when $\mathbf{G}(t) = \mathbf{G} = g\mathbf{g}$. Similarly the dephasing accrued by the spin, now in the position $\mathbf{r}$ due to diffusion, during the second gradient is $\phi_2 = \gamma \int_\Delta^{\Delta+\delta} \mathbf{G}(t) \cdot \mathbf{r} dt = \gamma\delta\mathbf{G} \cdot \mathbf{r}$. Since the second gradient is applied after the 180° RF pulse, which flips the spins around, the net phase shift accrued by a spin is $\phi = \phi_2 - \phi_1 = \gamma\delta\mathbf{G} \cdot (\mathbf{r} - \mathbf{r}_0)$. Of course, if the spins hadn't diffused and had remained static during the period Δ (between the gradients), then the net phase shift would have cancelled out. In other words the amount of net phase shift is proportional to the diffused distance $(\mathbf{r} - \mathbf{r}_0)$.

The NGP condition $\delta \ll \Delta$ can also be interpreted in the way Stejskal proposed it $\delta \to 0$, with $\delta\mathbf{G}$ finite. Although in practice the NGP condition can never be achieved, it provides a powerful insight into the process of measuring diffusion from NMR.

The complex signal generated by individual spins with a net phase shift ϕ is $\exp(i\phi) = \exp\left[i\gamma\delta\mathbf{G}\cdot(\mathbf{r}-\mathbf{r}_0)\right]$ [13,44]. However, the spin echo signal $E(\mathbf{G},\Delta)$ is the averaged net signal from the spin ensemble, or it is the expected value of the complex signal given the probability of spins starting at $\mathbf{r}_0$ and diffusing to $\mathbf{r}$ in the time Δ. This probability is the product of the probabilities $f(\mathbf{r}_0,0)$, of finding a spin initially at $\mathbf{r}_0$, and $P(\mathbf{r}|\mathbf{r}_0,\Delta)$, of a single spin starting at $\mathbf{r}_0$ and diffusing to $\mathbf{r}$ in time Δ. The product $f(\mathbf{r}_0,0)P(\mathbf{r}|\mathbf{r}_0,\Delta)$ introduces the random-walk model for the spin bearing particles diffusing from $\mathbf{r}_0$ to $\mathbf{r}$, and [13]:

$$E(\mathbf{G},\Delta) = \int f(\mathbf{r}_0,0)\int \exp\left[i\gamma\delta\mathbf{G}\cdot(\mathbf{r}-\mathbf{r}_0)\right]P(\mathbf{r}|\mathbf{r}_0,\Delta)d\mathbf{r}d\mathbf{r}_0. \qquad (6.14)$$

This indicates that in the absence of diffusion encoding gradients $E(\mathbf{0},t) = 1$. In practice $E(\mathbf{G},\Delta)$ is obtained by dividing the echo signal amplitude from a PGSE experiment with diffusion gradients by the echo signal amplitude from a Hahn spin echo experiment without gradients $E(\mathbf{G},\Delta) = S(\mathbf{G})/S_0$.

This leads to the q-space formalism by defining a reciprocal space q where [13] $\mathbf{q} := \gamma\delta\mathbf{G}/2\pi$. Inserting $\mathbf{q}$ in Eq. (6.14) gives the q-space signal:

$$E(\mathbf{q},\Delta) = \int f(\mathbf{r}_0,0)\int \exp\left[i2\pi\mathbf{q}\cdot(\mathbf{r}-\mathbf{r}_0)\right]P(\mathbf{r}|\mathbf{r}_0,\Delta)d\mathbf{r}d\mathbf{r}_0. \qquad (6.15)$$

Assuming the transition probability $P(\mathbf{r}|\mathbf{r}_0,\Delta)$ to be translationally invariant or that the movement of a spin is independent of the movements of the other spins and also of its own position and movements in the past—as in a random-walk, implies that $P(\mathbf{r}|\mathbf{r}_0,\Delta) = P(\Delta\mathbf{r},\Delta)$, which is the diffusion propagator. Also since in a random-walk the movements of all the particles are independent and identical, and since the complex signal and the diffusion propagator for a spin only depend on the spin displacement $\Delta\mathbf{r} = (\mathbf{r}-\mathbf{r}_0)$, it is useful to consider the *ensemble average propagator* (EAP), which describes the average probability of any spin in the ensemble diffusing by $\Delta\mathbf{r}$ during the time Δt [13]:

$$\overline{P}(\Delta\mathbf{r},\Delta t) = \int P(\Delta\mathbf{r},\Delta t)f(\mathbf{r}_0,0)d\mathbf{r}_0. \qquad (6.16)$$

Combining Eqs. (6.15) and (6.16) gives the main result of the q-space formalism [13]:

$$E(\mathbf{q},t) = \int \overline{P}(\Delta\mathbf{r},t)\exp\left(i2\pi\mathbf{q}\cdot\Delta\mathbf{r}\right)d\Delta\mathbf{r}, \qquad (6.17)$$

which establishes an inverse Fourier transform relationship between the EAP, henceforth denoted $P(\mathbf{r})$, and the normalized echo signal, henceforth denoted $E(\mathbf{q})$.

This Fourier relationship between the ensemble average diffusion propagator and the diffusion NMR signal ushers in the paradigm change that diffusion can be viewed more than just an intrinsic property, but also as a probe of the microstructure of the underlying medium. This becomes apparent when the medium is anisotropic and has a complex microstructure, which is the case in cerebral white matter where numerous fiber bundles criss-cross at a resolution much finer than that of dMRI.

6.4 From Diffusion MRI to Tissue Microstructure

Although Stejskal in Eq. (6.13) formulated the signal for anisotropic diffusion using a *diffusion tensor* (DT), the reason he fell short of providing a method for estimating the DT is perhaps because he was involved with dNMR. In such experiments it was generally possible to re-orient the experimental setup to align the primary anisotropy direction with the laboratory frame, sufficing it to measure the diffusion coefficient in only three directions [34].

This however changed with MRI and dMRI, when large anisotropic specimen that couldn't be rotated in the scanner began to be imaged. Imaging such specimen, e.g. cerebral white matter tissue, or the entire brain, revealed that the diffusion coefficient measured in such specimen depended upon the direction of the diffusion encoding gradient. In other words the dMRI signal decay was different along different gradient directions; or again such dMRI images revealed different contrasts as the diffusion encoding gradient was rotated. These dMRI images were called Diffusion Weighted Images (DWIs). DWIs were at first cryptic because while they revealed the diffusion coefficient, they clearly also indicated that the underlying tissue was highly anisotropic, but they did not provide a method for inferring the preferential directions of this anisotropy. The diffusion coefficients computed from these DWIs using Stejskal-Tanner's isotropic signal decay formulation $S = S_0 \exp(-bD)$ (Eq. 6.12) were called the *apparent diffusion coefficient* (ADC), since these changed in the highly anisotropic tissue depending on the direction of the diffusion encoding gradient (Fig. 6.5). This limitation of the DWI's, and of Stejskal-Tanner's isotropic signal formulation, shifted the interest from measuring only the diffusion coefficient to inferring the preferential diffusion anisotropy directions, or to using diffusion as a probe to infer the tissue's microstructure. This brought forth a whole new meaning to Stejskal's DT formulation, and it's measurement from dMRI, since its diagonalisation provided a local coordinate system that was a good indicator of the preferential diffusion anisotropy directions or the underlying medium's microstructure.

6.4.1 *Diffusion Tensor Imaging: The Simplest Model*

Diffusion tensor imaging (DTI) was introduced by Basser et al. [5, 6] in 1994, which for the first time provided a method for measuring the DT from dMRI and for inferring the local tissue microstructure from the DT. Starting from Stejskal's equation, Basser et al. defined the b-matrix, which also accounted for the imaging

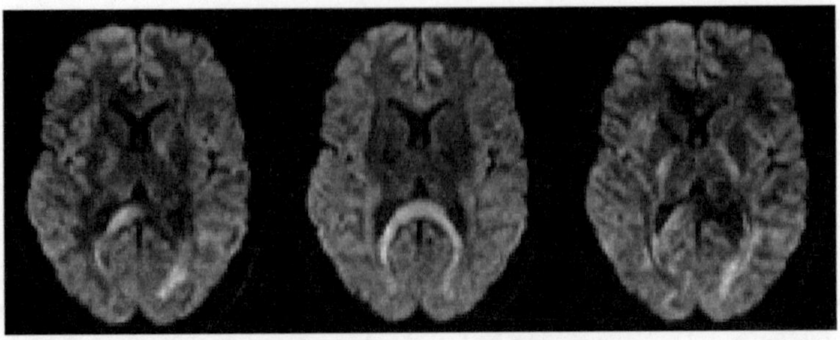

Fig. 6.5 Diffusion weighted images of the brain acquired along different gradient encoding directions showing different contrasts

gradients in addition to the pure diffusion encoding gradients [5,6]. They formulated the PGSE echo signal to be:

$$S = S_0 \exp\left(-tr(\mathbf{bD})\right), \tag{6.18}$$

where $tr(\mathbf{A})$ represents the trace of the matrix $\mathbf{A}$. This simplifies to Stejskal's formulation (Eq. 6.13) $S = S_0 \exp\left(-b\mathbf{g}^T\mathbf{D}\mathbf{g}\right)$ in the absence of the imaging gradients, or under the consideration that the imaging gradients are small compared to the diffusion encoding gradients, which is mostly true. Otherwise, the **b**-matrix has to be computed from the dynamics of the imaging and the diffusion encoding gradients.

DTI Estimation: **D** is a covariance tensor, therefore, it is symmetric and positive definite. This implies that there are six unknowns to be estimated from the DTI signal in Eq. (6.18). Therefore, at least six DWIs, acquired along linearly independent and non-coplanar gradient directions, and a non diffusion weighted or Hahn spin echo (S_0) image is required to measure the six unknown coefficients of **D**. The linearized version of Eq. (6.18) provides the simplest scheme for doing this [5, 6]:

$$\ln\left(\frac{S}{S_0}\right) = -b_{ij}D_{ij}. \tag{6.19}$$

In practice, often more than six DWIs are used to account for acquisition noise. In the case of N DWIs, the linearized equation for the signal is written in a matrix form:

$$\begin{bmatrix} -\ln(S_1/S_0) \\ -\ln(S_2/S_0) \\ -\ln(S_3/S_0) \\ \vdots \\ -\ln(S_N/S_0) \end{bmatrix} = \begin{bmatrix} b^1_{11} & 2b^1_{12} & 2b^1_{13} & b^1_{22} & 2b^1_{23} & b^1_{33} \\ b^2_{11} & 2b^2_{12} & 2b^2_{13} & b^2_{22} & 2b^2_{23} & b^2_{33} \\ b^3_{11} & 2b^3_{12} & 2b^3_{13} & b^3_{22} & 2b^3_{23} & b^3_{33} \\ \vdots & \vdots & \vdots & \vdots & \vdots & \vdots \\ b^N_{11} & 2b^N_{12} & 2b^N_{13} & b^N_{22} & 2b^N_{23} & b^N_{33} \end{bmatrix} \begin{bmatrix} D_{11} \\ D_{12} \\ D_{13} \\ D_{22} \\ D_{23} \\ D_{33} \end{bmatrix}, \tag{6.20}$$

$$\mathbf{X} = \mathbf{Bd}. \tag{6.21}$$

The easiest option for solving this is to use the least squares optimization $\mathbf{d}_{opt} = argmin_{\mathbf{d}}||\mathbf{X} - \mathbf{B}\mathbf{d}||^2$, which translates to the Moore-Penrose pseudo-inverse solution:

$$\mathbf{d} = (\mathbf{B}^T\mathbf{B})^{-1}\mathbf{B}^T\mathbf{X}.$$

Due to its linear form which only involves matrix manipulations, this solution is extremely rapid. However, it doesn't account for the signal noise or of the distortion to the noise it introduces while taking the logarithms of the signal in the linearization process. Due to DTI's popularity and maturity as a technique of probing tissue microstructures, a number of sophisticated solutions exist for measuring $\mathbf{D}$ from the dMRI signal. These range from Basser's original weighted least squares approach [5] which accounts for the logarithmic distortion of the signal noise, to non-linear optimization approaches that account for signal noise, spatial smoothing, and also for constraining the DT to be positive definite [18, 19, 29, 40, 45, 51, 70].

Microstructure from DTI: The consistency between the phenomenological approach and the q-space formalism, under the NGP condition, implies that the propagator describing the diffusion measured by DTI is the Gaussian PDF (Eq. 6.10). This is an oriented Gaussian parameterized by the DT $\mathbf{D}$, or its inverse. The orientation of the PDF can be deduced from the eigen-decomposition of the DT. The eigenvalues and eigenvectors of $\mathbf{D}$ form a local coordinate system that indicates the preferential diffusion direction orienting the Gaussian PDF. In other words it indicates the diffusion direction favoured by the microstructure of the medium. This preferential orientation of the microstructure can be visually represented by the ellipsoid represented by $\mathbf{D}$ whose implicit quadratic form is [6]:

$$\frac{\mathrm{x}^T\mathbf{D}^{-1}\mathrm{x}}{2t} = 1. \tag{6.22}$$

Since $\mathbf{D}$ is symmetric it can be diagonalized $\mathbf{D} = \mathbf{W}^{\mathbf{T}}\mathbf{\Lambda}\mathbf{W}$, where $\mathbf{W}$ are its orthonormal eigenvectors and $\mathbf{\Lambda}$ is a diagonal matrix whose diagonal elements are its eigenvalues. The canonical form of the diffusion ellipsoid defined by $\mathbf{D}^{-1}$ emerges in the coordinate frame of its eigenvectors:

$$\left(\frac{x'}{\sqrt{2\lambda_1 t}}\right)^2 + \left(\frac{y'}{\sqrt{2\lambda_2 t}}\right)^2 + \left(\frac{z'}{\sqrt{2\lambda_3 t}}\right)^2 = 1.$$

To infer the microstructure of the cerebral white matter from DTI, the fundamental assumption is that the coherent fiber bundle structures formed by the axons hinder the perpendicular diffusion of water molecules (spin bearing ^1H atoms) more than the parallel diffusion. Therefore, the elongation and orientation of the DT are good indicators of these coherent structures or fiber bundles locally. The eigenvector corresponding to the largest eigenvalue, the major eigenvector, indicates the main fiber direction, while the other eigenvectors and eigenvalues indicate diffusion anisotropy in the perpendicular plane (Figs. 6.6 and 6.7).

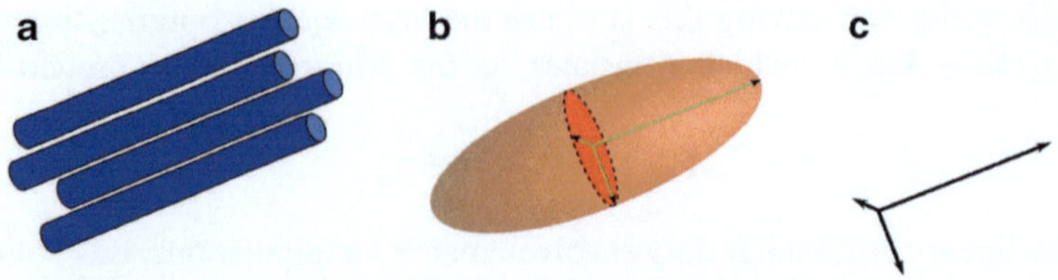

Fig. 6.6 Microstructure from DTI. Coherent fibers (**a**) can be inferred from the ellipsoid of the diffusion tensor (DT) (**b**). The DT also provides a local coordinate system (**c**) that indicates the preferential diffusion anisotropy directions

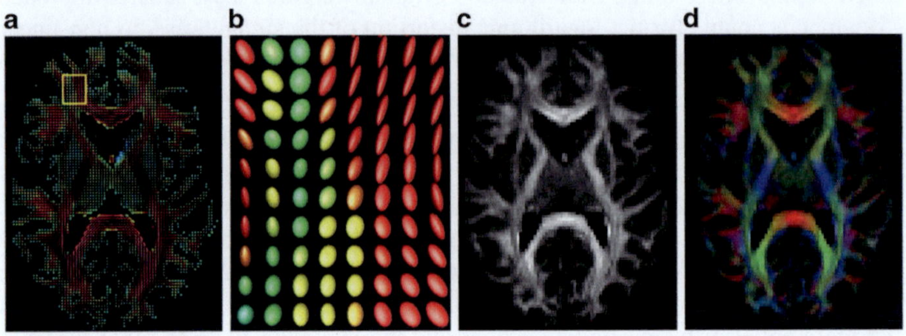

Fig. 6.7 Diffusion Tensor Imaging (DTI). (**a**) DTI ellipsoids from an axial slice of the brain. (**b**) Zoom of highlighted box in (**a**). (**c**) Fractional Anisotropy (FA). (**d**) Colour FA where the colour components RGB represent the unit major eigenvector of the diffusion tensor $\mathbf{e}_1 = [r, g, b]^T$ weighted by FA, or the colours indicate the preferential diffusion anisotropy directions

Scalar Measures: The microstructure of the medium or the white matter can be further characterized from a number of rotationally invariant scalar measures derived from the DT that reveal diffusive properties of the underlying tissue. A number of such scalar measures are of primary importance since they are often used as bio-markers.

- The first is *mean diffusivity* (MD) [4]:

$$\overline{\lambda} = tr(\mathbf{D})/3.$$

- *Fractional anisotropy* (FA) is the other measure which describes the amount of anisotropy presented by the microstructure (Fig. 6.7). It too is derived from the eigenvalues of $\mathbf{D}$ ($\lambda_1 \geq \lambda_2 \geq \lambda_3$) [4]:

$$\mathrm{FA} = \sqrt{\frac{3}{2}} \sqrt{\frac{(\lambda_1 - \overline{\lambda})^2 + (\lambda_2 - \overline{\lambda})^2 + (\lambda_3 - \overline{\lambda})^2}{\lambda_1^2 + \lambda_2^2 + \lambda_3^2}},$$

- The *Relative anisotropy* (RA) is given by [4]:

$$\mathrm{RA} = \frac{1}{\sqrt{3}} \frac{\sqrt{(\lambda_1 - \overline{\lambda})^2 + (\lambda_2 - \overline{\lambda})^2 + (\lambda_3 - \overline{\lambda})^2}}{\overline{\lambda}}.$$

- Westin et al. [73] proposed anisotropy measures to discern *linear* c_l, *planar* c_p and *spherical* c_s diffusion tensor shapes:

$$c_l = \frac{\lambda_1 - \lambda_2}{\sqrt{\lambda_1^2 + \lambda_2^2 + \lambda_3^2}}; \quad c_p = \frac{2(\lambda_2 - \lambda_3)}{\sqrt{\lambda_1^2 + \lambda_2^2 + \lambda_3^2}}; \quad c_s = \frac{3\lambda_3}{\sqrt{\lambda_1^2 + \lambda_2^2 + \lambda_3^2}}.$$

6.4.2 *Beyond DTI*

In DTI, the Gaussian assumption over-simplifies the diffusion of water molecules. While it is adequate for voxels in which there is only a single fiber orientation (or none), it breaks down for voxels in which there are more complex internal structures. This is an important limitation, since the resolution of dMRI acquisitions is between 1 mm^3 and 3 mm^3 while the physical diameter of fibers can be between 1 μm and 30 μm [8, 54]. Research groups currently agree that there is complex fiber architecture in most fiber regions of the brain [53]. In fact, it is currently thought that between one third and two thirds of imaging voxels in the human brain white matter contain multiple fiber bundle crossings [9]. This has led to the development of various High Angular Resolution Diffusion Imaging (HARDI) techniques [69] such as Q-Ball Imaging or Diffusion Spectrum Imaging for exploring the microstructure of biological tissues with greater accuracy.

HARDI samples q-space along as many directions as possible in order to reconstruct estimates of the true diffusion PDF—also referred as the EAP—of water molecules. This true diffusion PDF is model-free and can recover the diffusion of water molecules in any underlying fiber population. HARDI depends on the number of measurements N and the gradient strength (b-value), which directly affects acquisition time and signal to noise ratio in the signal.

6.4.2.1 Diffusion Spectrum Imaging: Recovering the Diffusion EAP

Diffusion spectrum imaging (DSI) introduced by Wedeen et al. [67, 71] in 2000, was the first dMRI method that applied the q-space formalism to measure or estimate the EAP in biological tissue. The forte of the q-space formalism, where diffusion could reveal more than just the intrinsic properties like the diffusion coefficient, and show how it could probe the complex microstructure of the underlying tissue, became quickly apparent. Crossing fiber microstructures were clearly revealed by the geometric forms of the measured EAPs, or their characteristics. Such EAPs clearly represented non-free diffusion and were non-Gaussian. DSI was based on the Fourier transform relationship between the signal and the EAP described in Eq. (6.17). Although in practice it couldn't satisfy the NGP condition required by the q-space formalism, DSI essentially imaged the q-space densely and reconstructed the EAP via a fast Fourier transform (FFT). In fact, $\delta \approx \Delta$ in the DSI experiments conducted in [67, 71, 72]. Nonetheless, the angular results produced by such DSI experiments clearly spoke in favour of q-space imaging—as seen in [58].

DSI, however, considered the modulus Fourier transform:

$$P(\mathbf{r}) = \int |E(\mathbf{q})| \exp\left(-i2\pi\mathbf{q}\cdot\mathbf{r}\right) d\mathbf{q},$$

in place of the true Fourier transform, justifying that in the case of pure diffusion the modulus Fourier transform is equal to the true Fourier transform. This was done to counter biological motion such as cardiac pulsation, which tended to contaminate the phase of the signal.

Initially visualizing iso-surfaces of the estimated EAP [71], DSI techniques later introduced the *orientation distribution function* (ODF), to emphasize the angular results, which indicated underlying fibers. Two ODFs were used, namely [72] (Fig. 6.8):

$$\Psi_{SA}(\mathbf{u}) = \int_{\mathbf{R}+} P(r\mathbf{u})r^2 dr, \quad \mathbf{u} = \mathbf{r}/|\mathbf{r}|, \tag{6.23}$$

and the one introduced by Tuch [67]:

$$\Psi_{T}(\mathbf{u}) = \int_{\mathbf{R}+} P(r\mathbf{u}) dr, \quad \mathbf{u} = \mathbf{r}/|\mathbf{r}|. \tag{6.24}$$

However, DSI had severe acquisition setbacks. To correctly estimate the EAP, the q-space had to be densely sampled, and also at very high b-values. The DSI sampling scheme was a Cartesian grid inside a sphere in q-space, where both the radius of the sphere—the maximum b-value, and the number of grid-points—the number of acquisitions played important roles in determining the accuracy of the estimated EAP. For example in [71, 72], the maximum b-value was in the range of 20,000 s/mm^2, and the number of acquisitions were more than 500. In comparison, DTI acquisitions are done for b-values of 1,000 s/mm^2, and only require a minimum of 6 acquisitions. These demanding requirements played unfavourably for DSI, since its clinical viability was near impossible.

However, DSI was the proof of concept for q-space imaging, which quickly became the popular approach for dMRI, and gave rise to a plethora of techniques for estimating complex EAPs or their characteristics like the ODF. These q-space techniques were developed to overcome the acquisition limitations of DSI.

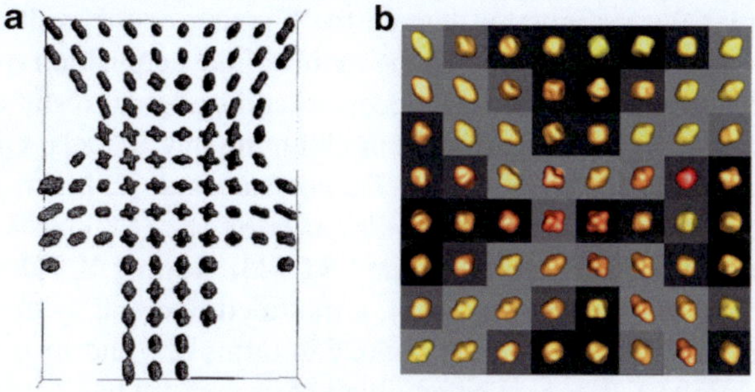

Fig. 6.8 Diffusion Spectrum Imaging & Q-Ball Imaging. (**a**) DSI diffusion PDFs from [71]. Corticospinal tract (orientations sup.-inf.) and pontine decussation (*left-right*). (**b**) ODFs, estimated from an analytical q-ball approach, such that they represent the angular marginal distributions of the true and unknown EAPs (biological rat phantom)

6.4.2.2 Q-Ball Imaging: Emphasizing the Anisotropic Diffusion Orientation Information

Q-Ball Imaging (QBI) was proposed by Tuch [67, 68] spurred by the facts that DSI had severe acquisition requirements, and that the DSI result of interest wasn't the estimated EAP itself, but rather its radial projection—the ODF, which emphasized angular details. His idea was to retrieve the same angular result with reduced acquisition requirements. His initial attempt was the model based multi-tensor approach which was stricken with instabilities induced by the assumed model. Therefore, he proposed QBI, a model free method that sampled q-space only on a sphere or q-shell with fixed q-radius with high angular resolution.

QBI like DSI is based on the q-space formalism and shows promising results, although like DSI, in practice it cannot satisfy the NGP condition [67]. However, QBI became a forerunner to a plethora of q-space methods that attempted to reconstruct the EAP or its characteristics from partial sampling of the q-space. QBI, itself maps spherical acquisitions in q-space to the ODF—a spherical function in real space.

QBI is based on the *Funk Radon transform* (FRT), which is a mapping from a sphere to a sphere $\mathcal{G} : S^2 \to S^2$. To a point on the sphere, called the pole, the FRT of a spherical function f, assigns the value of the integral of the spherical function along the equator on the plane that has for normal the vector connecting the centre of the sphere to the pole:

$$\mathcal{G}[f(\mathbf{u})](\mathbf{u}) = \int_{S^2} f(\mathbf{u})\delta(\mathbf{u}^T\mathbf{w})d\mathbf{w},$$

where $\mathbf{u}, \mathbf{w} \in S^2$. Using the Fourier slice theorem, Tuch was able to show that the FRT of the signal acquired on a q-sphere was equal to the ODF in Eq. (6.24) blurred by a zeroth-order Bessel function, where the blurring or the width of the Bessel function was inversely proportional to the radius of the acquisition q-sphere.

QBI, therefore, made it possible to reconstruct the angular result of DSI, i.e. the ODF, with fewer acquisitions and without assuming any models. QBI was further boosted by Anderson [2], Hess [32], and Descoteaux et al. [25], where an analytical solution was proposed, by using the *spherical harmonic* (SH) basis. It was shown that the SHs are the eigenfunctions of the FRT [25]. Letting Y_l^m denote the SH of order l and degree m ($m = -l, \cdots, l$), a modified real and symmetric SH basis is defined. For even order l, a single index j in terms of l and m is used such that $j(l,m) = (l^2 + l + 2)/2 + m$. The modified basis is given by:

$$Y_j = \begin{cases} \sqrt{2}\mathrm{Re}(Y_l^{|m|}), & \text{if } m < 0, \\ Y_l^m, & \text{if } m = 0, \\ \sqrt{2}(-1)^{m+1}\mathrm{Im}(Y_l^m), & \text{if } m > 0, \end{cases} \tag{6.25}$$

where $\mathrm{Re}(Y_l^m)$ and $\mathrm{Im}(Y_l^m)$ represent the real and imaginary parts of Y_l^m respectively. This modified basis is designed to be real, symmetric and orthonormal, and it is then possible to obtain an analytical estimate of the ODF in Eq. (6.24) with [25]:

$$\Psi_T(\mathbf{u}) = \sum_{j=1}^{L} \underbrace{2\pi P_{l(j)}(0) c_j}_{c'_j} Y_j(\mathbf{u}), \tag{6.26}$$

where $L = (l+1)(l+2)/2$ is the number of elements in the modified SH basis, c_j are the SH coefficients describing the input HARDI signal, $P_{l(j)}$ is the Legendre polynomial of order l that is associated with jth element of the modified SH basis and c'_j are the SH coefficients describing the ODF Ψ_T.

Aganj et al. [1] recently proposed an analytical solution to QBI using SHs to compute the ODF in Eq. (6.23), under a mono-exponential assumption of the signal. The ODF in Eq. (6.23) takes into account the solid angle factor during the radial integration, therefore, it is a true marginal density function of the EAP. This solution was also proposed by Vega et al. in [66]. The ODF in Eq. (6.24) proposed by Tuch on the other hand doesn't account for this solid angle, and therefore needs to be numerically normalized after estimation [68] (Fig. 6.8).

6.5 Computational Framework for Processing Diffusion MR Images

Diffusion MRI is a rich source of complex data in the form of images. Processing dMRI data poses a challenging problem since diffusion images can range from scalar images such as DWIs, where each voxel contains a scalar grey-level value, to tensor images such as in DTI, where each voxel contains a second order tensor, to more complex and generic function images such as in DSI or QBI, where each voxel contains a diffusion function such as the EAP in DSI or a spherical function such as the ODF in QBI, represented as coefficients in a particular basis of choice—such as the SH basis. Therefore, processing such higher dimensional images requires sophisticated mathematical and computational tools.

Processing diffusion images also forms an important part of the dMRI pipeline from acquisition to extraction of meaningful physical and medical information from the data. Operations such as regularization are important for denoising diffusion images as they render the tensor field in DTI or ODF field in QBI more coherent and therefore greatly improve the results of post-processing algorithms such as tractography. A rich body of literature for regularizing tensor fields in DTI can be found in [3, 15, 46, 50, 51]. The capacity to segment tensor images or ODF images makes it possible to identify and reconstruct white matter structures in the brain such as the corpus callosum, which is not possible from simple scalar MR images. However, the extension of such operations from scalar images to tensor fields or ODF fields requires the correct mathematical definitions for spaces of tensors (or EAPs or ODFs) with the appropriate metric. In this section we will present the tools required to process tensor images (DTI), which has seen extensive mathematical development recently. In particular we will present appropriate metrics for the space of symmetric positive definite matrices (or diffusion tensors) $\mathcal{S}ym_n^+$, an estimation algorithm for DTI that ensures that the DT is estimated in $\mathcal{S}ym_n^+$ using the Riemannian metric of $\mathcal{S}ym_n^+$ and a segmentation algorithm that uses the Riemannian metric of $\mathcal{S}ym_n^+$ to segment regions in a tensor field.

6.5.1 The Affine Invariant Riemannian Metric for Diffusion Tensors

Diffusion tensors are 3×3 symmetric matrices. However, since negative diffusion is non-physical these matrices are also required to be positive definite. In other words DTs belong to the space $\mathcal{S}ym_3^+$ of 3×3 symmetric positive definite matrices, which is a non-Euclidean space. Therefore an appropriate metric needs to be defined on this space which would render it into a Riemannian manifold and which would permit to constrain all operations naturally to $\mathcal{S}ym_3^+$ by using Riemannian geometry. $\mathcal{S}ym_3^+$ naturally inherits the Euclidean and the Frobenius metrics from the space of all matrices, however, $\mathcal{S}ym_3^+$ is neither complete nor closed under these metrics.

A number of works have recently proposed the affine invariant Riemannian metric for $\mathcal{S}ym_n^+$ which has been used extensively to compute on DTs [40, 45, 51]. In [51] the Riemannian metric is derived to be $g_{ij} = g(X,Y) = \langle X,Y\rangle_{\mathbf{S}} = tr\left(\mathbf{S}^{-\frac{1}{2}}X\mathbf{S}^{-1}Y\mathbf{S}^{-\frac{1}{2}}\right), \forall \mathbf{S} \in \mathcal{S}ym_3^+$ and with $X, Y \in T_{\mathbf{S}}\mathcal{S}ym_3^+ = \mathcal{S}ym_3$ the tangent space at $\mathbf{S}$. The geodesic distance between DTs induced by this metric can be computed to be:

$$\mathcal{D}_{Rm}(\mathbf{S}_1, \mathbf{S}_2) = \sqrt{\frac{1}{2} tr\left(\ln^2\left(\mathbf{S}_1^{-1/2}\mathbf{S}_2\mathbf{S}_1^{-1/2}\right)\right)}. \tag{6.27}$$

An equally well known metric for $\mathcal{S}ym_n^+$ is the Log-Euclidean metric [3]. Although it isn't affine invariant and only similarity invariant, computationally it is more efficient than the affine invariant Riemannian metric and resembles closely the latter. The distance between DTs induced by this metric is:

$$\mathcal{D}_{LE}(\mathbf{S}_1, \mathbf{S}_2) = \sqrt{\frac{1}{2} tr\left((\ln(\mathbf{S}_1) - \ln(\mathbf{S}_2))^2\right)}. \tag{6.28}$$

6.5.2 Estimation of DTs in $\mathcal{S}ym_3^+$ Using the Riemannian Metric

Using the appropriate Riemannian metric and geometry for $\mathcal{S}ym_3^+$ can constrain all operations to $\mathcal{S}ym_3^+$. For example using the Riemannian metric for DTI estimation can ensure that the DTs are positive definite or that no negative diffusion will be estimated even in the presence of noisy DWIs. This can be done by using the logarithmically transformed version of the Stejskal-Tanner equation (Eq. 6.13) and considering its explicit least square minimization $E(\mathbf{D}) = \frac{1}{2}\sum_{i=1}^{N}\left(\frac{1}{b}\ln\left(\frac{S_i}{S_0}\right) + \mathbf{g}_i^T\mathbf{D}\mathbf{g}_i\right)^2$. The gradient of this least square functional using the affine invariant Riemannian metric can be shown to be [40]:

$$\nabla E(\mathbf{D}) = \sum_{i=1}^{N}\left(\frac{1}{b}\ln\left(\frac{S_i}{S_0}\right) + \mathbf{g}_i^T\mathbf{D}\mathbf{g}_i\right)\cdot(\mathbf{D}\mathbf{g}_i)(\mathbf{D}\mathbf{g}_i)^T. \tag{6.29}$$

Furthermore the Euclidean gradient descent algorithm required for optimization has to be appropriately replaced by a Riemannian geodesic descent algorithm to respect the Riemannian geometry of $\mathcal{S}ym_3^+$.

6.5.3 Segmentation of a Tensor Field

The goal of segmenting a tensor field or an image of DTs is to compute the optimal 3D surface separating an anatomical structure of interest from the rest of the tensor image (Fig. 6.9). To do this we follow the method proposed in [23]. The idea will be to treat the tensor field as a field of Gaussian probability density functions and to utilize the affine invariant Riemannian metric on $\mathcal{S}ym_3^+$, which also forms a Riemannian metric in the space of Gaussian density functions, to compute the

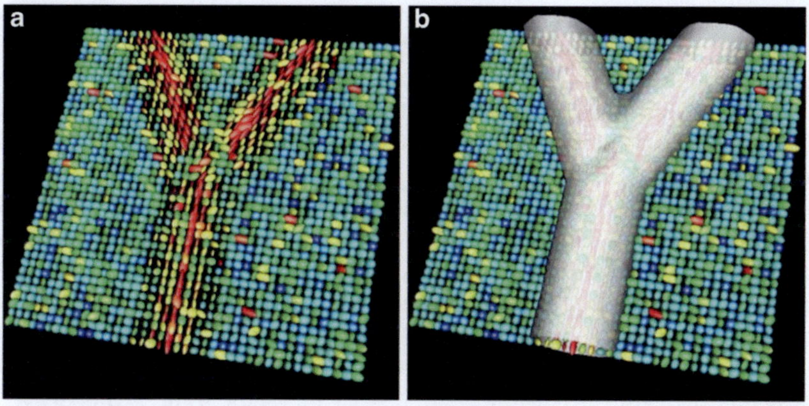

Fig. 6.9 Tensor field segmentation on a synthetic dataset simulating DTI [41]. (**a**) A slice from a $40 \times 40 \times 40$ dataset of synthetic diffusion tensors composed by a *divergent* tensor field and a background of isotropic tensors. Within the *Y* shape FA decreases as one gets further from the center-line. Noise was added to the original dataset. The colour of the tensors represent anisotropy with *red* indicating high anisotropy and *blue* indicating isotropy. (**b**) The segmented divergent *Y* shape using the level-set approach

segmentation boundary. Therefore, a DT at the point x in the image corresponds to the 3D Gaussian distribution $N(x, r)$.

Using the level-set approach and the optimal boundary Γ between the object of interest Ω_1 and the background Ω_2, the level-set $\phi : \Omega_1 \cup \Omega_2 \rightarrow \mathbb{R}$ can be defined as:

$$\begin{cases} \phi(x) = 0, & \text{if } x \in \Gamma \\ \phi(x) = \mathcal{D}_E(x, \Gamma), & \text{if } x \in \Omega_1 \\ \phi(x) = -\mathcal{D}_E(x, \Gamma), & \text{if } x \in \Omega_2 \end{cases} \tag{6.30}$$

where $\mathcal{D}_E(x, \Gamma)$ represents the Euclidean distance between x and Γ. Then according to the geodesic active regions model along with a regularity constraint on the interface, the optimal boundary Γ or the segmentation of the tensor field is obtained by minimizing the functional:

$$\begin{aligned} E(\phi, P_1, P_2) = \nu \int_{\Omega = \Omega_1 \cup \Omega_2} |\nabla H_\varepsilon(\phi)| dx - \int_\Omega H_\varepsilon(\phi) \ln(P_1(N(x, r))) dx \\ - \int_\Omega (1 - H_\varepsilon(\phi)) \ln(P_2(N(x, r))) dx, \end{aligned} \tag{6.31}$$

where $H_\varepsilon(\cdot)$ is a regularized version of the Heaviside function [23], and P_1 and P_2 are the probability distributions of the set of Gaussian distributions $N(x, r)$ in Ω_1 and Ω_2 respectively.

Equation (6.31) can be solved computationally by assuming the distributions P_1 and P_2 themselves to be Gaussians distributions. However, that would require the computation of the mean and the standard deviation of the set of 3D Gaussian distributions $N(x,r)$ constituting Ω_1 and Ω_2. This mean and standard deviation of a set of Gaussian distributions would require a metric to be defined on the space of Gaussian distributions. A number of examples are provided in [23]—the Euclidean metric, the Kullback-Leibler divergence, and the Riemannian metric. We reproduce here only the final example.

Using the affine invariant Riemannian metric on $\mathcal{S}ym_3^+$, which also forms a Riemannian metric on the space of 3D Gaussian distributions $N(x,r)$, it is possible to compute the mean distribution $\overline{N}(X,r)$ of a set of Gaussian distributions by a process of Riemannian geodesic descent—a modified gradient descent process. Similarly the empirical covariance matrix relative to the mean $\overline{N}$ is defined to be $\Lambda_{Rm} = \frac{1}{n-1}\sum_{i=1}^{n}\beta_i\beta_i^T$, with $\beta_i = \overline{N}\ln(N_i^{-1}\overline{N})$ which is the gradient of the squared geodesic distance $\nabla\mathcal{D}^2_{Rm}(N_i,\overline{N})$ in vector form. Using these it is possible to define a generalized Gaussian distribution over the space of 3D Gaussian distributions with a covariance matrix Λ_{Rm} of small variance $\sigma^2 = tr(\Lambda_{Rm})$:

$$P_{Rm}(N \mid \overline{N}, \Lambda_{Rm}) = \frac{1 + O(\sigma^3) + \varepsilon(\sigma/\eta)}{\sqrt{(2\pi)^{m(m+1)/2}|\Lambda_{Rm}|}} \exp\left(\frac{-\beta^T\gamma\beta}{2}\right) \; \forall N \in \mathcal{S}ym_3^+, \tag{6.32}$$

where $\gamma = \Lambda_{Rm}^{-1} - \mathcal{R}/3 + O(\sigma) + \varepsilon(\sigma/\eta)$ with $\mathcal{R}$ as the Ricci curvature tensor at $\overline{N}$, η as the injection radius at $\overline{N}$ and ε such that $\lim_{0^+} x^{-\beta}\varepsilon(x) = 0 \forall\beta \in \mathbb{R}^+$.

6.6 Tractography: Inferring the Connectivity

When DTI/DSI/QBI is performed on the brain, the DT/EAP/ODF—hereafter referred to as the *spherical diffusion function* (SDF), is a local indicator of coherent structures or fiber bundles in the cerebral white matter. However, the process of reconstructing the global structures of fiber bundles by connecting the local information is known as fiber tracing or *tractography*. Tractography graphically reconstructs the connectivity of the cerebral white matter by integrating along the direction indicated by the local geometry of the SDF. It is a modern tool that is unique in the sense that it permits an indirect dissected visualization of the brain in vivo and non-invasively [16]. The underpinnings of tractography are also based on the fundamental assumption of dMRI—the diffusion of water molecules is hindered to a greater extent perpendicular to coherent fiber bundle structures than parallel to these. Therefore, following the geometry of the local diffusion function and integrating along reveals the continuous dominant structure of the fiber bundle. However, in spite of the gain due to its non-invasive nature, tractography can only infer such structures indirectly. Therefore, tractography is acutely sensitive to the local geometry and the error is cumulative. The correct estimation of the local geometry is crucial.

Deterministic tractography is a well established tool that has seen considerable success in researching neurological disorders [20]. Deterministic tractography begins from a seed point and traces along the dominant fiber direction by locally connecting the "fiber" vectors or mathematically becoming tangent to these. Classically the major eigenvector of the diffusion tensor in DTI represented these "fiber" vectors [7,38,47]. However, since DTI is ambiguous and cannot accurately describe the fiber directions in regions with complex fiber configurations, DTI tractography, in spite of its successful usage, is known to be prone to errors. Hence the trend in recent years to extend tractography to complex shaped SDFs that describe the underlying fiber directions more accurately [26,67,72].

Probabilistic tractography was proposed to address the reliability of deterministic tractography which remains sensitive to a number of parameters. The concept and output of probabilistic tractography is, however, subtly different from determinist tractography. While the latter attempts to find the connectivity between two regions, the former measures the likelihood that two regions are connected, or it provides a connectivity confidence. Given the capabilities and ambiguities of dMRI acquisition and reconstruction schemes of today, due to partial voluming, noise, etc., probabilistic tractography provides a more complete statement. However, probabilistic tractography is also computationally more expensive than deterministic tractography.

6.6.1 Deterministic Tractography

Of the many deterministic tractography algorithms, the continuous *streamline tractography* is a widely used scheme. The continuous version of streamline tractography [7] defined for DTI, considers a fiber tract as a 3D space curve parametrized by its arc-length, $\mathbf{r}(s)$, and describes it by its Frenet equation:

$$\frac{d\mathbf{r}(s)}{ds} = \mathbf{t}(s) = \varepsilon_1(\mathbf{r}(s)), \tag{6.33}$$

where $\mathbf{t}(s)$ the tangent vector to $\mathbf{r}(s)$ at s is equal to the unit major eigenvector $\varepsilon_1(\mathbf{r}(s))$ of the diffusion tensor at $\mathbf{r}(s)$. This implies that fiber tracts are locally tangent to the dominant eigenvector of the diffusion tensor at every spatial position. The differential equation Eq. (6.33) along with the initial condition $\mathbf{r}(0) = \mathbf{r}_0$ means that starting from $\mathbf{r}_0$, a fiber can be traced by continuously integrating Eq. (6.33) along the direction indicated locally by the major eigenvector of the diffusion tensor at that point.

However, integrating Eq. (6.33) requires two things—first, a spatially continuous tensor (or SDF) field, and second, a numerical integration scheme. In [7], the authors proposed two approaches for estimating a spatially continuous tensor field from a discrete DTI tensor field, namely approximation and interpolation. They also proposed the Euler's method, the second order Runge-Kutta method, and the adaptive fourth order Runge Kutta method as numerical integration schemes.

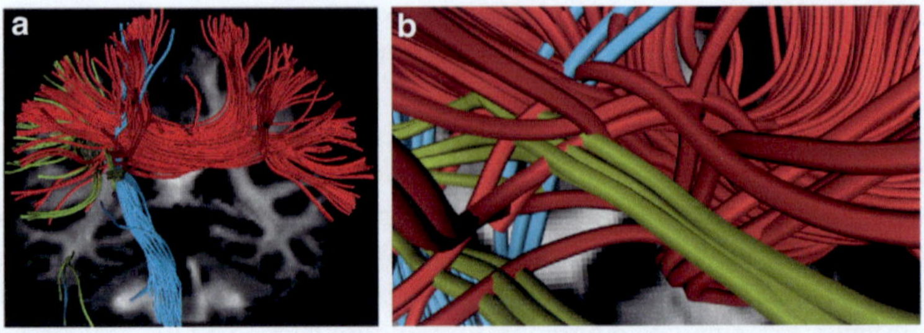

Fig. 6.10 Deterministic Tractography on ODFs. (**a**) Three fiber bundles are traced, starting from different seed regions. *Red*: seeds in the CC (*left-right*). *Blue*: seeds in the CST (superio-inferior). *Green*: seeds in the SLF (anterio-posterior). (**b**) Zoom into crossing region

Finally, for stopping they proposed four criteria—the tracts are within the image volume, the tracts are in regions with FA value higher than a threshold, the curvature of a tract is smaller than a threshold, and that a tract is better aligned with the major eigenvector in the next spatial location than any of the two other eigenvectors.

The streamline tractography algorithm can be adapted to SDFs with multiple maxima (EAP/ODF etc.) by modifying Eq. (6.33) to

$$\frac{d\mathbf{r}(s)}{ds} = \eta_{\theta_{min}}(\mathbf{r}(s)), \;\; \mathbf{r}(0) = \eta_{max}(0), \tag{6.34}$$

where $\eta_i(\mathbf{r}(s))$ are all the unit maxima vectors of the SDF at $\mathbf{r}(s)$, η_{max} is the unit maximum vector whose function value is the largest amongst all the η_i, and $\eta_{\theta_{min}}$ is the unit maximum vector in the current step that is most collinear to the unit maximum vector followed by the integration in the previous step. Equation (6.34) and the initial condition state that at the starting point the integrating begins along the dominant maximum direction, and at each consecutive step first all the maxima of the SDF are detected and the maximum direction most collinear to the maximum direction from the previous integration step is chosen to move forward in the integration (Fig. 6.10).

The maximum direction most collinear to the previously chosen maximum direction can be chosen by computing the dot product between the previously chosen maximum direction and all the maxima of the current step, and by considering the absolute values of the dot products. Hence, the maximum chosen for the next integration direction is the maximum whose absolute value of its dot product with the previously chosen maximum direction is the largest. Also to avoid erratic forward and backward steps during the integration, if the value of the dot product is negative, it is necessary to flip the unit maximum direction to point it consistently along the fiber direction, just like in [7].

Since streamline tractography requires a continuous field of SDFs for integrating Eq. (6.34), a continuous field of SDFs can be computed from a simple Euclidean interpolation of the SDFs when the SDFs are expressed in the SH basis. The

Euclidean interpolation in the SH basis is equivalent to the L2-norm interpolation of the SDFs, since the SHs form an orthonormal basis [24].

6.6.2 Probabilistic Tractography

General SDF-based (ODF etc.) probabilistic tractography have recently been published in the literature [9, 17, 33, 35, 48, 52, 57, 60] to generalize several existing DT-based methods. First, in [35] parametric spherical deconvolution is used as the SDF [65] and in [9] a mixture of Gaussian model is used to extend the probabilistic Bayesian DT-based tracking [10]. Related to these techniques, [33] uses a Bayesian framework to do global tractography instead of tracking through local orientations. In [52], Monte Carlo particles move inside the continuous field of q-ball diffusion ODF and are subject to a trajectory regularization scheme. In [48], an extension to their DT-based approach [49] is also proposed using a Monte Carlo estimation of the white matter geometry and recently, a Bingham distribution is used to model the peak anisotropy in the fiber distributions [60]. Finally, in [17], large number of M-FACT QBI streamlines are reconstructed and all pathways are reversed-traced from their end points to generate of map of connection probability. In this chapter, a new probabilistic algorithm is presented based on the ODF using a Monte Carlo random walk algorithm.

The new algorithm is an extension of the random walk method proposed in [36] to use the distribution profile of the fiber ODF. It starts off a large number of particles from the same seed point and lets the particles move randomly according to the local ODF estimate, F, and counts the number of times a voxel is reached by the path of a particle. This yields higher transitional probabilities along the main fiber directions. The random walk is stopped when the particle leaves the white matter mask.

For each elementary transition of the particle, the probability for a movement from the seed point x to the target point y in direction $\mathbf{u}_{xy}$ is computed as the product of the local ODFs in direction $\mathbf{u}_{xy}$, i.e.,

$$P(x \to y) = F(\mathbf{u}_{xy})_x \cdot F(\mathbf{u}_{xy})_y, \tag{6.35}$$

where $P(x \to y)$ is the probability for a transition from point x to point y, $F(\mathbf{u}_{xy})_x$ is the ODF at point x in direction xy (by symmetry, direction xy and yx are the same).

The transition directions in the local model are limited to 120 discrete directions corresponding to the angular sampling resolution of the acquired brain data and the step size of the particle step was fixed to 0.5 times the voxel size. A trilinear interpolation of the ODF was used for the subvoxel position and a white matter mask computed from a minimum FA value of 0.1 and a maximum ADC value of 0.0015 was used. A total of 100,000 particles were tested for each seed voxel. The connectivity of any given voxel with the seed voxel is estimated by the number of particles that reach the respective voxel, called a *tractogram*.

6.7 Clinical Applications

Diffusion MRI has now proved to be extremely useful for studying the normal and pathological human brain [11, 27] and for improving diagnosis and therapy of brain diseases. It has led to many applications in clinical diagnosis of neurological diseases and disorders, neuroscience applications in assessing connectivity of different brain regions and more recently, therapeutic applications, primarily in neurosurgical planning. Indeed, the sensitivity of diffusion measures to changes in the white matter have led to a wealth of clinical studies using second order diffusion tensor imaging as an in vivo clinical marker and dMRI has opened up a landscape of discoveries for neuroscience and medicine through research and clinical exploration of fiber connectivity and CNS diseases.

The dMRI models and tools presented in this chapter are at the heart of what is strongly needed to better recover, study and understand the anatomical connectivity of the normal and pathological human CNS. These dMRI methodological developments have already been applied in many places to better describe, characterize and quantify abnormalities within the CNS white matter and develop in vivo markers for diseases in clinical neuroscience.

The first and most successful application of dMRI since the early 1990s has been in acute brain ischemia, that emerged from the discovery of Moseley et al. that water diffusion immediately drops after the onset of an ischemic event, when brain cells undergo swelling through cytotoxic edema. This result was later confirmed by different groups using animal models and human patients with stroke.

Since then, many important CNS diseases have attracted the interest of dMRI researchers. Hence, DTI has appeared as a promising tool to look at brain maturation and development, especially to monitor the myelination process, as well as changes in connectivity in relation to functional disorders. Indeed, it has been shown by many studies that the degree of diffusion anisotropy in white matter increases during the myelination process, so that dMRI could be used to assess brain maturation in children, newborns or premature babies, as well as to characterize white matter disorders in children. DTI has now been used for more than 10 years to look, in vivo, at the disorganization of white matter in schizophrenia, which is thought to come from myelin abnormalities. Most neuroimaging studies of schizophrenia have used second order DTI so far, as a means to quantify the integrity of the white matter.

Alzheimer's and Parkinson's diseases, two of the most important neurodegenerative diseases, as well as others like multiple sclerosis have characteristic abnormalities in the microstructure of brain tissues that are not apparent and cannot be revealed reliably by standard imaging techniques. Diffusion MRI can reveal these co-lateral damages to the fibers of the CNS white matter that connect different brain regions.

We in the Athena Project Team, Inria, France, are currently focusing on the use of ultra high field MRI (3T and 7T) techniques for investigating the structural disorders of the brain stem and of the central deep nuclei in the field of Parkinson's syndrome. This work aims to identify the biomarkers of the pathology by benefitting from the latest developments in the domain of ultra high fields and from the latest image processing innovations in the field of diffusion imaging based on high angular resolution models. Determining the biomarkers for the various forms of Parkinson's disease using high field systems will surely open the way for developing new diagnosis tools at lower field strengths and thus make them available to a clinical environment. Deep brain nuclei are important structures that are involved in a large array of behaviors, such as locomotion, eye-movement, or sleep. Damage to these structures leads to movement disorders such as Parkinson's disease. To date, except for the larger striato-pallidal complex, there are no reliable imaging markers of small deep nuclei. With dMRI and refinements of neuroimaging methods and higher field magnets, imaging of these nuclei has become possible. Together with our collaborators at the Center of Neuroimaging Research (CENIR, Paris, France) and CEA Neurospin (Saclay, France), we are currently conducting a series of experiments in normal volunteers and in patients with basal ganglia pathology to characterize deep brain structures and study the structural disorders of the brainstem in the case of Parkinsonian syndromes. More precisely, we are involved in the data analysis part with the objective to detect the anatomical connectivity of the brainstem structures and their connectivity to the brain and we hope to find new neuroimaging markers of deep brain nuclei that could be used for the diagnosis of Parkinsonian syndromes at an early stage. Possible extension and improvements of the tractography algorithms presented here would be necessary to obtain a satisfactory spatial resolution for identifying the anatomical network involved in Parkinson's disease and improve the characterization of lesions of deep brain structures.

An important application where dMRI is expected to significantly impact in the close future is Traumatic Brain Injury (TBI), which is the damage caused to the brain due to external mechanical force, such as rapid acceleration or deceleration, falls, motor vehicle accidents, impact or penetration by a projectile. The worst injuries can lead to permanent brain damage or death. Because a sudden and violent trauma to the head can cause injury to and shearing of the white matter fibers, it's indeed possible to use dMRI to examine the integrity of white matter that is especially vulnerable to TBI. This opens the way to exciting and challenging problems to quantify and qualify structural changes in white matter.

To this date, a large number of dMRI clinical studies of TBI only uses simple scalar diffusion measurements such as FA and/or MD to characterize the structural abnormalities present along a given fiber pathway to identify pathologies and compare patients with healthy controls. This clearly opens the road to many exciting and challenging problems to examine with more elaborate diffusion models the white matter's integrity and to better quantify and qualify structural changes in white matter.

Through our collaborations with clinical partners and our development of innovative tensor and HARDI processing methods, we are convinced that we can advance further our ability to better understand the architecture of the CNS and help to prove that dMRI can provide a relevant source of useful information, such as in vivo markers of diseases in clinical neuroscience.

Efficient and better tools are therefore more than welcome for undertaking a more systematic study of these abnormalities. For instance, HARDI's improved signal-to-noise ratio can image fiber paths with extraordinary angular detail, identifying more relevant anatomical features, connections and disease biomarkers not seen with conventional MRI.

Moving from quantitative comparisons of scalar values derived from second order diffusion tensors in voxels or along existing pathways, to more relevant measures based on higher order diffusion models and demonstration of qualitative changes in white matter anatomy is extremely challenging.

Therefore, key challenges to tackle include defining new biomarkers related to the quantities, scalar or not, that could be retrieved from High Order Tensors, HARDI, Q-Ball or the EAP to better describe and characterize the biological tissue being analyzed.

Overall, one of the main objectives now is to apply the new methodological developments to improve dMRI as a clinical tool and make it more apt in characterizing and quantifying abnormalities within CNS white matter and to develop better neuroimaging markers for the diagnosis of CNS neurodegenerative diseases. There is a consensus in the neuroimaging field that more elaborate models and HARDI measures will certainly advance research in a large number of CNS diseases.

We are confident that the recent development of novel mathematical analysis tools for dMRI and HARDI will certainly result in fundamental advances for research on stroke, multiple sclerosis, amyotrophic lateral sclerosis, Alzheimer's and Parkinson's diseases, HIV/AIDS, neurosurgery, tumor growth modeling or neuropsychiatric disorders like schizophrenia. Moreover, our understanding of the development of the human brain, the effect of aging or the organization of anatomo-functional networks has already started to greatly benefit from the unprecedented insight that dMRI provides into the brain's microstructure.

6.8 Conclusion

In this chapter we presented an overview of the mathematical tool and framework of dMRI. Diffusion MRI is unique in its capacity at inferring the microstructure of the cerebral white matter in vivo and non-invasively, albeit in an indirect fashion. We first presented a quick perusal of the brain—its general structure and organization, the tissues constituting it, and in highlight, it's major neuronal pathways interconnecting its various regions. These are central for understanding the diffusion of water molecules in the brain, and for understanding the usefulness of dMRI in studying the brain. The Brownian motion of the water molecules contained in the brain is hindered in the white matter by its fibrous structures in a particular fashion. While the diffusion of water molecules is greatly hindered perpendicular to these structures, the diffusion parallel to these structures is relatively less affected. In other words the diffusing water molecules probe the white matter's microstructure. Therefore, since dMRI is sensitive to the diffusion of water molecules, it is used to measure the constrained or anisotropic diffusion of water molecules in the white matter, to infer its major axon fiber bundles non-invasively.

Next we presented the fundamentals of the NMR phenomenon, the diffusion NMR experiment, and reviewed three important diffusion MRI reconstruction algorithms. The NMR experiment can recover several different physical properties from samples which contain spin bearing particles by simply applying a set of magnetic fields and gradients. This forms the core of the non-invasive nature of MRI. However, NMR can only examine a tiny region of a sample or a single spin ensemble and cannot image an entire biological specimen. This is made possible by the spatial encoding technique of MRI, which allows to spatially encode various juxtaposed regions or spin ensembles where NMR can be applied independently. This is done in MRI again using magnetic gradients. Therefore, this allows MRI to examine entire biological specimen, like the brain or the body, in vivo and non-invasively.

One of the properties that NMR can be sensitized to is the Brownian motion of the spin bearing particles in a sample. Therefore, NMR can be used to measure the diffusion properties of a sample by modelling the diffusion of the spin bearing particles in the sample. Since diffusion has been historically modelled in two different ways, namely the Fick's phenomenological laws of diffusion and Einstein's random walk model of Brownian motion, the diffusion NMR signal is also modelled in two ways, namely the Stejskal-Tanner formulation and the q-space formalism.

DTI was the first dMRI technique that was proposed to infer the tissue microstructure. It is the most commonly used technique since its mathematical framework is simple, it has few acquisition requirements and has a number of powerful and practical applications. However, it is limited in regions with microstructural heterogeneity. Many higher order techniques have been therefore proposed recently in dMRI to overcome this limitation of DTI. Of these we presented DSI and QBI, and in particular the ODF.

Diffusion MRI data represents images that contain complex mathematical objects. Recently the computational framework of mathematical tools required to process such images has been vastly improved. We presented the appropriate metrics, in particular the Riemannian metric for $\mathcal{S}ym_n^+$, an estimation algorithm and a segmentation framework using this metric for DTI.

DTI, ODFs and other general SDFs represent the local microstructure of the cerebral white matter in each voxel. As the final mathematical tools we presented tractography algorithms which spatially integrate anisotropy information to reconstruct more global structures such as white matter fiber tracts. Tractography is a unique tool which permits one to indirectly dissect and visualize the brain's white matter in vivo and non-invasively.

Finally we concluded the chapter with an overview of major clinical applications to highlight and emphasize the usefulness and strengths of dMRI.

6.9 Online Resources

medInria

http://med.inria.fr

medInria is a multi-platform medical image processing and visualization software, and it's free. Through an intuitive user interface, medInria offers from standard to cutting-edge processing functionalities for your medical images such as 2D/3D/4D image visualization, image registration, or dMRI processing and tractography. medInria was initially developed by the Asclepios Project Team, Inria, France. A new version of medInria is being jointly developed by the Asclepios, Athena, Parietal and the Visages Inria Project Teams.
Key Features: Log-Euclidean metric, HARDI/ODF, Fiber Tracking.

BrainVISA/Anatomist

http://brainvisa.info/

BrainVISA/Anatomist is developed by the Institut Fédératif de Recherche no.49, France.

- BrainVISA is a software, which embodies an image processing factory. A simple control panel allows the user to trigger sequences of treatments on series of images. These treatments are performed by calls to command lines provided by different laboratories. These command lines, hence, are the building blocks on which are built the assembly lines of the factory.
- Anatomist is a visualization software, which main originality is a generic module dedicated to structural data, namely sets of objects linked one another into a graph structure. These objects may be cortical folds inferred from T1 weighted MR data, fiber bundles inferred from MR diffusion weighted data, activated clusters inferred from Statistical Parametric Maps, etc. . . This module includes a nomenclature control panel, which can drive several brains simultaneously. Anatomist provides also some tools to easily map Statistical Parametric Maps on 3D renderings of the brain, inflated meshes of the cortical surface, etc. . . . Finally, a manual drawing toolbox can be used for various purpose.

Key Features: toolboxes for T1-MRI, dMRI (DTI, QBI), fMRI, MEG/EEG, etc.

FMRIB Software Library (FSL)

http://www.fmrib.ox.ac.uk/fsl/index.html

FSL is a comprehensive library of analysis tools for fMRI, MRI and DTI brain imaging data. FSL is written mainly by members of the Analysis Group, fMRIB, Oxford, UK. FSL runs on Apple and PCs (Linux and Windows), and is very easy to install. Most of the tools can be run both from the command line and as GUIs ("point-and-click" graphical user interfaces).
Key Features: Diffusion toolbox contains tools for low-level diffusion parameter reconstruction and probabilistic tractography, including crossing-fiber modelling. It is also capable of tract-based spatial statistics—voxel-wise analysis of multi-subject diffusion data.

Camino Diffusion MRI Toolkit

http://web4.cs.ucl.ac.uk/research/medic/camino/pmwiki/pmwiki.php

Camino is a free, open-source, object-oriented software package for analysis and reconstruction of dMRI data, tractography and connectivity mapping. It is developed by the Microstructure Imaging Group, University College London, UK.

Key Features: DTI, multi-tensor model, QBI, MESD/PAS-MRI, deterministic/probabilistic tractography, synthetic data generation and more.

MRI Studio an Image Processing Program

https://www.mristudio.org/

MRI Studio is an image processing program running under Windows. It is suitable for such tasks as tensor calculation, color mapping, fiber tracking, and 3D visualization. Most of operations can be done with only a few clicks. DTI Studio is being developed through the support of the Laboratory of Brain Anatomical MRI and Center for Imaging Science at Johns Hopkins University, USA.

Key Features: DTI, Fiber-tracking and editing, 3D visualization, Region of Interesting (ROI) drawing and statistics, image registration.

3D Slicer

http://www.slicer.org/

Slicer, or 3D Slicer, is a free, open source software package for visualization and image analysis. 3D Slicer is natively designed to be available on multiple platforms, including Windows, Linux and Mac Os X. 3D Slicer provides image registration, processing of DTI (diffusion tractography), an interface to external devices for image guidance support, and GPU-enabled volume rendering, among other capabilities. It is developed by the Surgical Planning Laboratory, Brigham and Women's Hospital and the Harvard Medical School, USA.

Key Features: Functionality for segmentation, registration and 3D visualization of multi-modal image data, advanced image analysis algorithms for DTI and fMRI, supports standard image file formats.

MRtrix: MR tractography including crossing fibers

http://www.nitrc.org/projects/mrtrix/

MRtrix provides a set of tools to perform diffusion-weighted MRI white matter tractography in the presence of crossing fibers, using Constrained Spherical Deconvolution, and a probabilistic streamlines algorithm. These applications have been written from scratch in C++, using the functionality provided by the GNU Scientific Library, and gtkmm. The software is currently capable of handling DICOM and AnalyseAVW image formats, amongst others. The source code is distributed under the GNU General Public License. MRtrix is being developed by the Brain Research Institute (BRI), Melbourne, Australia.

Key Features: Detection of crossing fibers using Constrained Spherical Deconvolution, probabilistic streamline fiber tracking.

The Fiber Cup Phantom Database
http://www.lnao.fr/spip.php?rubrique79
The Fiber Cup was originally intended to be a tractography contest at the MICCAI conference help in London in 2009. Today, the Fiber Cup is still open and is intended to provide a common dataset with known ground truth along with an evaluation methodology to compare and challenge tractography algorithms. It was developed at Neurospin, CEA, France.
Key Features: Provides a MR phantom containing a plethora of crossing, kissing, splitting and bending fiber configurations, acquisitions done at three b-values providing dMRI data on three q-ball shells.

Acknowledgements This work was partially supported by the ANR project NucleiPark and the France-Parkinson Association.

List of Acronyms

ADC	Apparent Diffusion Coefficient
CC	Corpus Callosum
CNS	Central Nervous System
CR	Corona Radiata
CST	Corticospinal Tract
dMRI	Diffusion Magnetic Resonance Imaging
dNMR	Diffusion Nuclear Magnetic Resonance
DT	Diffusion Tensor
DTI	Diffusion Tensor Imaging
DSI	Diffusion Spectrum Imaging
DWI	Diffusion Weighted Image
EAP	Ensemble Average Propagator
EPI	Echo Planar Imaging
FA	Fractional Anisotropy
FRT	Funk Radon Transform
HARDI	High Angular Resolution Diffusion Imaging
ILF	Inferior Longitudinal Fasciculus
MD	Mean Diffusivity
NGP	Narrow Gradient Pulse
NMR	Nuclear Magnetic Resonance
ODF	Orientation Distribution Function
PD	Parkinson's disease
PDF	Probability Density Function
PGSE	Pulsed Gradient Spin Echo
PNS	Peripheral Nervous System
QBI	Q-Ball Imaging
RA	Relative Anisotropy

SDF	Spherical Diffusion Function
SH	Spherical Harmonics
SLF	Superior Longitudinal Fasciculus
TBI	Traumatic Brain Injury

References

1. I. Aganj, C. Lenglet, G. Sapiro, E. Yacoub, K. Ugurbil, and N. Harel. Reconstruction of the orientation distribution function in single and multiple shell Q-ball imaging within constant solid angle. *Magnetic Resonance in Medicine*, 64(2):554–566, 2010.
2. A.W. Anderson. Measurements of fiber orientation distributions using high angular resolution diffusion imaging. *Magnetic Resonance in Medicine*, 54:1194–1206, 2005.
3. V. Arsigny, P. Fillard, X. Pennec, and N. Ayache. Log-Euclidean metrics for fast and simple calculus on diffusion tensors. *Magnetic Resonance in Medicine*, 56(2):411–421, August 2006. PMID: 16788917.
4. P. J. Basser. Inferring microstructural features and the physiological state of tissues from diffusion-weighted images. *NMR in Biomedicine*, 8:333–344, 1995.
5. P. J. Basser, J. Mattiello, and D. LeBihan. Estimation of the effective self-diffusion tensor from the NMR spin echo. *Journal of Magnetic Resonance*, B(103):247–254, 1994.
6. P. J. Basser, J. Mattiello, and D. LeBihan. MR diffusion tensor spectroscopy and imaging. *Biophysical Journal*, 66(1):259–267, 1994.
7. P. J. Basser, S. Pajevic, C. Pierpaoli, J. Duda, and A. Aldroubi. In vivo fiber tractography using DT-MRI data. *Magnetic Resonance in Medicine*, 44(4):625–632, October 2000.
8. C. Beaulieu. The basis of anisotropic water diffusion in the nervous system - a technical review. *NMR in Biomedicine*, 15:435–455, 2002.
9. T. E. J. Behrens, H. Johansen-Berg, S. Jbabdi, M. F. S. Rushworth, and M. W. Woolrich. Probabilistic diffusion tractography with multiple fibre orientations. what can we gain? *NeuroImage*, 34(1):144–155, 2007.
10. T.E.J. Behrens, M.W. Woolrich, M. Jenkinson, H. Johansen-Berg, R.G. Nunes, S. Clare, P.M. Matthews, J.M. Brady, and S.M. Smith. Characterization and propagation of uncertainty in Diffusion-Weighted MR Imaging. *Magnetic Resonance in Medicine*, 50:1077–1088, 2003.
11. D. Le Bihan, J.-F. Mangin, C. Poupon, C.A Clark, S. Pappata, N. Molko, and H. Chabriat. Diffusion tensor imaging: concepts and applications. *J Magn Reson Imaging.*, 13(4):534–46, 2001.
12. F. Bloch. Nuclear induction. *Physical Review*, 70:460–474, 1946.
13. P. T. Callaghan. *Principles of nuclear magnetic resonance microscopy*. Oxford University Press, Oxford, 1993.
14. H. Y. Carr and E. M. Purcell. Effects of diffusion on free precession in nuclear magnetic resonance experiments. *Physical Review*, 94:630–638, May 1954.
15. C. A. Castano-Moraga, C. Lenglet, R. Deriche, and J. Ruiz-Alzola. A Riemannian Approach to Anisotropic Filtering of Tensor Fields. *Signal Processing [Special Issue on Tensor Signal Processing]*, 87(2):263–276, 2006.
16. M. Catani, R.J. Howard, S. Pajevic, and D.K. Jones. Virtual in vivo interactive dissection of white matter fasciculi in the human brain. *Neuroimage*, 17:77–94, 2002.
17. Yi P. Chao, Chia Y. Yang, Kuan H. Cho, Chun H. Yeh, Kun H. Chou, Jyh H. Chen, and Ching P. Lin. Probabilistic anatomical connection derived from QBI with MFACT approach. In *2007 Joint Meeting of the 6th International Symposium on Noninavsive Functional Sourse Imaging of the Brain and Heart and the International Conference on Functional Biomedical Imaging*, 2007.

18. C. Chefd'hotel, D. Tschumperlé, R. Deriche, and O. Faugeras. Constrained flows on matrix-valued functions : application to diffusion tensor regularization. In *Proceedings of ECCV'02*, June 2002.
19. C. Chefd'hotel, D. Tschumperlé, R. Deriche, and O. Faugeras. Regularizing flows for constrained matrix-valued images. *J. Math. Imaging Vis.*, 20(1–2):147–162, January 2004.
20. O Ciccarelli, M Catani, H Johansen-Berg, C Clark, and A Thompson. Diffusion-based tractography in neurological disorders: concepts, applications, and future developments. *Lancet Neurol*, 7(8):715–27, 2008.
21. J. Crank. *The Mathematics of Diffusion*. Oxford University Press, 2nd edition, 1975.
22. J.J. Dejerine. *Anatomie des Centres Nerveux*. Paris, Rueff & Cie., 1901.
23. R. Deriche, D. Tschumperlé, C. Lenglet, and M. Rousson. Variational Approaches to the Estimation, Regularization and Segmentation of Diffusion Tensor Images. In Chen & Faugeras Paragios, editor, *Mathematical Models of Computer Vision: The Handbook*. Springer, 2005.
24. M. Descoteaux. *High Angular Resolution Diffusion MRI: From Local Estimation to Segmentation and Tractography*. PhD thesis, Universite de Nice - Sophia Antipolis, February 2008.
25. M. Descoteaux, E. Angelino, S. Fitzgibbons, and R. Deriche. Regularized, Fast, and Robust Analytical Q-Ball Imaging. *Magnetic Resonance in Medicine*, 58:497–510, 2007.
26. M. Descoteaux, R. Deriche, T. R. Knosche, and A. Anwander. Deterministic and Probabilistic Tractography Based on Complex Fibre Orientation Distributions. *IEEE Transactions on Medical Imaging*, 28(2):269–286, 2009.
27. Q. Dong, R.C. Welsh, T.L. Chenevert, R.C. Carlos, P. Maly-Sundgren, D.M. Gomez-Hassan, and S.K. Mukherji. Clinical applications of diffusion tensor imaging. *Journal of Magnetic Resonance Imaging*, 19:6–18, 2004.
28. A. Einstein. Investigations on the theory of the Brownian movement. *Ann. der Physik*, 1905.
29. P.T. Fletcher and S. Joshi. Principal geodesic analysis on symmetric spaces: Statistics of diffusion tensors. In *Proc. Computer Vision Approaches to Medical Image Analysis*, Prague, May 2004.
30. H. Gray. *Gray's Anatomy of the Human Body*. LEA and FEBIGER, 1918, 1918.
31. E.L. Hahn. Spin echoes. *Physical Review*, 80:580–594, 1950.
32. C.P. Hess, P. Mukherjee, E.T. Han, D. Xu, and D.B. Vigneron. Q-ball reconstruction of multimodal fiber orientations using the spherical harmonic basis. *Magnetic Resonance in Medicine*, 56:104–117, 2006.
33. S. Jbabdi, M.W. Woolrich, J.L.R. Andersson, and T.E.J. Behrens. A bayesian framework for global tractography. *NeuroImage*, 37:116–129, 2007.
34. H. Johansen-Berg and T. E.J. Behrens. *Diffusion MRI, From quantitative measurement to in vivo neuroanatomy*. Elsevier, 1st edition, April 2009.
35. E. Kaden, T. R. Knosche, and A. Anwander. Parametric spherical deconvolution: Inferring anatomical connectivity using diffusion mr imaging. *NeuroImage*, 37:474–488, 2007.
36. M.A. Koch, D.G. Norris, and M. Hund-Georgiadis. An investigation of functional and anatomical connectivity using magnetic resonance imaging. *NeuroImage*, 16:241–250, 2002.
37. P.C. Lauterbur. Image formation by induced local interactions: examples employing nuclear magnetic resonance. *Nature*, 242:190–191, 1973.
38. M. Lazar, D.M. Weinstein, J.S. Tsuruda, K.M. Hasan, K. Arfanakis, M.E. Meyerand, B. Badie, H.A. Rowley, V.Haughton, A. Field, and A.L. Alexander. White matter tractography using diffusion tensor deflection. In *Human Brain Mapping*, volume 18, pages 306–321, 2003.
39. D. LeBihan and E. Breton. Imagerie de diffusion *in vivo* par résonnance magnétique nucléaire. *CR Académie des Sciences*, (301):1109–1112, 1985.
40. C. Lenglet. *Geometric and Variational Methods for Diffusion Tensor MRI Processing*. PhD thesis, Universite de Nice - Sophia Antipolis, December 2006.
41. C. Lenglet, M. Rousson, R. Deriche, and O. Faugeras. Statistics on the manifold of multivariate normal distributions: Theory and application to diffusion tensor mri processing. *Journal of Mathematical Imaging and Vision*, 25(3):423–444, October 2006.
42. P. Mansfield. Multi-planar image formation using nmr spin echoes. *Journal of Physics C*, 10:55–58, 1977.

43. K.D. Merboldt, W. Hanicke, and J. Frahm. Self-diffusion nmr imaging using stimulated echoes. *J. Magn. Reson.*, 64:479–486, 1985.
44. L. Minati and W. P. Węglarz. Physical foundations, models, and methods of diffusion magnetic resonance imaging of the brain: A review. *Concepts in Magnetic Resonance Part A*, 30A(5):278–307, 2007.
45. M. Moakher. A differential geometric approach to the geometric mean of symmetric positive-definite matrices. *SIAM J. Matrix Anal. Appl.*, 26(3):735–747, April 2005.
46. M. Moakher and M. Zéraï. The Riemannian Geometry of the Space of Positive-Definite Matrices and Its Application to the Regularization of Positive-Definite Matrix-Valued Data. *Journal of Mathematical Imaging and Vision*, 40(2):171–187, 2011.
47. S. Mori, B.J. Crain, V.P. Chacko, and P.C.M. Van Zijl. Three-dimensional tracking of axonal projections in the brain by Magnetic Resonance Imaging. *Annals of Neurology*, 45(2):265–269, February 1999.
48. G. J. M. Parker and D. C. Alexander. Probabilistic anatomical connectivity derived from the microscopic persistent angular structure of cerebral tissue. *Philosophical Transactions of the Royal Society, Series B*, 360:893–902, 2005.
49. G.J.M. Parker and D.C Alexander. Probabilistic monte carlo based mapping of cerebral connections utilising whole-brain crossing fibre information. In *IPMI*, pages 684–695, 2003.
50. O. Pasternak, N. Sochen, and Y. Assaf. Variational Regularization of Multiple Diffusion Tensor Fields. In Joachim Weickert, Hans Hagen, Gerald Farin, Hans-Christian Hege, David Hoffman, Christopher R. Johnson, and Konrad Polthier, editors, *Visualization and Processing of Tensor Fields*, Mathematics and Visualization, pages 165–176. Springer Berlin Heidelberg, 2006. 10.1007/3-540-31272-2_9.
51. X. Pennec, P. Fillard, and N. Ayache. A Riemannian Framework for Tensor Computing. *International Journal of Computer Vision*, 66(1):41–66, January 2006. A preliminary version appeared as Inria Research Report 5255, July 2004.
52. M. Perrin, C. Poupon, Y. Cointepas, B. Rieul, N. Golestani, C. Pallier, D. Riviere, A. Constantinesco, D. Le Bihan, and J.-F. Mangin. Fiber tracking in q-ball fields using regularized particle trajectories. In *Information Processing in Medical Imaging*, pages 52–63, 2005.
53. C. Pierpaoli, P. Jezzard, P.J. Basser, A. Barnett, and G. Di Chiro. Diffusion Tensor MR imaging of human brain. *Radiology*, 201:637–648, 1996.
54. C. Poupon. *Détection des faisceaux de fibres de la substance blanche pour l'étude de la connectivité anatomique cérébrale.* PhD thesis, Ecole Nationale Supérieure des Télécommunications, December 1999.
55. K. Pribam and P. MacLean. Neuronographic analysis of medial and basal cerebral cortex. *J. of Neurophysiology*, 16:324–340, 1953.
56. E.M. Purcell, H.C. Torrey, and R.V. Pound. Resonance absorption by nuclear magnetic moments in a solid. *Physical Review*, 69:37–38, 1946.
57. P. Savadjiev, J. Campbell, M. Descoteaux, R. Deriche, G. B. Pike, and K. Siddiqi. Disambiguation of complex subvoxel fibre configurations in high angular resolution fibre tractography. In *Joint Annual Meeting ISMRM-ESMRMB*, Berlin, Germany, 19–25th May 2007, to appear. International Society of Magnetic Resonance in Medicine.
58. J. D. Schmahmann, D. N. Pandya, R. Wang, G. Dai, H. E. D'Arceuil, A. J. de Crespigny, and Van J. Wedeen. Association fibre pathways of the brain: parallel observations from diffusion spectrum imaging and autoradiography. *Brain*, 130:630–653, 2007.
59. N.R. Selden, D.R. Gitelman, N. Salamon-Murayama, T.B. Parrish, and M.M. Mesulam. Trajectories of cholinergic pathways within the cerebral hemispheres of the human brain. *Brain*, 121:2249–2257, 1998.
60. K. K. Seunarine, P. A. Cook, M. G. Hall, K. V. Embleton, G. J. M. Parker, and D. C. Alexander. Exploiting peak anisotropy for tracking through complex structures. *Computer Vision, IEEE International Conference on*, 0:1–8, 2007.
61. E. O. Stejskal. Use of Spin Echoes in a Pulsed Magnetic-Field Gradient to Study Anisotropic, Restricted Diffusion and Flow. *The Journal of Chemical Physics*, 43(10):3597–3603, 1965.

62. E. O. Stejskal and J. E. Tanner. Spin diffusion measurements: Spin echoes in the presence of a time-dependent field gradient. *The Journal of Chemical Physics*, 42(1):288–292, 1965.
63. D.G. Taylor and M.C. Bushell. The spatial mapping of translational diffusion coefficients by the nmr imaging technique. *Phys. Med. Biol.*, 30:345–349, 1985.
64. H. C. Torrey. Bloch equations with diffusion terms. *Phys. Rev.*, 104(3):563–565, November 1956.
65. J. D. Tournier, F. Calamante, D.G. Gadian, and A. Connelly. Direct estimation of the fiber orientation density function from diffusion-weighted MRI data using spherical deconvolution. *NeuroImage*, 23:1176–1185, 2004.
66. A. Tristan-Vega, C. F. Westin, and S. Aja-Fernandez. A new methodology for the estimation of fiber populations in the white matter of the brain with the Funk-Radon transform. *NeuroImage*, 49:1301–1315, 2010.
67. D. Tuch. *Diffusion MRI of Complex Tissue Structure*. PhD thesis, Harvard University and Massachusetts Institute of Technology, 2002.
68. D. Tuch. Q-ball imaging. *Magnetic Resonance in Medicine*, 52(6):1358–1372, 2004.
69. D.S. Tuch, J.W. Belliveau, T.G. Reese, and V.J. Wedeen. High angular resolution imaging of the human brain. In *Proceedings of the International Society for the Magnetic Resonance in Medecine*, Philadelphia, April 1999.
70. Z. Wang, B. C. Vemuri, Y. Chen, and T. H. Mareci. A constrained variational principle for direct estimation and smoothing of the diffusion tensor field from complex DWI. *IEEE Transactions on Medical Imaging*, 23(8):930–939, 2004.
71. V. J. Wedeen. Mapping fiber orientation spectra in cerebral white matter with Fourier transform diffusion MRI. In *In Proc. of the 8th Annual Meeting of ISMRM*, page 82, 2000.
72. V. J. Wedeen, P. Hagmann, W.-Y. I. Tseng, T. G. Reese, and R. M. Weisskoff. Mapping complex tissue architecture with diffusion spectrum magnetic resonance imaging. *Magn. Reson. Med.*, 54(6):1377–1386, 2005.
73. C.F. Westin, S.E. Maier, H. Mamata, A. Nabavi, F.A. Jolesz, and R. Kikinis. Processing and Visualization for Diffusion Tensor MRI. *Medical Image Analysis*, 6(2):93–108, June 2002.
74. T. H. Williams, N. Gluhbegovic, and J. Jew. *The human brain: dissections of the real brain*. Virtual Hospital, 1997.
75. M.P. Young, G. Burns, and J.W. Scannell. *The Analysis of Cortical Connectivity*. Landes Bioscience, October 1995.

Part III
Modeling in Neuroscience

Chapter 7
Single-Trial Analysis of Bioelectromagnetic Signals: The Quest for Hidden Information

Maureen Clerc, Théodore Papadopoulo, and Christian Bénar

7.1 Introduction

7.1.1 Electric Activity in the Brain

Electric signals propagate within the brain along nerve fibers (axons) as a series of action potentials (APs). The corresponding primary current can be approximated by a pair of opposite current dipoles, one at the depolarization and one at the repolarization front, and this quadrupolar source moves along the axon as the activation propagates. The separation of the two dipoles depends on the duration of the AP and on the conduction velocity of the fiber. For a cortical axon with a conduction speed of 5 m/s, the opposite dipoles would be about 5 mm apart.

In synapses, the chemical transmitter molecules change the ion permeabilities of the postsynaptic membrane and a postsynaptic potential (PSP) and current are generated. In contrast to the currents associated with an action potential, the postsynaptic current can be adequately described by a single current dipole oriented along the dendrite. The electric and magnetic fields of a current dipole fall off more slowly with distance (as $1/r^2$) than the fields associated with the quadrupolar AP currents (as $1/r^3$). Local field potentials (LFPs) correspond to a spatial average of the individual activities: because summation of currents flowing in neighboring fibers is more effective for synaptic currents (lasting up to tens of milliseconds), than for APs (about 1 ms-long), LFPs mostly correspond to summations of PSPs.

M. Clerc (✉) · T. Papadopoulo
Inria Sophia Antipolis Méditerranée, Athena project-team, 2004 Route des Lucioles, 06902, Sophia Antipolis, France
e-mail: maureen.clerc@inria.fr; theodore.papadopoulo@inria.fr

C. Bénar
Institut des Neurosciences des Systèmes -INS, UMR 1106 INSERM, Aix-Marseille Université, Faculté de Médecine La Timone, 27 Bd Jean Moulin, 13385 Marseille Cedex 05
e-mail: christian.benar@univmed.fr

F. Cazals and P. Kornprobst (eds.), *Modeling in Computational Biology and Biomedicine*, DOI 10.1007/978-3-642-31208-3_7,

There are several scales at which bioelectricity can be described and measured. At a microscopic scale, microelectrodes are placed inside or in very close vicinity to neurons (see Chap. 8): at this scale, it is possible to measure APs as well as LFPs which correspond to the local electric potential within the cortex as generated by the nearby neurons. At the mesoscopic scale (below a square millimeter), intracortical recordings (stereotaxic EEG, SEEG) only mesure the local field potential. Non-invasive measurements of the electric potential via EEG or the magnetic field via MEG are made on the scalp, and the spatial extent of brain activity to which these measurements can be related does not yet meet a consensus, but lies between a square millimeter and a square centimeter. Similarly to SEEG, MEG and EEG mostly see PSPs: this results both from the faster decrease of the AP field with distance and from the summation argument used for LFPs.

The remainder of this chapter mainly focusses on SEEG, MEG and EEG and thus on PSPs.

7.1.2 Measuring Brain Activity

Several well-established techniques allow to non-invasively probe the brain in function. Some are sensitive to the metabolic activity of the brain, and others to its electric activity. Functional Magnetic Resonance Imaging (fMRI), which is sensitive to the ratio between oxygenated and de-oxygenated haemoglobin, is a metabolic imaging modality. It is an established technique for mapping regions involved in cognitive tasks. Because of its spatial encoding, one can estimate from fMRI the activity within the brain volume, with millimeter resolution. Unfortunately, the recorded signal, and thereby the detection of active areas, relies on the slow hemodynamic response to neuronal activity (taking several seconds to reach its peak). Therefore, the temporal sequence of activated regions is difficult to estimate through this technique. In contrast, electroencephalography (EEG) is a more direct reflection of neuronal function. It records variations of electric potential with millisecond resolution, at the time scale of neuronal synaptic activity. The EEG measures variations of electric potential on the scalp, produced by the pyramidal neurons of the gray matter. These brain sources actually produce an electromagnetic field, the magnetic component of which can be measured by magnetoencephalography (MEG). MEG and EEG thus measure two complementary consequences of the same brain sources.

Localizing active brain areas from EEG requires to solve a difficult inverse problem [1]. The resulting spatial resolution within the brain is of the order of a centimeter for EEG and MEG. A synopsis of most current brain exploration techniques is presented in Fig. 7.1, allowing to compare their respective time and spatial resolutions, and their degree of invasiveness. MEG, EEG and fMRI rank among the least invasive techniques. Local field potentials are measured by intra-cerebral electrodes: because their recording sites are only a few millimeters away from the sources of activity, they do not suffer from convolution effects

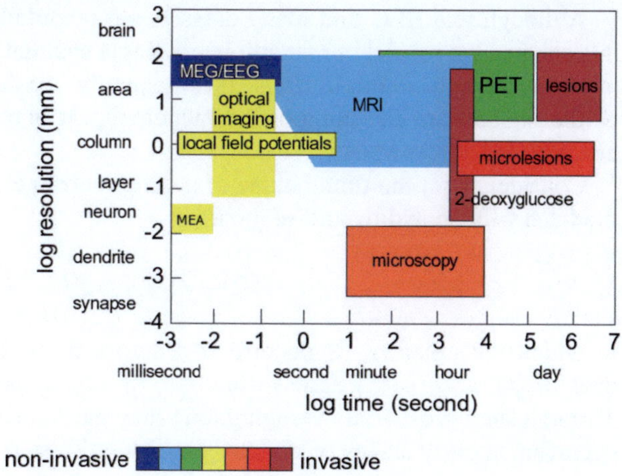

Fig. 7.1 Comparison of several brain imaging techniques, in terms of their spatial resolution (*vertical axis*), time resolution (*horizontal axis*), and invasiveness (*color-code*)

between different sources as in EEG. Chapter 6 deals with diffusion MRI data, a type of MR imaging which bring complementary structural connectivity information to EEG and MEG.

This chapter is devoted to the analysis of MEG and EEG signals, in order to extract meaningful information from them. The signals measured by MEG and EEG belong to one of two categories: spontaneous, i.e., endogeneously produced by the brain, or evoked, i.e., triggered by an outside stimulus. In the latter case, the EEG is referred to as Evoked Response Potential (ERP). Since the seminal work of Lehmann et al. on microstates [16], much effort is being devoted in the community in order to be able to analyze single-trial measurements, or to segment continuous strands of data into pieces within which the signals enjoy similar properties.

7.1.3 Bioelectromagnetic Signal Analysis

In EEG and MEG signals, the information of interest has a very low signal-to-noise ratio (SNR), because of high ongoing brain activity, not necessarily related to the object of the study. In fact, the notion of "noiseless signal" does not really exist. In evoked studies, a way to improve the SNR is to measure brain activity across many repetitions of the same experiment. Each of the repeated measurements is referred to as a trial, and multitrial analyses allow to deal with the global dataset.

A simple procedure to apply to a multitrial dataset is to translate each trial in time so that zero corresponds to the onset of the stimulation, and to average the measurements across trials. The resulting data is called Averaged Evoked Response Potentials in the case of EEG. Averaged ERP typically consist of a series of waves, whose latencies and amplitudes can be extracted.

Although real EEG and MEG datasets are recorded over multiple channels, in this chapter, we restrict our attention to a single channel, and we focus on extracting information from multiple trials. It is generally considered that the relevant part of the signals are the components which repeat across trials, whereas noise is uncorrelated across trials.

Consider $s_k(t)$ the time-course of a (single-channel) signal measured at trial k, modeled with an additive noise model

$$s_k(t) = x(t) + n_k(t), \tag{7.1}$$

in which the noise n_k is decorrelated across trials. Directly averaging the signals $\{s_1(t), \ldots, s_K(t)\}$ across trials $\frac{1}{K}\sum_{k=1}^{K} s_k(t)$ provides an estimate of $x(t)$. Through such cross-trial averaging, one may gain information about the processes occurring at early stages of sensory processing (visual evoked potentials, auditory evoked potentials, somatosensory evoked potentials), which are stereotypical, and do not vary much across trials.

Unfortunately, the simple additive model (7.1) is quite far from the truth in most operational settings: the "relevant" part of the data $x(t)$ can generally not be assumed to be constant across trials. There are several sources of variability, that are represented in Fig. 7.2:

- When $x(t)$ includes oscillatory components, their phase is generally variable across trials, and cross-trial averaging tends to cancel out the oscillations;
- The latency, i.e., the time between the stimulus onset and the response, is often variable, and the resulting averages are blurred estimators of the actual responses.

Rather than the naive additive model (7.1), multitrial datasets should be modeled as the sum of trial-dependent components $x_k(t)$ and noise:

$$s_k(t) = x_k(t) + n_k(t). \tag{7.2}$$

The challenge is then to distinguish between the relevant part of the signals, x_k, and the noise, n_k, when both vary across trials. Some constraints must be set on x_k and n_k in order for their separation to be possible.

7.1.4 Chapter Overview

Public

This chapter is mainly targeted at researchers in cognitive and clinical neuroscience and in signal and image processing, who are interested in processing brain signals. Most of the material covered hereafter is presented in sufficient detail for graduate-level students with a good background in signal processing and linear algebra.

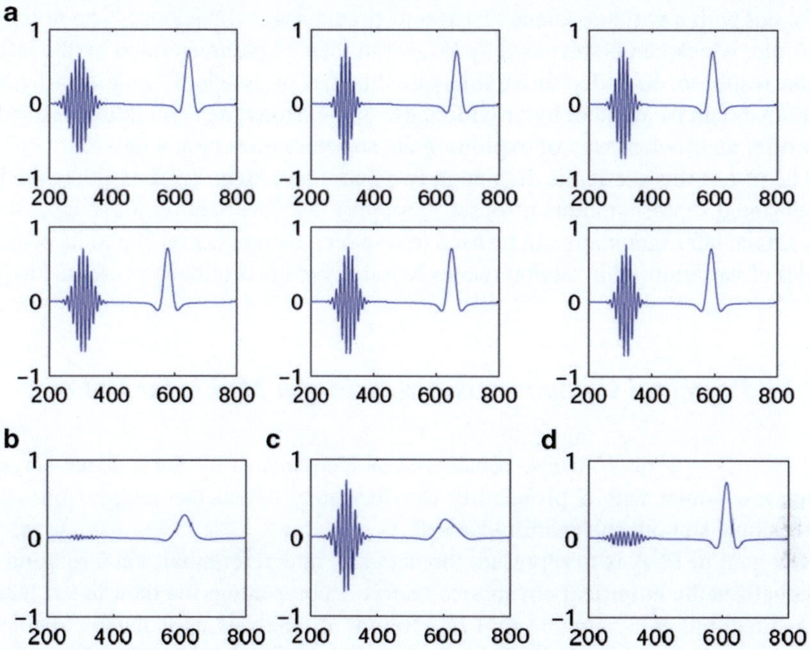

Fig. 7.2 Illustration of the difficulties due to averaging signals across trials. (**a**) six trials of simulated EEG, comprising two events that are subject to variablity in latency, amplitude and frequency. (**b**), (**c**) and (**d**): Averaged signal with various time alignments. (**b**) no time alignment, (**c**) time alignment on the *left* event, (**d**) time alignment on the *right* event

Outline

This chapter deals with the analysis of multitrial MEG or EEG dataset, and presents two classes of approaches: data-driven, and model-driven. In Sect. 7.2 we present a data-driven approach for dimensionality reduction which allows to reorder the trials, and subsequently simplifies their analysis. In Sect. 7.3 we present a multitrial version of Matching-Pursuit, which models the signals of interest as linear combinations of atoms from a predefined dictionary.

7.2 Data-Driven Approaches: Non-linear Dimensionality Reduction

Considering a dataset described by the additive model (7.2), statistical methods such as Principal Components Analysis (PCA) can be used to explore the structure of the trial-dependent activity $x_k(t)$, if the additive noise $n_k(t)$ is not too large.

We consider multitrial datasets that lie on a noisy 1-D manifold. This often occurs in multitrial ERP recordings, in which similar neural activations occur across trials, but with a variable latency between stimulation and response. The noisy 1-D manifold, which can be revealed by PCA, can then be parameterized by the latency of the response. In order to capture this "degree of freedom" manifold-learning techniques can be applied: by providing low-dimensional representation of the data, they offer an efficient way of exploiting the structure present in a dataset.

The rest of this section is divided in two parts: first, Sect. 7.2.1 explores the low-dimensional representations afforded through the PCA. Section 7.2.2 next shows how Laplacian Eigenmaps can be used to reorder a dataset according to its principal modes of variation. This section relates research results originally presented in [10].

7.2.1 Principal Components Analysis of a Multitrial Dataset

Let, $(x_i)_{i=1,\ldots,K}$ be K trials, considered as elements of a metric space $(\mathcal{X}, d_{\mathcal{X}})$, which are drawn with a probability distribution p which has support on a low-dimensional smooth sub-manifold $\mathcal{M}$ of $\mathcal{X}$.

The goal of PCA is to represent the data in a new referential, via a rotation that diagonalizes the empirical covariance matrix. Representing the data in the leading PCA directions is a valuable tool in exploratory analysis as it makes visible the structure present in the data. Chapters 3 and 5 of this book apply PCA to other biomedical data.

To illustrate this, a dataset containing 1,000 time series was simulated, each with 512 time samples. The time series mainly differ by their latency. Figure 7.3 presents this dataset: in (a), nine out of 1,000 time series are plotted; (b) displays a raster plot, i.e., an image whose 1,000 lines consist of the time series, in color-scale; (c) and (d) show 2D and 3D PCA projections of the dataset , in which each dot represents one of the time series. Fig. 7.3d clearly displays how the dataset is organized along a 1D structure.

It is possible to exploit this 1D structure in view of reordering the time-series according to their latency. This reordering is equivalent to finding a parameterization of the curve in Fig. 7.3d. This challenge is addressed in the following sections by a nonlinear embedding method called the Graph Laplacian.

7.2.2 Nonlinear Embedding via the Graph Laplacian

Given a set of points $(x_i)_i$, each representing a time series, the aim is to recover the structure of $\mathcal{M}$ via an embedding function f that maps the $(x_i)_i$ into a low-dimensional Euclidian space $\mathbb{R}^n$. The embedding function f provides a low-dimensional representation of the dataset and also a parameterization of the manifold. When $\mathcal{M}$ has a 1D structure, the first coordinate of f can be used to reorder the points (i.e., in our context, the time series), provided that f satisfies a "*regularity*" constraint, sometimes referred to as a minimal distortion property: if

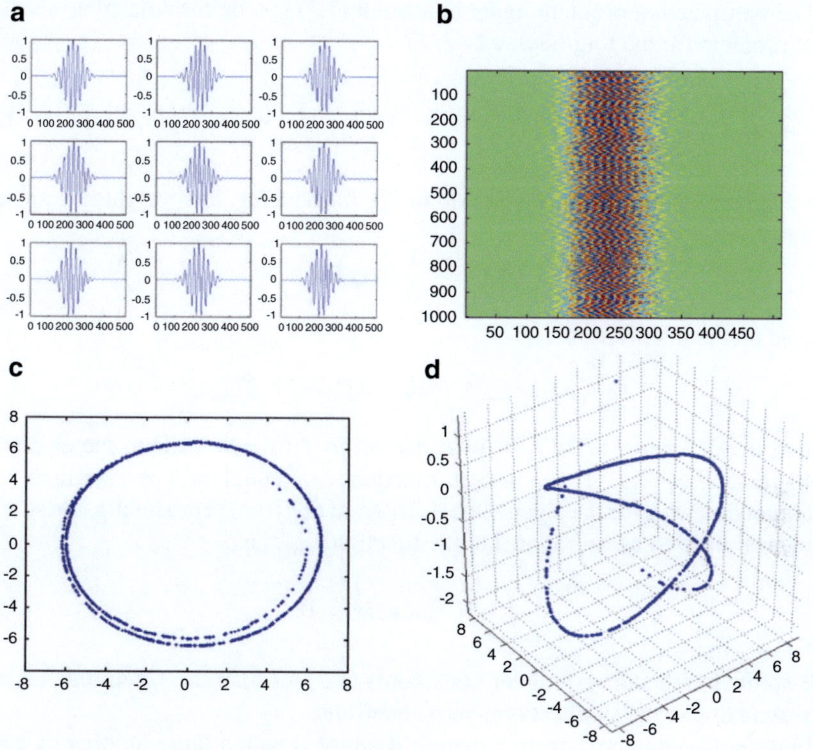

Fig. 7.3 PCA analysis of a set of 1,000 jittered synthetic time series of 512 time samples. (**a**) Nine time series out of 1,000. (**b**) Original raster plot. (**c**) 2D PCA projection. (**d**) 3D PCA projection

two points x and z are close according to the manifold $\mathcal{M}$, then $f(x)$ and $f(z)$ must also be close in $\mathbb{R}^n$. To express this regularity constraint in mathematical terms, notice that for $n = 1$, a Taylor expansion provides the following inequality [2]

$$|f(z) - f(x)| \leq d_{\mathcal{M}}(x,z)||\nabla f(x)|| + o(d_{\mathcal{M}}(x,z)) \ ,$$

where ∇f stands for the gradient of f and $d_{\mathcal{M}}$ is the geodesic distance on the manifold between points x and z. The notation $g(z) = o(d_{\mathcal{M}}(x,z))$ means that $\frac{g(z)}{d_{\mathcal{M}}(x,z)}$ tends to 0 as z tends towards x.

In order to obtain an embedding that satisfies the *"regularity"* constraint, Laplacian-based methods control the smoothness of f globally by minimizing $\int_{\mathcal{M}} ||\nabla f(x)||^2 p(x)dx$, provided that

$$\int_{\mathcal{M}} ||f(x)||^2 p(x)dx = 1. \tag{7.3}$$

The latter condition removes the scaling indeterminacy for the function f.

The optimization problem under constraint (7.3) can be formulated as a saddle-point problem for the Lagrangian $\mathbf{L}(f,\lambda)$:

$$\mathbf{L}(f,\lambda) = \int_{\mathcal{M}} ||\nabla f(x)||^2 p(x)dx + \lambda\left(1 - \int_{\mathcal{M}} ||f(x)||^2 p(x)dx\right). \qquad (7.4)$$

The Lagrangian (7.4) can be rewritten by introducing the weighted Laplacian operator

$$\Delta f = -\frac{1}{p}\mathrm{div}(p\nabla f),$$

defined on the manifold $\mathcal{M}$:

$$\mathbf{L}(f,\lambda) = \langle \Delta f, f\rangle_{\mathcal{M}} + \lambda(1 - \langle f, f\rangle_{\mathcal{M}}).$$

Setting the derivative of $\mathbf{L}(f,\lambda)$ with respect to f to zero leads to the eigenvalue problem $\Delta f = \lambda f$. The constant function f_{cst} equal to 1 everywhere is an eigenfunction of Δ for the eigenvalue 0. To avoid this trivial embedding, the solution f is constrained to be orthogonal to the function f_{cst}, i.e.,

$$\int_{\mathcal{M}} f(x)p(x)dx = 0. \qquad (7.5)$$

The optimal embedding f under constraints (7.3)–(7.5) is the eigenfunction of Δ corresponding to the smallest non-zero eigenvalue.

In order to compute f from a manifold sampled with a finite number of points, the operator Δ needs to be approximated. Graph Laplacian methods approximate Δ through the Laplacian of a particularly designed graph. Let $\mathcal{G} = (\mathcal{V},\mathcal{E})$ be an undirected graph where $\mathcal{V}$ are the nodes $(x_i)_{i=1,\ldots,N}$ and $\mathcal{E}$ are the edges. A weight w_{ij} is associated to every edge $(i,j) \in \mathcal{E}$, leading to a weighted graph $\mathcal{G}$. The Laplacian L of the graph is a matrix defined by $L = D - W$ where W is the matrix composed of elements w_{ij} and D is diagonal with $D_{ii} = \sum_j w_{ij}$. The random-walk Laplacian, L_{rw}, is a normalized version of L defined by $L_{rw} = D^{-1}L = I - D^{-1}W$ where I is the identity.

A weighting matrix W allowing a good approximation of Δ must now be defined. This is done with the help of a similarity measure k which is a non-increasing function of the distance $d_{\mathcal{X}}$. Here, k is a Gaussian kernel with standard deviation σ:

$$k(x_i,x_j) = e^{-d_{\mathcal{X}}(x_i,x_j)^2/\sigma^2}.$$

In [12], the authors proved that the random-walk Laplacian of $\mathcal{G}$ converges almost surely to, the Laplacian operator Δ when the sample size N goes to infinity, if

$$w_{ij} = \frac{k(x_i,x_j)}{(d(x_i)d(x_j))^{1/2}}, \qquad (7.6)$$

where $d(x) = \sum_{i=1}^{N} k(x,x_i)$ is an estimator of the probability density function over the manifold $\mathcal{M}$.

Algorithm 10 Laplacian-based n-dimensional embedding

1: Set $d_{\mathcal{X}}$, σ and n the dimension of the embedding.
2: Compute K with $K(i,j) = e^{-\frac{d_{\mathcal{X}}(x_i,x_j)^2}{\sigma^2}}$.
3: Compute D_K with $D_K(i,i) = \sum_{j=1}^{N} K(i,j)$ and $D_K(i,j) = 0$ if $i \neq j$.
4: Compute $W = D_K^{-1/2} K D_K^{-1/2}$ (Eq. 7.6).
5: Compute D with $D(i,i) = \sum_{j=1}^{N} W(i,j)$ and $D(i,j) = 0$ if $i \neq j$.
6: Find the $n+1$ first generalized eigenvectors f_k solution of $(D - W)f_k = \lambda_k D f_k$, $k = 0, \ldots, n$.
7: The coordinates of point x_i in $\mathbb{R}^n$ are $(f_1(i), f_2(i), \ldots, f_n(i))$.

The embedding function f can therefore be obtained by computing the eigenvectors f_q of L_{rw}, which satisfy:

$$(I - D^{-1}W)f_q = \lambda_q f_q.$$

Eigenvectors of L_{rw} are generalized eigenvectors of L since

$$Lf_q = \lambda_q D f_q \ .$$

It can easily be proved that 0 is a trivial eigenvalue of L, and also that L is a symmetric positive matrix, which implies, since $d(x_i) \geq 0$ for all i, that the generalized eigenvalues λ_q are positive and can be ordered: $0 = \lambda_0 \leq \lambda_1 \leq \cdots \leq \lambda_q \leq \lambda_{q+1} \leq \cdots \leq \lambda_{N-1}$. The embedding function f into $\mathbb{R}^n$ is then:

$$f(x_i) = (f_1(i), \ldots, f_n(i)), \tag{7.7}$$

where $f_q(i)$ is the ith component of f_q.

Algorithm 10 presents the different steps of the approach. Note that, since L is symmetric, its first eigenvectors can be computed efficiently with an iterative method. It is possible to set to 0 the w_{ij} below a threshold, leading to sparse matrices, and reducing the computational cost of matrix-vector multiplications at each iteration.

Manifold learning methods are "data driven". They capture the structure of the dataset, provided that the chosen distance $d_{\mathcal{X}}$ is appropriate. When dealing with time series, many distance functions can be used. In practice, $d_{\mathcal{X}}$ does not need to be a distance, in the mathematical sense: instead it can measure the difference between features of interest for two elements of the dataset. One may even design $d_{\mathcal{X}}$ to be *blind* to features of the data that are irrelevant for the application at hand.

7.2.3 Application to the Reordering of EEG Times Series

The purpose of the previous section was to compute an embedding function, which now allows the parameterization of the dataset according to the first embedding

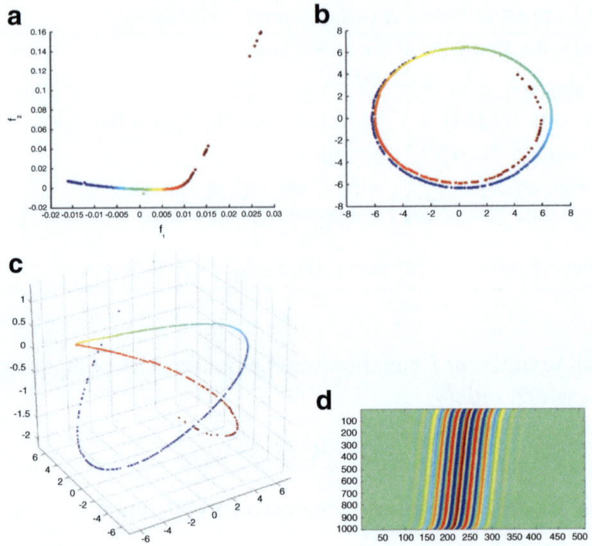

Fig. 7.4 Illustration of reordering by Graph Laplacian embedding, on the synthetic dataset already presented in Fig. 7.3. (**a**) f_1 on the *horizontal axis* and f_2 on the *vertical axis*, i.e. the first two coordinates of the Laplacian-embedding function in Eq. (7.7). The *color* codes for the f_1 coordinate. (**b**) 2D PCA projection of the dataset, color-coded as the corresponding points in (**a**); (**c**) 3D PCA projection, with the same *color-code*. In (**d**), the time series are reordered according to the *color-code* (or f_1 coordinate)

coordinate. In the context of biosignal analysis, this parameterization is used to reorder the dataset, allowing subsequent adapted signal processing, such as realignment.

Once computed, the Laplacian embedding f provides a parameterization of the manifold $\mathcal{M}$: if $\mathcal{M}$ has a noisy 1D structure, the first coordinate, f_1, orders the elements along the manifold. This is now illustrated with the noiseless synthetic dataset already presented in Sect. 7.2.1 and Fig. 7.3. It is simulated with $T = 512$ and $N = 1{,}000$ trials (represented as N points from a 512-dimensional manifold $\mathcal{M}$). The embedding was performed using the Euclidian distance and a Gaussian kernel. The embedded points are represented in Fig. 7.4a. It can be observed that the embedding unfolds the manifold structure. The ordering provided by f_1 can be encoded with a color, which is thus associated to each point of the PCA representation. The 2D and 3D PCA colored point clouds are presented in Fig. 7.4b, c: observe that the color changes gradually along the one-dimensional structure. Figure 7.4d presents the reordering of the raster plot which was presented in Fig. 7.3b.

We now apply the nonlinear embedding method to a real dataset coming from an auditory oddball experiment, which consists in alternating frequent tones and rare ("target") tones. This protocol is known to elicit a positive EEG deflection to the rare tones, referred to as the "P300" or "P3" wave, more prominent on the midline electrodes and occurring at a latency around 300 ms [4]. The data is recorded from the central electrode *Cz*, sampled at a rate of 256 Hz and processed

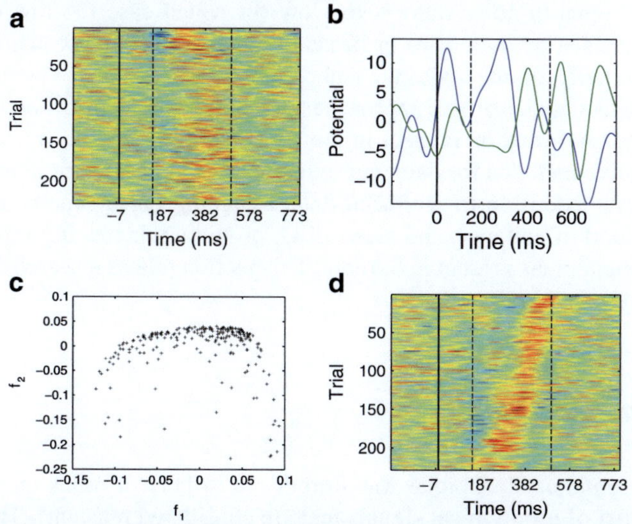

Fig. 7.5 Reordering results on EEG oddball time series. The *solid vertical line* in the raster plots corresponds to the stimulus onset. The *vertical dashed lines* provide the limits on the time window used to reorder the data. This time window is manually set around the largest fluctuation of the evoked response. Embedding was performed with $r = 2$ and $\sigma = 0.05$. (**a**) Raster plot of raw time series; (**b**) Two sample time series; (**c**) 2D embedding of the time series; (**d**) Raster plot reordered using the first coordinate f_1 of the embedding

with a high-pass filter at 0.5 Hz (Butterworth zero phase filter, time constant 0.3183 s, 12 dB/oct) and a low-pass filter at 8 Hz (Butterworth zero phase filter, 48 dB/oct). The positive deflection of the P300 wave, in the 3–5 Hz range, is preserved. Figure 7.5a presents a raster plot of the data set. The random nature of the time latency of the P300 is apparent, as first observed in [15].

The time series were embedded into a two dimensional space (Fig. 7.5), after restricting the time series to a time interval around the largest fluctuation of the evoked response. This interval was manually defined by visual inspection of the dataset. It can be noticed that the points are clustered along an elongated 1D structure, as in the synthetic dataset presented in Fig. 7.4. The first coordinate can therefore be used to correctly parameterize the manifold (Fig. 7.5c) and to order the time series. By observing the reordered raster plot in Fig. 7.5d, it appears that the first coordinate of f has correctly captured the latency ordering.

7.3 Model-Driven Approaches: Matching Pursuit and Its Extensions

The above-presented data-driven approaches require that the data belong to a low dimensional manifold. However, this is often not the case in practise either because the signal-to-noise ratio is too low (in which case the low dimensional manifold is hidden by the noise) or because signals are complex, which increases the dimensionality of the required embedding space. In such cases, it may be preferable to use model-driven approaches which leverage a priori on the structure of the relevant information present in the signal. In this section, we will review a class of iterative methods for data approximation, called Matching Pursuit, and we will present its extension to multitrial data. Matching pursuit methods have been used in the field of epilepsy, and sleep EEG, providing interesting information on oscillatory phenomena present in the data. This section relates research that has been originally presented in [3].

7.3.1 Matching Pursuit

The *matching pursuit* technique was introduced in [17]. It relies on a *dictionary* of *atoms* (a list of prototypical signals that are considered relevant). The algorithm iteratively decomposes the signal into a sequence of atoms. At each iteration, it seeks the atom of the dictionary that best matches the signal. The atom contribution is then subtracted from the signal before starting the next iteration of the algorithm. The method is repeated until the remaining residual is considered negligible. Consequently, the algorithm mainly relies on: (1) a choice of a dictionary, (2) a procedure to combine dictionary elements into a signal and a means to extract the most significant atom out of a signal.

In our case, the dictionary $\mathcal{D}$ is a set of known predefined functions $\psi_{\mathbf{p}_i}(t), i = 1 \ldots M$ (the atoms), where $\mathbf{p}_i$ is a set of parameters defining the function $\psi_{\mathbf{p}_i}$. Given $\mathcal{D}$ and multiple recordings (trials) $s_k(t)$ of a given brain activity, classical matching pursuit (independently used on each recording) aims at decomposing the signals $s_k(t)$ into a low number P of components chosen in $\mathcal{D}$:

$$s_k(t) = \sum_{i=1}^{P} a_{ik} \psi_{\mathbf{p}_i(k)}(t) + n_k(t) \ , \tag{7.8}$$

where $n_k(t)$ is the noise on trial k and a_{ik} is the amplitude associated to the atom $\psi_{\mathbf{p}_i}(t)$ in $s_k(t)$. Note that in theory, the noise $n_k(t)$ should encompass any activity that is not present in all trials, but when the decomposition is done independently for each trial, there is no way to guarantee this and thus noise represents any residual signal that cannot be modelled with P atoms $\psi_{\mathbf{p}_i(k)}$.

Contrarily to usual bases in functional spaces, dictionaries $\mathcal{D}$ are most often overcomplete (the predefined functions are highly redundant so that decompositions are not unique), but should have a good descriptive power, i.e., a good ability to represent any allowed input signal with a low number of components P leading to sparse representations. Another important property of dictionaries is their interpretability, i.e., the possibility to attach some meaningful semantics to each of their atoms.

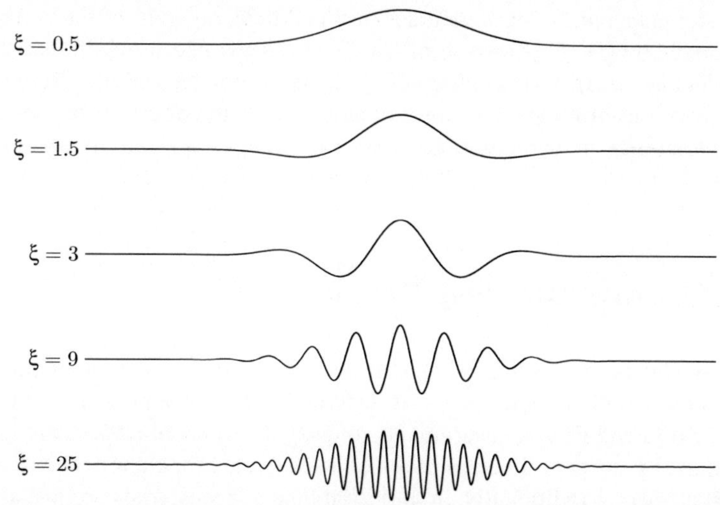

Fig. 7.6 Gabor atoms for different values of the oscillation parameter, obtained by varying f with a constant σ (Eq. 7.9). A low oscillation parameter produces a transient wave, and a high value a sustained oscillation

7.3.2 A Dictionary Tuned for MEG and EEG

There are two types of activity that can appear in MEG or EEG measurements: oscillatory activity such as gamma or alpha bursts, and transient, phasic activity. In order to model both types of events, an over-complete dictionary of Gabor wavelets is chosen. In the time domain, this dictionary is defined by:

$$\psi_{t_0,f,\sigma}(t) = \frac{1}{(\pi\sigma^2)^{\frac{1}{4}}} e^{2i\pi f(t-t_0)} e^{-\frac{(t-t_0)^2}{2\sigma^2}}. \qquad (7.9)$$

An element of this dictionary is a Gaussian function centered at t_0, modulated by a sinusoidal signal with frequency f. These atoms are thus both oscillatory (the "evoked potential" type of activity being associated to low frequencies) and localized (their temporal extent is bounded by the Gaussian with σ standard deviation). The parameter t_0 allows for time invariance (meaning that the representation of a translated signal is the translation of the original signal representation). Each $\psi_{t_0,f,\sigma}$ is a wavelet centered around time t_0 and oscillating at a frequency f. The duration of the wavelet is given by σ but it is also interesting to specify it by its number of oscillations which is given by $\xi = 2\pi f\sigma$. Figure 7.6 shows various such wavelets, all centered around a same time instant, and for different values of ξ. Note that the initial factor in (7.9) ensures that all atoms are normalized to have a unit energy (in L^2 norm).

As it is equivalent, in the remainder of this article, an atom of the dictionary is parameterized by (t_0, f, ξ) instead of (t_0, f, σ). This is because ξ directly qualifies the transient ($\xi < 2$) vs. oscillatory ($\xi \geq 2$) nature of the activity. Both types of parameterizations provide for a convenient interpretation of events as they account for the variability in the duration (ξ or σ), latency (t_0) and frequency (f) of activities.

7.3.3 Consensus Matching Pursuit

Consensus Matching Pursuit (CMP) is an evolution of the standard matching pursuit algorithm presented in Sect. 7.3.1. It is designed to take advantage of the multitrial nature of MEG and EEG measurements. Indeed, as events of interest are supposed to be evoked by the experimental protocol, such events should repeat in every trial of the experiment. Traditionally, in multitrial data analysis, trials are just averaged, which is simple and easy. Unfortunately, as explained in Sect. 7.1.3, working with averaged data has two major drawbacks:

1. Average activities do not give access to the variability of the events of interest across trials;
2. The events detected in the averaged signals are often deformed both in their amplitudes and durations, as illustrated in Fig. 7.2b–d.

To overcome these difficulties, CMP completely avoids the use of averaging and instead uses a voting scheme. Each individual signal is first represented into a time-frequency-ξ map (an extension of the more classical time-frequency map). Such maps provide for each atom (i.e. each possible value of (t_0, f, ξ)) its associated amplitude obtained by convolving $s_k(t)$ with the atom. As atoms vary smoothly, with (t_0, f, ξ) the map is also smooth, and its peaks (local maxima of the map with respect to the variables (t_0, f, ξ)) indicate atoms that locally best represent the signal.

In practise, the dictionary is discretized: the time step is given by the sampling of the signal, and the ranges of the parameters in the frequency and ξ dimensions derive from a priori information on the range of interest. Hereafter, discretization of ξ is set to $0.8; 1.5; 3; 5; 7; 9; 13; 25$. For a given ξ and a particular window length, only a certain range of frequencies are acceptable, as the time-support of the atom (given by σ) must be smaller than the signal time window. Moreover, we choose not to analyze high frequencies ($f > 15\,Hz$) for small oscillation parameters ($\xi < 2$): this stems from the assumption that high-frequency activity must be oscillatory. In the maps, frequencies that were not computed were set to zero.

To extract the most sensible atoms that repeat across trials, the peaks of the time-frequency-ξ map are extracted for each trial. Each peak then votes in a voting map common to all trials with a weight given by the amplitude associated with the peak (a vote is a smooth function around the peak position). Once the votes of all the peaks of all the trials have been accumulated, the highest peak of the voting

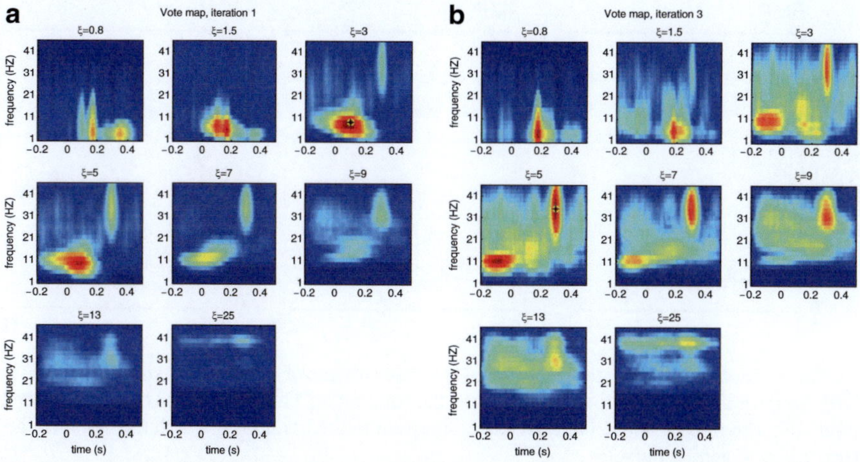

Fig. 7.7 The voting map at two iterations of the algorithm for a given (synthetic) trial. *Deep blue* areas correspond to disallowed parameter values (Sect. 7.3.3). The *black dot* in each map indicates the highest peak of the map, i.e., the best "common atom" extracted at that iteration: (**a**) Result after one iteration; (**b**) Result after three iterations. Please note that each iteration is normalized independently

map is extracted as the next most significant "consensus atom" to extract for the matching pursuit algorithm. As this "consensus atom" is not necessarily a peak in the individual trial maps, the peak closest to this "consensus atom" is computed and the corresponding atom is recorded for each individual map. Its contribution is subtracted from the signal to finish the iteration of the matching pursuit algorithm. Figure 7.7 illustrates two iterations of the algorithm for one trial of a synthetic dataset.

After P iterations of CMP, the signals have been decomposed as:

$$s_k(t) = \sum_{i=1}^{P} a_{ik} \psi_{\mathbf{p}_i(k)}(t) + n_k(t) \; . \tag{7.10}$$

Because of the voting map, the CMP decomposition guarantees a cross-trial coherence between atom parameters $p_i(k)$.

7.3.4 Experiments with Real Data

Other adaptations of the matching pursuit algorithm have been devised to deal with EEG data. Evoked Matching Pursuit (EMP) is a standard matching pursuit algorithm applied on a signal obtained by averaging all the individual trials. Closer to the Consensus Matching Pursuit, Induced Matching Pursuit (IMP) is a matching pursuit algorithm applied on a multi-trial dataset obtained by averaging the time-frequency

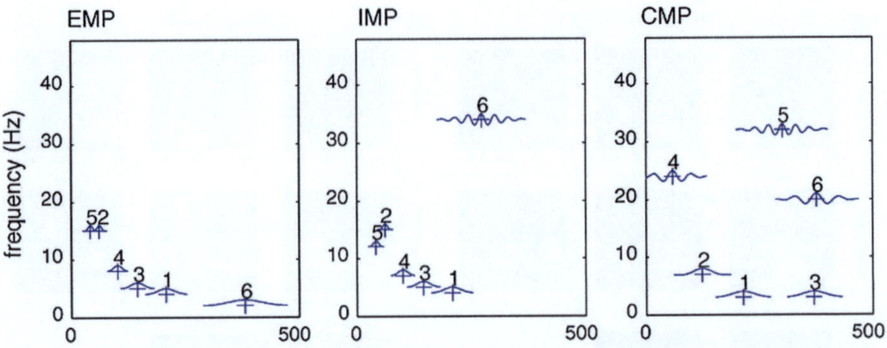

Fig. 7.8 Comparison of the atoms recovered by three variants of matching pursuit: EMP, IMP and CMP. Atoms are shown a time-frequency map, the parameter ξ being illustrated by a small wavelet in the plot. The indices above the wavelets correspond to the iteration at which the atom has been extracted by matching pursuit

power representations of all the individual trials. It was originally proposed by Gribonval et al. for multichannel data [11]. Figure 7.8 compares the atoms obtained on a real dataset with the three variants of matching pursuit: CMP (the variant described in this article), EMP and IMP.

The real dataset was measured during a visual task which is a variation of the protocol introduced in [21]. The original protocol consisted in observing a succession of illusory triangles (curved and non-curved), real triangles and no-triangle stimuli. This results in a mixture of event-related potentials and high-frequency oscillatory activity. The variation consists in presenting the triangles pointing either up or down. The subject is instructed to press a button as a response to the curved illusory triangles. Intracerebral EEG was acquired (sampled at 1 kHz) from an epileptic patient implanted in the occipital region (stereotaxic EEG, SEEG) for the sole clinical purpose of presurgical evaluation. The studied channel is located in the occipital cortex, for which a standard time-frequency has shown there is high-frequency activity. For evoked potential (frequencies < 10 Hz), the three methods identified rather similar atoms. Both IMP and CMP identify an oscillation around 30 Hz, which was not visible with EMP. Furthermore, CMP detects one oscillation at 400 ms and 20 Hz. IMP missed this 20 Hz component because it required more atoms to represent the averaged signal under 20 Hz. For EMP and IMP, these missed components appear only if the number of extracted components is increased. This shows that CMP provides a sparser description of the analyzed signal (or equivalently a more accurate description for a limited of extracted components).

7.4 Success Stories

This section presents examples where single-trial analysis has allowed the extraction of unique information that was not available on the averaged data.

7.4.1 Co-variations Between EEG and fMRI Signals

Although functional MRI and EEG have very different characteristics and resolutions (Sect. 7.1), EEG single-trial analysis can make a bridge between these complementary measures. The general idea is to use parameters extracted from the EEG at the level of single trials, and test whether the fMRI signal from each brain region is co-varying with the EEG -derived information (parametric analysis). The existence of such co-variation can indicate that the region as defined by fMRI is at the source of the EEG signal, or that both phenomena are modulated by the same processes (e.g., attention). Parameters can represent the amplitude or latency of a given wave on the evoked activity. This method has been used successfully with two well-known evoked potentials: the error negativity [8], and the oddball-related P300 [4,9] (Fig. 7.9). A recent study has pushed even further the principle and used information arising from high frequency gamma oscillations [18].

7.4.2 Distinction Between Latency and Amplitude Effects in Evoked Potentials

As mentioned in the previous section, a classical marker used in EEG evoked activity is the amplitude of a given wave. For example, the amplitude of the P300 (a positive wave peaking around 300 ms after presentation of an unexpected stimulus) has been used as a marker in many clinical applications [19]. However, the amplitude of a wave on the average signal can in principle be influenced by two factors at the single trial level (Fig. 7.10):

- The actual amplitude of the waves;
- An increased jitter in latency across trials.

There was a classical view in the literature that the amplitude of the P300 (response to rare events) depends on the number of frequent stimuli preceding the rare event. However, in [13], it-has been shown that in fact this could be attributed to ISI (inter-stimulus interval) effects: when considering the single-trial amplitude, there is no effect of latency. This can also be seen on Fig. 7.9, where trials are sorted by increasing reaction time. Long intervals between rare events tend to induce a faster response: with cumulating frequent events, the probability of having a rare event increases, and the subject is more prepared to respond. Therefore, the ISI is inversely correlated with the reaction time. In addition, the P300 latency is very highly correlated with the reaction time (as can be visualized on the sorted trials figure, see also [4]). As a consequence, long ISI correspond to low variability in latency of P300, and therefore to higher amplitude on average, similarly to what is illustrated in Fig. 7.10.

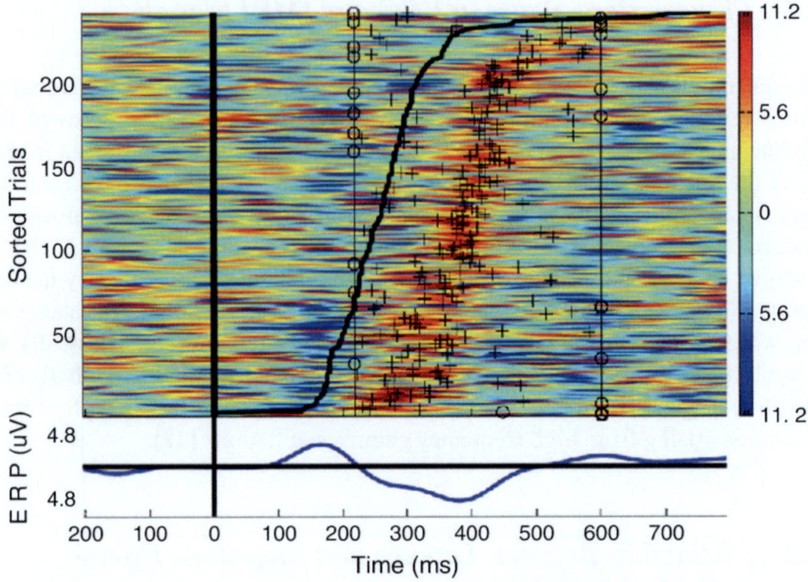

Fig. 7.9 Raster plot of single-trial P300 on EEG recorded in the fMRI scanner, sorted by response time (response time is indicated by a *black line*). Single-trial P300 latencies are shown with a cross. The event related potential (average across trials) is shown below the raster plot. This information was used in [4] in order to detect regions where fMRI signals and P300 parameters are co-varying

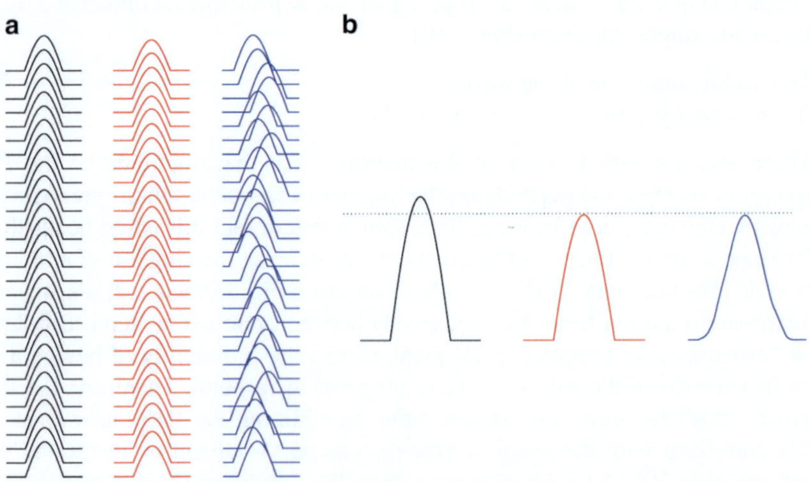

Fig. 7.10 Illustration of latency and amplitude effects on simulated waves (**a**) 30 trials with different conditions 1: fixed amplitude and latency 2: amplitude fluctuations 3: latency fluctuations. (**b**) Corresponding event-related potentials (i.e., averages across trials). Conditions 2 and 3 have a similar amplitude decrease with respect to condition 1, even though the underlying mechanisms are very different

7.4.3 *Habituation and Learning Effects*

In classical analysis of ERPs, averages are performed across large sets of trials and subjects, ignoring possible fluctuations arising from e.g. attention-related processes, learning and changes of strategy. Another interesting application of single-trial analysis is to open the way for a fine characterization of such fluctuations. In [7], the authors showed that it is possible to characterize habituation effects with single-trial analysis in a simple stimulation protocol. Quian Quiroga and Van Luijtelaar [20] have shown that different waves of the evoked potentials have different profiles of habituation, some with slow variations and other waves with habituation taking place in a few trials only. This suggests that these waves are related to different processes.

7.5 Conclusion

In this chapter, we presented the analysis of multitrial electrophysiology datasets coming from neuroelectromagnetic recordings by EEG and MEG. Such measurements present several characteristics: the absence of ground-truth data, and a high level of noise, which can be defined as the part of the data which is uncorrelated across trials. Multitrial recordings are compulsory, in order to extract meaningful information from the data. Moreover, the information of interest is subject to inter-trial variability. This chapter has focussed on two families of data processing methods that are applied in this context: data-driven methods, in a section on non-linear dimensionality reduction, and model-driven methods, in a section on Matching Pursuit and its extensions. The importance of correctly capturing the inter-trial variability is underlined in the last section which presents three case-studies in clinical and cognitive neuroscience. Furthermore, the rapidly growing field of Brain Computer Interfaces is driving research on the online interpretation of EEG signals, and we can therefore expect much progress on single-trial biosignal analysis in the years to come.

7.6 Online Resources

7.6.1 Datasets

BCI Competitions

http://www.bbci.de/competition/

This site provides EEG/MEG/ECoG datasets for the purpose of comparing signal processing and classification methods for Brain Computer Interfaces.

EEG/ERP datasets

http://sccn.ucsd.edu/~arno/fam2data/publicly_available_EEG_data.html

Various freely downloadable datasets with many subjects and up to 2,500 trials.

EEG datasets catalogue

http://www.brainsignals.de/

A list of freely available brain signal datasets (EEG/MEG/fMRI).

7.6.2 Software

MPTK

http://mptk.irisa.fr/

The Matching Pursuit Tool Kit (MPTK) provides a fast implementation of the Matching Pursuit algorithm for the sparse decomposition of multichannel signals. It comprises a C++ library, standalone command line utilities, and some scripts for running it and plotting the results through Matlab.

EEGLAB

http://sccn.ucsd.edu/eeglab/

EEGLAB is an interactive Matlab toolbox dedicated to continuous and event-related EEG and MEG. It contains various tools to visualize and process such signals.

OpenViBE

http://openvibe.inria.fr/

OpenViBE is an open source easy-to-use Brain Computer Interface platform. It can be used to run online experiments, to present stimuli, to acquire EEG signals, and to process them online.

BCI 2000

http://bci2000.org/

BCI2000 is a general-purpose system for brain-computer interface (BCI) research. It can also be used for data acquisition, stimulus presentation, and brain monitoring applications.

OpenMEEG

http://openmeeg.gforge.inria.fr

OpenMEEG is an open source program that solves the MEG/EEG forward problem, providing state-of-the-art accuracy. It comprises a C++ library, standalone command line utilities, and plugins for python and Matlab.

FieldTrip/TurboFieldTrip

http://fieldtrip.fcdonders.nl/

FieldTrip is a Matlab software toolbox for MEG and EEG analysis. The FieldTrip software is released as open source under the GNU general public license. The software includes algorithms for simple and advanced analysis of MEG and EEG data, such as time-frequency analysis, source reconstruction using dipoles, distributed sources and beamformers and non-parametric statistical testing. It supports the data formats of all major MEG systems and of the most popular EEG systems, and new formats can be added easily.

Brainstorm

http://neuroimage.usc.edu/brainstorm/

Brainstorm is a collaborative open-source Matlab application dedicated to magnetoencephalography (MEG) and electroencephalography(EEG) data visualization, processing and cortical source estimation.

7.6.3 Additional Information

MEG

http://megcommunity.org/

A site dedicated to MEG research. Among other things, the site references datasets, courses and software as well as some introductory material on MEG and its applications.

Wikipedia MEG article

http://en.wikipedia.org/wiki/Magnetoencephalography

A general introduction to MEG.

Wikipedia EEG article

http://en.wikipedia.org/wiki/Electroencephalography

A general introduction to EEG.

Acknowledgements The authors wish to thank Franck Vidal and Boris Burle for useful discussions. This article relates some work published with Alexandre Gramfort, Renaud Keriven and Bruno Torrésani. This work is partially funded by the French ANR project MultiModel.

List of Acronyms

CMP	Consensus Matching Pursuit
EEG	electro-encephalography
EMP	Evoked Matching Pursuit
ERP	Evoked Response Potential
fMRI	functional Magnetic Resonance Imaging
ICA	Independent Components Analysis
IMP	Induced Matching Pursuit
ISI	inter-stimulus interval
MEG	Magneto-encephalography
MRI	Magnetic Resonance Imaging

PCA Principal Component Analysis
P300 P300 wave, a positive deflection of the electroencephalogram occuring 300 ms after a stimulus
SNR Signal to Noise Ratio

References

1. S. Baillet, J.C. Mosher, and R.M. Leahy. Electromagnetic brain mapping. *IEEE Signal Processing Magazine*, 18(6):14–30, 2001.
2. M. Belkin and P. Niyogi. Laplacian Eigenmaps for Dimensionality Reduction and Data Representation. *Neural Computation*, 15:1373–1396, 2003.
3. C.G. Bénar, T. Papadopoulo, B. Torrésani, and M. Clerc. Consensus matching pursuit for multi-trial eeg signals. *Journal of Neuroscience Methods*, 180:161–170, 2009.
4. C.G. Bénar, D. Schön, S. Grimault, B. Nazarian, B. Burle, M. Roth, J.M. Badier, P. Marquis, C. Liegeois-Chauvel, and J.L. Anton. Single-trial analysis of oddball event-related potentials in simultaneous EEG-fMRI. *Human Brain Mapping*, 28:602–613, 2007.
5. B. Burle, C. Roger, S. Allain, F. Vidal, and T. Hasbroucq. Error negativity does not reflect conflict: a reappraisal of conflict monitoring and anterior cingulate cortex activity. *J. of Cogn. Neurosci.*, 20(9):1637–55, 2008.
6. P. Comon. Independent component analysis - a new concept? *Signal Processing*, 36, 1994.
7. J. de Munck, F. Bijma, P. Gaura, C. Sieluzycki, M. Branco, and R. Heethaar. A maximum-likelihood estimator for trial-to-trial variations in noisy MEG/EEG data sets. *IEEE Trans. Biomed. Eng.*, 51(12):2123–28, 2004.
8. S. Debener, M. Ullsperger, M. Siegel, K. Fiehler, D. von Cramon, and A. Engel. Trial-by-trial coupling of concurrent electroencephalogram and functional magnetic resonance imaging identifies the dynamics of performance monitoring. *Neuroscience*, 2005.
9. T. Eichele, K. Specht, M. Moosmann, M. Jongsma, R. Quian Quiroga, H. Nordby, and K. Hugdahl. Assessing the spatiotemporal evolution of neuronal activation with single-trial event-related potentials and functional MRI. *Proc Natl Acad Sci U.S.A.*, 2005.
10. A. Gramfort, R. Keriven, and M. Clerc. Graph-based variability estimation in single-trial event-related neural responses. *IEEE Trans. Biomed. Engin.*, 56(5):1051–1061, 2010.
11. R. Gribonval, H. Rauhut, K. Schnass, and P. Vandergheynst. Atoms of all channels, unite! Average case analysis of multi-channel sparse recovery using greedy algorithms. *The Journal of Fourier Analysis and Applications*, 14(5):655–687, 2008.
12. M. Hein, J. Audibert, and U. von Luxburg. Graph Laplacians and their convergence on random neighborhood graphs. *The Journal of Machine Learning Research*, 8:1325–1370, 2007.
13. A. Holm, P. Ranta-aho, M. Sallinen, P. Karjalainen, and K. Müller. Relationship of P300 single-trial responses with reaction time and preceding stimulus sequence. *Int. J. Psychophysiol.*, 2006.
14. T.P. Jung, S. Makeig, M. Westerfield, J. Townsend, E. Courchesne, and T.J. Sejnowski. Analysis and visualization of single-trial event-related potentials. *Human Brain Mapping*, 14:166–185, 2001.
15. M. Kutas, G. McCarthy, and E. Donchin. Augmenting mental chronometry: the P300 as a measure of stimulus evaluation time. *Science*, 197:792–795, August 1977.
16. D. Lehmann and W. Skrandies. Spatial analysis of evoked potentials in man - a review. *Progr Neurobiol*, 23(3):227–250, 1984.
17. S. Mallat and Z. Zhang. Matching pursuit with time-frequency dictionaries. *IEEE Trans. on Signal Processing*, 41(12):3397–3414, 1993.

18. C. Mulert, V. Kirsch, R. Pascual-Marqui, R.W. McCarley, and K.M. Spencer. Long-range synchrony of gamma oscillations and auditory hallucination symptoms in schizophrenia. *International Journal of Psychophysiology*, 79(1):55–63, January 2011. Special Issue: Correlations between gamma-band oscillations and human behaviour.
19. J. Polich. Clinical application of the p300 event-related brain potential. *Physical Medicine & Rehabilitation Clinics of North America*, 15(133), 2004.
20. R. Quian Quiroga and E. van Luijtelaar. Habituation and sensitization in rat auditory evoked potentials: A single-trial analysis with wavelet denoising. *International Journal of Psychophysiology*, 43(2):141–153, 2002.
21. C. Tallon-Baudry, O. Bertrand, C. Delpuech, and J. Pernier. Stimulus specificity of phase-locked and non-phase-locked 40 Hz visual responses in human. *J. Neurosci.*, 16(13):4240–4249, 1996.
22. F. Vidal, B. Burle, M. Bonnet, J. Grapperon, and T. Hasbroucq. Error negativity on correct trials: a reexamination of available data. *Biol. Psychol.*, 64(3):265–82, 2003.

Chapter 8
Spike Train Statistics from Empirical Facts to Theory: The Case of the Retina

Bruno Cessac and Adrian G. Palacios

8.1 Introduction

Given a stimulus from the external world (e.g., visual scene, sound or smell) biological sensors at the periphery of the nervous system are able to transduce the physical manifestations of this stimulus (light emission, air pressure variations, chemical concentrations) into sequences of action potentials (spike trains), which propagate through the nervous system. Then, the brain is able to *analyze* those spike trains and infer crucial information on the nature of the stimulus. Critical—yet unsolved—questions in neuroscience are How is the physical signal encoded by the nervous system? How does the brain analyze the spike trains? What are the underlying computational *coding* principles? At the current stage of scientific knowledge, answering those questions is still a challenge for biology and computational neuroscience.

Among sensory systems the retina provides functionality such as detection of movement, orientation, temporal and spatial prediction, response to flash omissions and contrast, that were up to recently viewed as the exclusive duty of higher brain centers [24]. The retina is an accessible part of the brain [15] and a prominent system to study the neurobiology and the underlying computational capacity of the neural coding. As a matter of fact, there is currently a wide research activity in understanding how the retina encodes visual information. However, basic questions are still open, such as: Are the ganglion cells (which send spikes from the eyes to the

B. Cessac (✉)
Inria Sophia Antipolis Méditerranée, Neuromathcomp project-team, 2004 Route des Lucioles, Sophia Antipolis 06902, France
e-mail: bruno.cessac@inria.fr

A.G. Palacios
CINV-Centro Interdisciplinario de Neurociencia de Valparaiso, Universidad de Valparaiso, Harrington 287, Valparaiso 2360102, Chile
e-mail: adrian.palacios@uv.cl

F. Cazals and P. Kornprobst (eds.), *Modeling in Computational Biology and Biomedicine*, DOI 10.1007/978-3-642-31208-3_8, © Springer-Verlag Berlin Heidelberg 2013

brain via the optical nerve), independent signal-encoders or are neural correlations important for coding a visual scene, and how to interpret them?

8.1.1 Chapter Overview

Public

This chapter addresses to readers having a master degree in Mathematics, Physics or Biology.

Outline

In this chapter, we present a state of the art about neural coding in the retina considered from the point of view of statistical physics and probability theory. As a consequence, this chapter contains both recent biological results and mathematical developments. The chapter is organized as follows. In Sect. 8.2 we introduce the current challenge of unraveling the neural code via spike trains statistics analysis. Such an analysis requires elaborated mathematical tools introduced in Sect. 8.3. We mainly focus on the so-called Gibbs distributions. This concept comes from statistical physics but our presentation departs from the classical physics courses since it is based on transition probabilities of Markov process. This way, as we show, allows to handle non-stationary dynamics, and is adapted to statistical analysis, of data as well as neural networks models. As an illustration, we present, in Sect. 8.4, two "success stories" where spike train statistics analysis has allowed to make a step further in our understanding of information encoding by the retina. In the same section, we also present an example of a rigorous spike train analysis in a neural network and compare the spike trains probability distribution to the models currently used on the experimental side.

8.2 Unraveling the Neural Code in the Retina via Spike Train Statistics Analysis

8.2.1 Retina Structure and Functions

8.2.1.1 Retina Structure

The vertebrate retina is a tightly packed neural tissue, exhibiting a rich diversity of neurons. It is structured in three cells nuclei layers and two plexiform synaptic

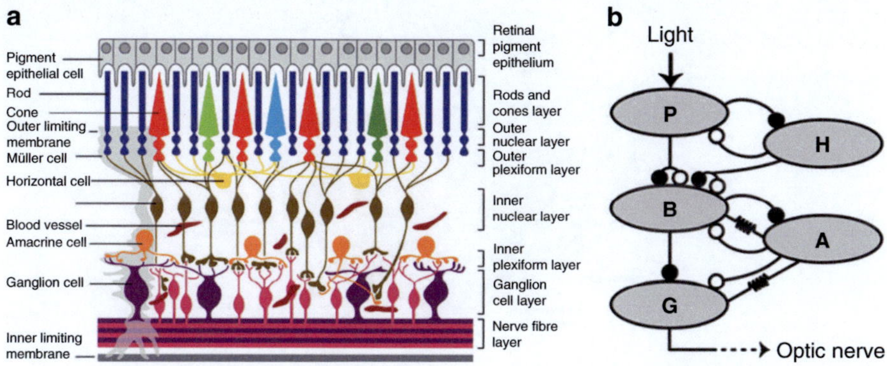

Fig. 8.1 Processing steps of the visual stream. (**a**) The cellular organization of the retina (From Expert Reviews in Molecular Medicine by Cambridge University Press 2004); (**b**) Main connectivity structure between retina cells types (Reproduced from [24]). Chemical synapses are represented by *circles*, either excitatory (*closed circles*) or inhibitory (*open circles*). Electric synapses (*gap junctions*) are marked by resistor symbols

layers [37, 77] (Fig. 8.1a). The outer nuclear layer (ONL) contains the rods and cones *photoreceptors* (P) somata; the inner nuclear layer (INL) contains *bipolar* (B), *horizontal* (H) and *amacrine* cells (A). Finally, the most internal nuclear layer is composed with ganglion cells (G) and displaced amacrine cells. The outer plexiform layer (OPL) corresponds to synaptic contacts between P, B and H cells. The inner plexiform layer (IPL) corresponds to synaptic contacts between B, A and G cells.

The retina is about 300–500 μm thick, depending on species, and has about 100–130 millions of photoreceptors, 10–12 millions of bipolar, horizontal and amacrines cells and 0.4–1.6 millions of G cells. Together with this high and compact number of cells there is a very large number of synapses present in dendrites and axons terminal, that has been roughly estimated to one billion of synapses [58]. The retina is also rich in terms of the variability of neurotransmitters, where rods, cones, and bipolar cells liberate glutamate, horizontal and amacrines cells can liberate gaba, glycine, serotonin, acetylcholine, dopamine among others. Together with the richness in chemical slow synapses circuits the retina has a variety of electrical ("gap-junctions"), fast synapses endowing the retina with specific functional circuits.

Single photons are converted by photoreceptors into a graded change in the resting potential, resulting in a neurotransmitter liberation (glutamate) into the synaptic region connecting photoreceptors with B and H cells. Those cells make synapses with G and A cells. Therefore, photons fluxes generate a chain of changes in the resting potential of B,H,A, and G cells with consequence the emission of action potentials ("spikes") by G cells. They are the principal neural encoders through the integration of neural signals. The retina output, formed by spike train sequence, is carried by different types of G cells through the optical nerve to the brain higher visual structures: e.g., lateral geniculate nucleus (LGN) or visual cortex layers (Fig. 8.2).

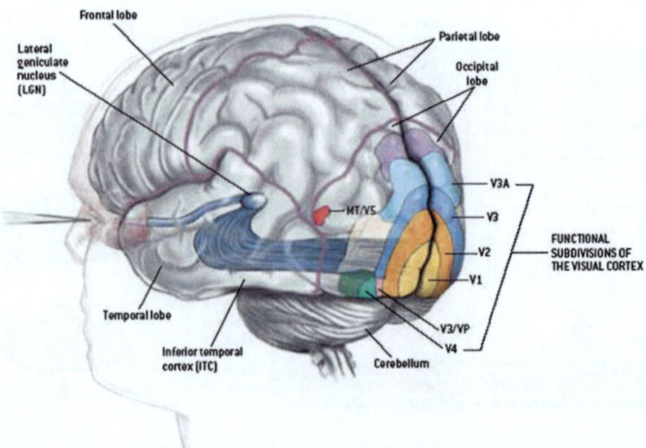

Fig. 8.2 Visual pathway in the human brain. The principal projection of the eye, which is formed by the optic nerve is carried to a first synapses in the lateral geniculate nucleus (*LGN*) in the thalamus and then for a second synapses to the main cortical visual area V1, from where many other projections target secondary cortical areas (*V2*, etc) (Reproduced from [32])

8.2.1.2 Retina Circuits and Receptive Fields

As a result of its stratified, horizontal and vertical structure, and of the various type of synaptic connections (electrical fast synapses $\sim$0.1 ms for short distance; chemical slow synapses $\sim$10 ms for long distances) between the different type of neurons (P,H,B,A,G) a large number of "circuits" are present in the retina. The main connectivity structure of the retina is shown in Fig. 8.1b. This circuitry results in the capacity of specific G cells to respond to specific stimuli in the visual field.

The *receptive field* (RF) of a sensory neuron is a region of space where the presence of a stimulus modifies the activity of that neuron. In the retina this change of activity is precisely the result of the transduction chain, from photoreceptor to G cells, converting photons into spike trains. As a consequence, one also defines the RF of a G cell as the input from all of the photoreceptors which synapse with it (via B, H, A) cells.

The RF of a cell can have different forms, depending on the network of neurons connected to this cell. A prominent example, is the *antagonist ON-OFF center-surround arrangement*. An ON-center cell is stimulated when the center of its receptive field is exposed to light, and is inhibited when the surround is exposed to light. OFF-center cells behaves exactly the opposite way.

The physiological properties of G cells are determined at the center and the surround of their RF by the input of ON or OFF B cell. First, photoreceptors make synapses with ON (excitatory) B cells and OFF (inhibitory) B cells according to their response to light. Figure 8.3a explains in a schematic way how this property results from the connectivity between P, B, and H cells. In the example, the illumination of the photoreceptors in the center of the RF results in a depolarization of ON B cells so in an increase of spikes rate in the respective connected G cells.

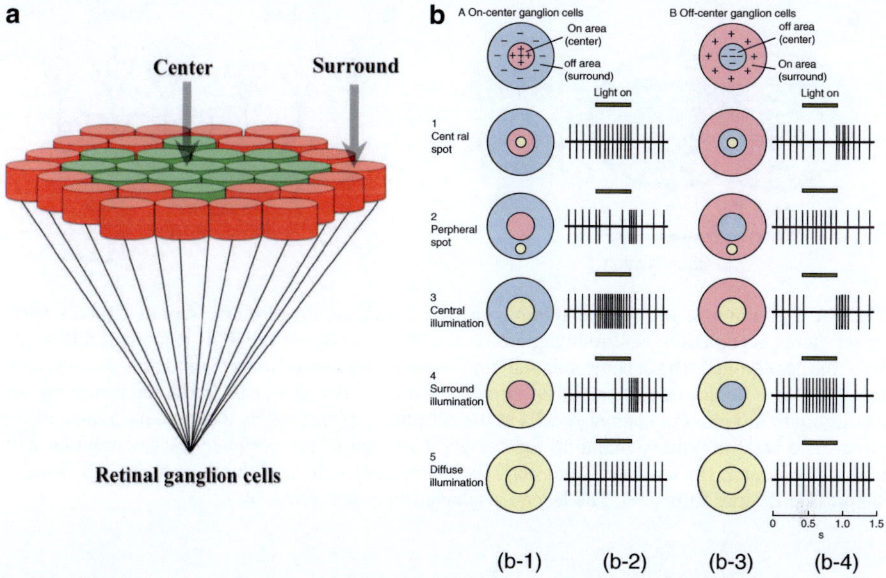

Fig. 8.3 Center-surround antagonism (**a**) Illumination of a piece of retina (Created by A.G. Palacios). (**b**) ON-center and OFF-center RF (Figure from [1]). The first line shows center-surround architecture of the cell while lines 2–6 shows a typical response of the G cells and the illumination pattern leading to that response. *(b-1)* Center-surround architecture of an ON-center cell and illumination pattern. *(b-2)* Time duration of the stimulus and spike response of the cell. Time is in abscissa. *(b-3)* and *(b-4)* Same as columns *(b-1)* and *(b-2)* for an OFF-center cell. Case 1 *left* (*right*) is a ON-center (OFF-center) G cell where a light spot (*yellow*) in the center of the RF generates an increase (decrease) of spike firing. In case 2 a spot stimulus in the surround generates a decrease (increase) of the spike rate. In case 3, 4 an increase in the size of the stimuli leads a sharper response. In case 5 a diffuse stimulus covering the center-periphery has no effect on the spike firing rate

On the opposite, the illumination of the photoreceptors at the periphery of the RF results in a hyper-polarization of OFF B cells so in a decrease of spikes rate in the respective connected G cells. In more general terms, as a consequence of this architecture (Fig. 8.3a), a G cell connected to that B cell fires spikes at the maximal rate when the center of the RF is illuminated and when the surround is dark (Fig. 8.3(b-2), case 3). On the opposite it fires no spike at all when the center of the RF is dark and the surround is illuminated (Fig. 8.3(b-4), case 4).

Figure 8.3b summarizes the different patterns of illuminations—G cell response in terms of spike firing and the functional implication of RF organization. For example, a full, uniform, illumination of the RF leads to a regular spiking activity with no difference between ON-OFF and OFF-ON cells (Fig. 8.3b, case 5).

As consequence of dynamical and complex interaction (spatial and temporal) opposite functions for, e.g., color, contrast, intensity are likewise found for a single G cell, depending where the stimulus is present in their RF.

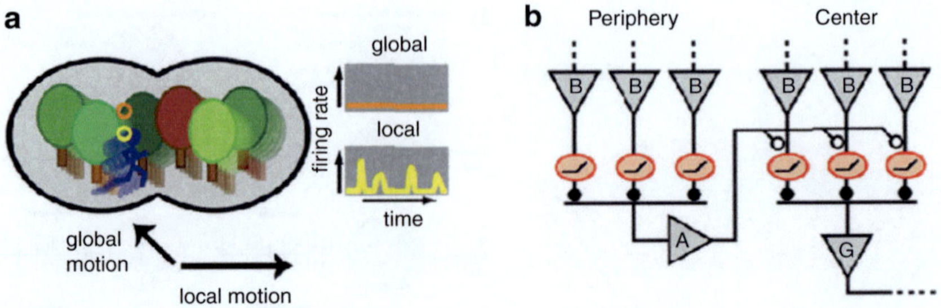

Fig. 8.4 Detection of differential motion. (**a**) An object-motion-sensitive G cell remains silent under global motion of the entire image but fires when the image patch in its RF moves differently from the background. (**b**) Scheme summarizing the circuitry behind this computation. Rectification (see [13] for a description of rectification mechanism.) of B cell signals in the RF center creates sensitivity to motion. Polyaxonal A cells in the periphery are excited by the same motion-sensitive circuit and send inhibitory inputs to the center. If motion in the periphery is synchronous with that in the center, the excitatory transients will coincide with the inhibitory ones, and firing is suppressed (Figure from [24]. The legend is adapted from this reference)

It has been long believed that retina was mainly acting as an image transducer, absorbing photons and producing electrical signals or acting as a temporal and spatial linear filter. It was also believed that the retina doesn't perform any pre-processing of the image before sending spike trains to the brain. More recently, researchers pointed out that retina, in some species, is "smarter" than previously believed and is able to detect salient features or properties in a image such as approaching motion, motion detection and discrimination, texture and object motion, creating predictive or anticipatory coding thanks to "specialized" G cells (see, e.g., [24] for a review). The specificity of these population of cells for the detection of differential motion results largely from the circuit they belong to. An example is shown in Fig. 8.4 (detection of differential motion) where A cells play a prominent role.

8.2.2 Multi-electrodes Array Acquisition

The pioneering work of Hubel and Wiesel based on anatomy and single cell recording on brain visual areas was very useful. However at that time, little was known about the properties of the retinal neural network. Similarly, today, the anatomical description of different types of G cells is a well known piece of literature, in contrast to their collective neural response that is partly missing. To overcome limitations of single-electrodes recording and to access to the coding response of a population of neurons, multi-electrodes (MEA) devices are used in physiology (for references on MEA see [69]). MEA devices are formed by an array of isolated electrodes (64–256, separated from 30 to 200 μm each, see Fig. 8.5).

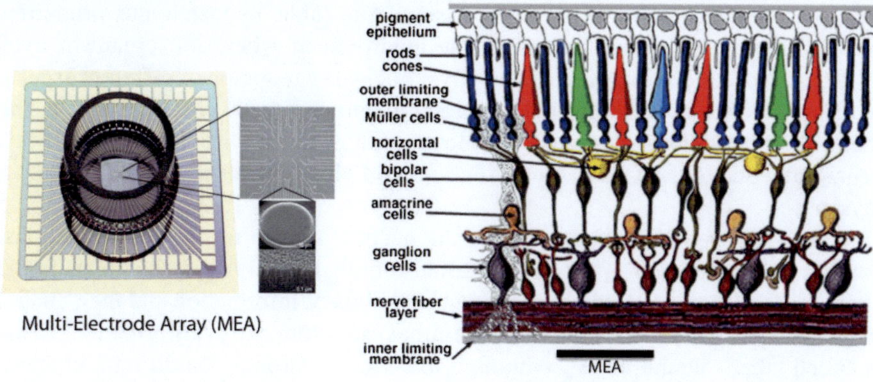

Fig. 8.5 Multi-electrode arrays. (*Left*) "Multichannelsystems MEA" from http://www.multichannelsystems.com; (*Right*) Schematic view of the implementantion of a MEA on the retina (not scaled)

When in contact with a small piece of neural tissue, a MEA is able to record the simultaneous activity (spike and/or field potential) from, e.g., 10–150 G cells. The final goal is to produce from the MEA signal a *raster plot* of G cells activity, namely a graph with time in abscissa and a neuron labeling in ordinate such that a vertical bar is drawn each "time" a neuron emits a spike. This poses an important challenge for signal processing: to sort out from a complex (spatial and temporal) neural signal superposition recording the contribution of each cell. With the recent increase in the number of electrodes of MEA devices, the necessity of adequate spike sorting algorithms turns out to be critical. Recently the Berry's lab at Princeton has developed an efficient method, enabling to sort out, from a 256 MEA experiment, about 200 different G cells (personal communication).

MEA devices constitute an excellent tool to track the physiological properties of G cells [45] as well as their coding capacity [29, 30]. Before the introduction of MEA devices, the neural coding properties of single G cells was studied using intra or extra cellular electrodes, giving a limited sense of their collective role. In that respect, the work of Meister et al. [40] using MEA devices was pioneer. With simple stimulus, like checkerboard random white noise, and spike sorting algorithms these authors were able to determinate the number of spiking cells and their respective RF. They have shown that concerted G cells are critical, not only for retina development, but for the neural coding processing.

8.2.3 Encoding a Visual Scene

The distribution and fluctuations of visual signals in the environment can be treated as a statistical problem [43, 67]. Natural scenes (a digital image or movie from a natural scenario) differ in their particular statistical structure and therefore the encoding capacity of a visual system should be able to match the properties and distribution of visual signals in the environment where the organism lives [3, 20, 21, 70, 72]. The anatomical and physiological segregation of different aspects of a visual scene in separate spatial, temporal and chromatic channels start at the retina and rely on local "circuits" [3]. However, how the precise articulation of this neural network contributes to local solutions and global perception is still largely a mystery.

G cells, as well as most neurons in the nervous system respond to excitations, coming from other neurons or from external stimuli, by emitting spike trains. In the contemporary view [48], spikes are quantum or bits of information and their spatial (neuron-dependent) and temporal (spike times) structure carry "information": This is called "the" "neural code". Although this view is strongly based on a contemporary analogy with computers, spike trains *are not* computer-like binary codes. Indeed, an experiment reproduced several times (e.g., an image presented several times to the retina) does not reproduce exactly the same spike train, although some regularity is observed. As a consequence, current attempts to deciphering the neural code are based on statistical models.

8.2.4 The Ganglion Cells Diversity

The recent use of MEA in retina has lead to the description of a diversity of G cells type and to the question about their actual early visual capacity. The vertebrate retina has in fact 15–22 anatomically different class of G cells making it a much more complex functional neural network than expected [24, 37, 49].

The three most frequent G cells in the retina can be classified from their morphology in: parasol (primates but α or Y in cats and rabbits) corresponding to 3–8 % of the total number of G cells; midget (β or X in cats and rabbits) corresponding to 45–70 %; and bi-stratified G cells. In physiological terms parasol (Y) cells can be classified as brisk-transient, and midget (X) as brisk-sustained. They can have an ON or OFF function.

Although only a reduced fraction of the existing G cells [37, 38, 62] has been studied in detail [24], their diversity raises questions such as: How do G cells encode an image? Which features from a natural visual scene are they coding? Are G cells independent or collective encoders?

An interesting approach has been advanced by Balasubramanian and Sterling [5]. The authors propose that the retina organization should use simple coding principles to carry the maximum of information at low energetic cost. However, as the authors point out, the statistic distribution (e.g., color, contrast) for natural images is *not* Gaussian. Therefore, the classical Gaussian estimator for Shannon information:

$$I = \frac{1}{2} \log_2(1 + SNR),$$

where SNR is the signal to noise ratio is not appropriate. Instead, "pixels" in natural images are highly correlated and the general form of statistical entropy (see Eq. (2) in [5]) is required to calculate the spike capacity of G cells to carry information. In that respect, the coding capacity for different G cells has been estimated (see, e.g., Eq. (5) in [5]). The larger capacity for information transmission comes from, e.g., "sluggish" G cells (32 %); local-edge (16 %), brisk-transient (9 %).

8.2.5 Population Code

This term refers to the computational capacity of a neural assembly (or circuit) to solve specific questions [4, 47]. Assuming that living systems have evolved to optimize the population code, how is this optimum reached in the retina? Are G cells sensors independent encoders or, on the opposite, are neural correlations important for coding? In an influential article, Nirenberg et al. [41] suggest that G cells act as independent encoder. However, orchestrated spikes train from G cells were reported by pioneer work of Rodieck [50] and Mastronarde [39]. Mastronarde shows that G cells responses tend to fire together and dynamically adapt to light or dark background [39]. This suggests that they act in a correlated way. However, this approach is by itself incomplete, since different sources of correlation were not clearly considered [44, 61]. On the other hand, MEA can now record many G cells from small pieces of retina ($<500\,\mu$m) [14, 17, 40] and help us to asses the importance, and origin, of neural synchrony for the neural coding. For example, in darkness, salamander G cells shows three types of synchrony depending on the time laps: (1) a common photoreceptor source through B cells (*broad correlation*: 40–100 ms) (2) A cells and G cells connected through gap junctions (*medium*: 10–50 ms) (3) gap junction between G cells ($narrow$:<1 ms) [6].

At present and although a large bunch of experimental facts enlighten our knowledge about the retina structure as well as its activity, basic questions on the way how a visual scene is encoded by spike trains remain still open. This is largely due to (1) the complex structure of the retina; (2) its large number of cells; (3) the lack of sufficiently accurate statistical models and methods to discriminate competing hypotheses. Apparently elementary questions such as determining whether correlations are significant from the analysis of MEA recordings requires in fact the use of smart statistical analysis techniques, based on "statistical models" defined by a set of a priori hypothesis as we see in the next section.

8.3 Spike Train Statistics from a Theoretical Perspective

In this section we develop the mathematical framework to analyze spike train statistics. The collective neuron dynamics, which is generally submitted to noise, produces spike trains with randomness though some statistical regularity can be observed. Spike trains statistics is assumed to be summarized by an hidden probability μ characterizing the probability of spiking patterns. One current goal in experimental analysis of spike trains is to approximate μ from data. We describe here several theoretical tools allowing to handle this question. Our presentation is based on the notion of transition probabilities. In this context we introduce Gibbs distributions, which is one of the main theoretical concept of this chapter. Gibbs distributions are usually considered in the stationary case where they are obtained from the maximal entropy principle. Their definition via transition probabilities, adopted in this chapter, affords the consideration of Gibbs distribution in the more general context of non-stationary dynamics with possibly infinite memory.

8.3.1 Spike Statistics

8.3.1.1 Raster Plots

We consider a network of N neurons. We assume that there is a minimal time scale $\delta > 0$ corresponding to the minimal resolution of the spike time, constrained by biophysics and by measurements methods (typically $\delta \sim 1$ ms) [8, 9]. Without loss of generality (change of time units) we set $\delta = 1$, so that spikes are recorded at integer times. One then associates to each neuron k and each integer time n a variable $\omega_k(n) = 1$ if neuron k fires at time n and $\omega_k(n) = 0$ otherwise. A *spiking pattern* is a vector $\omega(n) \stackrel{\text{def}}{=} [\,\omega_k(n)\,]_{k=1}^{N}$ which tells us which neurons are firing at time n. We note $\mathcal{A} = \{\,0,1\,\}^N$ the set of spiking patterns. A *spike block* is a finite ordered list of spiking patterns, written:

$$\omega_{n_1}^{n_2} = \{\,\omega(n)\,\}_{\{n_1 \le n \le n_2\}},$$

where spike times have been prescribed between the times n_1 to n_2 (i.e., $n_2 - n_1 + 1$ time steps). The *range* of a block is $n_2 - n_1 + 1$, the number of time steps from n_1 to n_2. The set of such blocks is $\mathcal{A}^{n_2 - n_1 + 1}$. Thus, there are 2^{Nn} possible blocks with N neurons and range n. For example, $N = 3$ neurons and $n = 2$ time steps the possible blocks are:

$$\begin{pmatrix} 0\,0 \\ 0\,0 \\ 0\,0 \end{pmatrix};\ \begin{pmatrix} 0\,1 \\ 0\,0 \\ 0\,0 \end{pmatrix};\ \begin{pmatrix} 1\,0 \\ 0\,0 \\ 0\,0 \end{pmatrix};\ \begin{pmatrix} 1\,1 \\ 0\,0 \\ 0\,0 \end{pmatrix};\ \ldots \begin{pmatrix} 1\,1 \\ 1\,1 \\ 1\,0 \end{pmatrix};\ \begin{pmatrix} 1\,1 \\ 1\,1 \\ 1\,1 \end{pmatrix}.$$

We call a *raster plot* a bi-infinite sequence $\omega \stackrel{def}{=} \{\omega(n)\}_{n=-\infty}^{+\infty}$, of spiking patterns. This notion corresponds to its biological counterpart (Sect. 8.2.2) with the obvious difference that experimental raster plots are finite. The consideration of infinite sequences is more convenient on the mathematical side but, at several places, we discuss the effects of having finite experimental rasters on spike statistics estimation. The set of raster plots is denoted $\mathcal{X} = \mathcal{A}^{\mathbb{Z}}$.

8.3.1.2 Transition Probabilities

The probability that a neuron emits a spike at some time n depends on the history of the neural network. However, it is impossible to know explicitly its form in the general case since it depends on the past evolution of all variables determining the neural network state. A possible simplification is to consider that this probability depends *only* on the spikes emitted in the past by the network. In this way, we are seeking a family of transition probabilities of the form $P\left[\omega(n) \,\middle|\, \omega_{n-D}^{n-1}\right]$, the probability that the firing pattern $\omega(n)$ occurs at time n, given a past spiking sequence ω_{n-D}^{n-1}. Here D is the *memory depth* of the probability, i.e., how far in the past does the transition probability depend on the past spike sequence. We use here the convention that $P\left[\omega(n) \,\middle|\, \omega_{n-D}^{n-1}\right] = P\left[\omega(n)\right]$ if $D = 0$ (memory-less case).

Transition probabilities depend on the neural network characteristics such as neurons conductances, synaptic responses or external currents. They give information on the dynamics that takes place in the observed neural networks. Especially, they have a *causal* structure where the probability of an event depends on the past. This reflects underlying biophysical mechanisms in the neural networks which are also causal. The explicit computation of transition probabilities can be done in some model-examples (Sect. 8.4.4). From them, one is able to characterize statistical properties of rasters generated by the network, as we now develop.

8.3.1.3 Markov Chains

Transition probabilities with a finite memory depth D define a "Markov chain", i.e., a random process where the probability to be in some state at time n (here a spiking pattern $\omega(n)$) depends only upon a finite past (here on $\omega(n-1), \ldots, \omega(n-D)$). Markov chains have the following property. Assume that we know the probability of occurrence of the block ω_m^{m+D-1},

$$P\left[\omega_m^{m+D-1}\right] = P\left[\omega(m+D-1), \omega(m+D-2), \ldots, \omega(m)\right]. \qquad (8.1)$$

Then, by definition, the probability of the block ω_m^{m+D} is:

$$\begin{aligned} P\left[\omega_m^{m+D}\right] &= P\left[\omega(m+D), \omega(m+D-1), \ldots, \omega(m)\right] \\ &= P\left[\omega(m+D) \,|\, \omega(m+D-1), \ldots, \omega(m)\right] \\ &\quad \times P\left[\omega(m+D-1), \ldots, \omega(m)\right]. \end{aligned}$$

Thus:

$$P\left[\omega_m^{m+D}\right] = P\left[\omega(m+D) \,\middle|\, \omega_m^{m+D-1}\right] P\left[\omega_m^{m+D-1}\right],$$

and, by induction, the probability of a block ω_m^n, $\forall n, n-m \geq D$ is given by:

$$P\left[\omega_m^n\right] = \prod_{l=m+D}^{n} P\left[\omega(l) \,|\, \omega_{l-D}^{l-1}\right] P\left[\omega_m^{m+D-1}\right]. \tag{8.2}$$

Thus, knowing the probability of occurrence of the block ω_m^{m+D-1} one can infer the probability of forthcoming blocks by the mere multiplication of transition probabilities.

Given the (joint) probability $P\left[\omega_m^n\right]$ the (marginal) probability of sub-blocks can be easily obtained, since for $m \leq n_1 \leq n_2 \leq n$,

$$P\left[\omega_{n_1}^{n_2}\right] = \sum_{m,n}^{*(n_1,n_2)} P\left[\omega_m^n\right], \tag{8.3}$$

where $\sum_{m,n}^{*(n_1,n_2)}$ means that we sum up over all possible spiking patterns in the interval $\{m, n\}$ excluding the interval $\{n_1, n_2\}$ (i.e., we sum up over all possible values of $\omega(m), \ldots, \omega(n_1 - 1), \omega(n_2 + 1), \ldots, \omega(n)$).

As a consequence, from (8.2) to (8.3), the probability of the spike block ω_{n-D}^n, of range D, is:

$$P\left[\omega_{n-D}^n\right] = \sum_{m,n}^{*(n-D,n)} \prod_{l=m+D}^{n} P\left[\omega(l) \,|\, \omega_{l-D}^{l-1}\right] P\left[\omega_m^{m+D-1}\right]. \tag{8.4}$$

Knowing the probability of an initial block of range D (here ω_m^{m+D-1}) one infers from this equation the probability of subsequent blocks of range D. Equation (8.4) can also be expressed in terms of vector-matrices multiplication, and the main properties of the Markov chain can be deduced from linear algebra and matrices spectra theorems [64]. For compactness we shall not use this possibility here though, (see [74] for further details).

However, this equation shows that the "future" of the Markov chain (the probability of occurrence of blocks) depends on an initial condition (here $P\left[\omega_m^{m+D-1}\right]$), which is a priori undetermined. Moreover, there are a priori infinitely many possible choices for the initial probability.

8.3.1.4 Asymptotic of the Markov Chain

Assume now that $n - m \to +\infty$ in Eq. (8.4) and more precisely that $m \to -\infty$. Practically, this limit corresponds to considering that the system began to exist in a distant past (defined by the initial condition of the Markov chain) and that it has evolved long enough, i.e., over a time larger than relaxation times in the system, so that it has reached sort of an "adult" age where its structure is essentially fixed. Note that this does not exclude adaptation processes, e.g., if the transition probabilities depend explicitly on time. Mathematically, the limit $n - m \to +\infty$ corresponds to studying the asymptotic of the Markov chain and related questions are: Is there a limit probability $P\left[\omega_{n-D}^{n}\right]$? Does it depend on the initial condition $P\left[\omega_{m}^{m+D-1}\right]$, $m \to -\infty$?

Let us first consider the easiest case where transition probabilities are invariant under time translation. This means that for each possible spiking pattern $\alpha \in \mathcal{A}$, for all possible "memory" blocks $\alpha_{-D}^{-1} \in \mathcal{A}^D$ and $\forall n$, $P\left[\omega(n) = \alpha \,\middle|\, \omega_{n-D}^{n-1} = \alpha_{-D}^{-1}\right] = P\left[\omega(0) = \alpha \,\middle|\, \omega_{-D}^{-1} = \alpha_{-D}^{-1}\right]$. We call this property *stationarity* referring rather to the physics literature than to the Markov chains literature (where this property is called *homogeneity*). If, additionally, all transition probabilities are strictly (positivity) then there is a unique probability μ, called the *asymptotic probability of the chain*, such that, whatever the initial choice of a probability $P\left[\omega_{m}^{m+D-1}\right]$ in (8.4) the probability of a block ω_{n-D}^{n} converges to $\mu\left[\omega_{n-D}^{n}\right]$ as m tends to $-\infty$. One says that the chain is *ergodic* (Note that positivity of all transition probabilities is a sufficient but not necessary condition for ergodicity [64]). In this sense, dynamics somewhat "selects" the probability μ, since, whatever the initial condition $P\left[\omega_{m}^{m+D-1}\right]$, it provides the statistics of spikes observed after a sufficiently long time. Additionally, μ has the following property: for any time $n_1, n_2, n_2 - n_1 \geq D$,

$$\mu\left[\omega_{n_1}^{n_2}\right] = \prod_{l=n_1+D}^{n_2} P\left[\omega(l) \,\middle|\, \omega_{l-D}^{l-1}\right] \mu\left[\omega_{n_1}^{n_1+D-1}\right]. \tag{8.5}$$

Let us return to the problem of choosing the initial probability in Eq. (8.4). If one wants to determine the evolution of the Markov chain after a initial observation time n_1 one has to fix the initial probability $P\left[\omega_{n_1}^{n_1+D-1}\right]$ and to use (8.4) (where m is replaced by n_1) and there is an indeterminacy in the choice of $P\left[\omega_{n_1}^{n_1+D-1}\right]$. This indeterminacy is released, though, if the system has started to exist in the infinite past. Then, $P\left[\omega_{n_1}^{n_1+D-1}\right]$ has to be replaced by $\mu\left[\omega_{n_1}^{n_1+D-1}\right]$ and Eq. (8.4) becomes:

$$\mu\left[\omega_{n_2-D}^{n_2}\right] = \sum_{n_1,n_2}^{*(n_2-D,n_2)} \prod_{l=n_1+D}^{n_2} P\left[\omega(l) \,\middle|\, \omega_{l-D}^{l-1}\right] \mu\left[\omega_{n_1}^{n_1+D-1}\right]. \tag{8.6}$$

In this way, taking the limit $m \to -\infty$ for an ergodic Markov chain, resolves the indeterminacy in the initial condition.

Positivity and stationary assumptions may not hold. If positivity is violated then several situations can arise: several asymptotic probability distributions,can exist, depending on the choice of the initial probability $P\left[\omega_{m}^{m+D-1}\right]$; it can also be that no asymptotic probability exist at all. If stationarity does not hold, as it is the case e.g., for a neural network with a time-dependent stimulus, then one can nevertheless define a probability μ selected by dynamics. In short, this is a probability μ on the set of raster plots $\mathcal{A}^{\mathbb{Z}}$ which still obeys (8.5) but without the conditions of stationarity (transition probabilities are not time-translation invariant [16]). In this case, which is realistic when dealing with living systems submitted, e.g., to time-dependent stimuli, the statistics of spikes is time-dependent. For example, the probability that a neuron emits a spike at time n depends on n, while it is not the case when dynamics is stationary.

8.3.1.5 Gibbs Distributions

Assume that $P\left[\omega(n) \mid \omega_{n-D}^{n-1}\right] > 0$ for all $n \in \mathbb{Z}$. Then, a probability distribution μ that obeys (8.5) is called a *Gibbs distribution*, and the function

$$\phi_n\left(\omega_{n-D}^{n}\right) \stackrel{def}{=} \log P\left[\omega(n) \mid \omega_{n-D}^{n-1}\right], \tag{8.7}$$

is called a (normalized) *Gibbs potential*. The advantage of this definition of Gibbs distribution is that it holds for time-dependent transition probabilities contrarily to the classical definition from the maximal entropy principle (Sect. 8.3.2.8). Moreover, in the case (8.7) the Gibbs potential depends explicitly on time (index n). This definition also extends to system with infinite memory (Sect. 8.4.4) although Eq. (8.5) has to be modified [16].

The Gibbs potential depends on the block ω_{n-D}^{n-1} and on the spiking pattern $\omega(n)$, thus, finally, this is a function of the block ω_{n-D}^{n} of range $D+1$. The term "normalized" refers to the fact that the potential in (8.7) is the logarithm of a transition probability. Below, we give example of Gibbs distributions where the potential is not normalized: this is an arbitrary function of the block ω_{n-D}^{n}. We call $R = D+1$ the *range* of the potential. A Gibbs potential can have an infinite range ($D \to +\infty$ in our setting).

The condition $P\left[\omega(n) \mid \omega_{n-D}^{n-1}\right] > 0$ (positivity) for all $n \in \mathbb{Z}$ ensures that there is a one-to-one correspondence between a Gibbs potential and a Gibbs distribution. If this condition is relaxed, i.e., some transitions are forbidden, then several Gibbs distribution can be associated with a Gibbs potential. This corresponds to a first-order phase transition in statistical physics [22]. In the infinite range case, the existence and uniqueness of a Gibbs distribution associated with this potential requires additional assumptions to the positivity of transition probabilities [16].

From (8.2), we have $\forall n - m \geq D$:

$$\mu\left[\omega_m^n \mid \omega_m^{m+D-1}\right] = \exp \sum_{l=m+D}^{n} \phi_l\left(l, \omega\right). \tag{8.8}$$

This form reminds the Gibbs distribution on spin lattices in statistical physics where one looks for lattice translation-invariant probability distributions given specific boundary conditions. Given a potential of range D the probability of a spin block depends on the states of spins in a neighborhood of size D of that block.

Thus, the conditional probability of this block given a fixed neighborhood is the exponential of the energy characterizing physical interactions within the block as well as with the boundaries. Here, spins are replaced by spiking patterns; space is replaced with time which is mono-dimensional and oriented: there is no dependence in the future. Boundary conditions are replaced by the dependence in the past.

8.3.2 Determining the "Best" Markov Chain to Describe an Experimental Raster

We now show how the formalism of the previous section can be used to analyze spike trains statistics in experimental rasters.

8.3.2.1 Observables

We call *observable* a function which associates to a raster plot a real number. Typical choices of observables are $\omega_{k_1}(n_1)$ which is 1 if neuron k_1 fires at time n_1 and which is 0 otherwise; $\omega_{k_1}(n_1)\,\omega_{k_2}(n_2)$ which is 1 if neuron k_1 fires at time n_1 and neuron k_2 fires at time n_2 and which is 0 otherwise, and so on. Another example is $\omega_{k_1}(n_1)\,(1-\omega_{k_2}(n_2))$ which is 1 is neuron k_1 fires at time n_1 *and neuron* k_2 *is silent at time* n_2. This example stresses that observables can consider as well events where some neurons are silent. One can also consider more general forms of observables, e.g., non linear functions of spike blocks (see for example Eqs. (8.38) and (8.42) below).

It is a general result from Hammersley and Clifford [25] that any function of spike blocks can be uniquely decomposed as a linear combination of what we call *monomials* in this chapter, namely a function of the form $\omega_{k_1}(n_1)\,\omega_{k_2}(n_2)\ldots\,\omega_{k_m}(n_m)$ which is equal to 1 if and only if neuron k_1 fires at time n_1, ..., neuron k_m fires at time n_m in the raster ω. Thus monomials attribute the value '1' to characteristic spike events.

8.3.2.2 Probabilities and Averages

Let μ be a probability on the set of rasters (typically the Gibbs distribution introduced above). Mathematically, the knowledge of μ is equivalent to knowing the probability $\mu\left[\,\omega_m^n\,\right]$ of any possible spike block. For an observable f we denote $\mu\left[\,f\,\right] \stackrel{\text{def}}{=} \int f d\mu$ the average of f with respect to μ. If f is only a function of finite blocks ω_m^n then:

$$\mu\left[\,f\,\right] = \sum_{\omega_m^n} f(\omega_m^n)\,\mu\left[\,\omega_m^n\,\right], \tag{8.9}$$

where the sum holds on all possible (2^{n-m+1}) values of ω_m^n. For example the average value of $f(\omega) = \omega_k(n)$ is given by $\mu\,[\,\omega_k(n)\,] = \sum_{\omega_k(n)} \omega_k(n)\,\mu\,[\,\omega_k(n)\,]$ where the sum holds on all possible values of $\omega_k(n)$ (0 or 1). Thus, finally

$$\mu\,[\,\omega_k(n)\,] = \mu\,[\,\omega_k(n) = 1\,], \tag{8.10}$$

which is the probability of firing of neuron k at time n. This quantity is called the *instantaneous firing rate*. Likewise, the average value of $\omega_{k_1}(n)\,\omega_{k_2}(n)$ is the probability that neuron k_1 and k_2 fire at the same time n: this is a measure of *pairwise synchronization* at time n.

8.3.2.3 Empirical Averages

In experiments, raster plots have a finite duration T and one has only access to a finite number $\mathcal{N}$ of those rasters, denoted $\omega^{(1)}, \ldots, \omega^{(\mathcal{N})}$. From these data one computes empirical averages of observables. Depending on the hypotheses made on the underlying system there are several ways of computing those averages.

A classical (though questionable assumption as far as experiments are concerned) is *stationarity*: The statistics of spike is time-translation invariant. In this case the *empirical average* reduces to a *time average*. We denote $\pi_\omega^{(T)}\,[\,f\,]$ the time average of the function f computed for the raster ω of T. For example, when $f(\omega) = \omega_k(n)$, $\pi_\omega^{(T)}\,[\,f\,] = \frac{1}{T}\sum_{n=0}^{T-1} \omega_k(n)$, which provides an estimation of the firing rate of neuron k (it is independent of time from the stationarity assumption). If f is a monomial $\omega_{k_1}(n_1)\ldots\omega_{k_m}(n_m)$, $1 \le n_1 \le n_2 \le n_m < T$, then $\pi_\omega^{(T)}\,[\,f\,] = \frac{1}{T-n_m}\sum_{n=0}^{T-n_m} \omega_{k_1}(n_1+n)\ldots\omega_{k_m}(n_m+n)$, and so on. Why using the cumbersome notation $\pi_\omega^{(T)}\,[\,f\,]$? This is to remind the reader that such empirical averages are *random variables*. They fluctuate from one raster to another i.e., $\pi_{\omega^{(1)}}^{(T)}\,[\,f\,] \neq \pi_{\omega^{(2)}}^{(T)}\,[\,f\,]$ for distinct rasters $\omega^{(1)}, \omega^{(2)}$. Moreover, those fluctuations depend on T.

Assume now that all empirical rasters have all been generated by an hidden Markov chain and additionally that this chain is ergodic with a Gibbs distribution μ. Then, all those rasters obey $\pi_{\omega^{(r)}}^{(T)}\,[\,f\,] \to \mu\,[\,f\,]$, $r = 1, \ldots, \mathcal{N}$, as $T \to +\infty$, whatever f: the time average converges (almost-surely) to the average with respect to the hidden probability μ (this is one of the definitions of ergodicity). As a consequence the fluctuations of the time-average about the exact mean $\mu\,[\,f\,]$ tends to 0, typically like $\frac{K_f}{\sqrt{T}}$, where K_f is a constant depending on f. This is the celebrated central limit theorem stating moreover that fluctuations about the mean are Gaussian [23]. We come back to this point in Sect. 8.3.2.4.

The remarkable consequence of ergodicity (which implies stationarity) is that the empirical average can be estimated from one raster only. Now, if we have $\mathcal{N}$ rasters available we can use them to enlarge artificially the sample size, e.g., computing empirical average by $\frac{1}{\mathcal{N}} \sum_{r=1}^{\mathcal{N}} \pi_{\omega^{(r)}}^{(T)} [\, f \,]$. This also allows the computation of error bars as well as more elaborated statistical estimation techniques [48].

What if the stationarity assumption is violated? Then, the average of f depends on time and one computes the empirical average from the $\mathcal{N}$ rasters. We denote $\pi^{(\mathcal{N})} [\, f(n) \,]$ the average of f at time n, performed over $\mathcal{N}$ rasters. For example when $f(\omega) = \omega_k(n)$, $\pi^{(\mathcal{N})} [\, f(n) \,] = \frac{1}{\mathcal{N}} \sum_{r=1}^{\mathcal{N}} \omega_k^{(r)}(n)$ is the sample-averaged probability that neuron k fires at time n . If all rasters are described by the same probability (the Gibbs distribution which is also defined in the non-stationary case), then $\pi^{(\mathcal{N})} [\, f(n) \,] \to \mu [\, f(n) \,]$ as $\mathcal{N} \to +\infty$.

8.3.2.4 Example of Empirical Average: Estimating Instantaneous Pairwise Correlations

Assume that spikes are distributed according to an hidden probability μ supposed to be stationary for simplicity. The instantaneous pairwise correlations of neurons k, j with respect to μ is:

$$C\,(\,k, j\,) = \mu\,[\,\omega_k(0)\,\omega_j(0)\,] - \mu\,[\,\omega_k(0)\,]\,\mu\,[\,\omega_j(0)\,]\,. \tag{8.11}$$

Since μ is stationary the index 0 can be replaced by any time index (time-translation invariance of statistics).

Assume now that we have a raster ω distributed according to μ. An estimator of $C\,(\,k, j\,)$ is:

$$C_\omega^{(T)}\,(\,k, j\,) = \pi_\omega^{(T)}\,[\,\omega_k(0)\omega_j(0)\,] - \pi_\omega^{(T)}\,[\,\omega_k(0)\,]\,\pi_\omega^{(T)}\,[\,\omega_j(0)\,]\,. \tag{8.12}$$

It converges to $C\,(\,k, j\,)$ as $T \to +\infty$.

The events "neuron k fires at time 0" ($\omega_k(n) = 1$) and "neuron j fires at time 0" ($\omega_j(n) = 1$) are *independent* if $\mu\,[\,\omega_k(0)\,\omega_j(0)\,] = \mu\,[\,\omega_k(0)\,]\,\mu\,[\,\omega_j(0)\,]$, thus $C\,(\,k, j\,) = 0$. (Note that independence implies vanishing correlation but the reverse is not true in general. Here the two properties are equivalent thanks to the binary $0, 1$ form of the random variables $\omega_k(0), \omega_j(0)$).

Assume now that the observed raster has been drawn from a probability where these events are independent, but the experimentalist who analyzes this raster does not know it. To check independence he computes $C_\omega^{(T)}\,(\,k, j\,)$ from the experimental raster ω. However, since T is finite, $C_\omega^{(T)}\,(\,k, j\,)$ will not be exactly 0. More precisely, from the central limit theorem the following holds. The probability that the random variable $\left| C_\omega^{(T)}\,(\,k, j\,) \right|$ is larger than ϵ, is well approximated (for large T and small ϵ) by $e^{-\frac{\epsilon^2}{2}\frac{T}{K}}$. K can be exactly computed (Sect. 8.3.1.5). In the simplest case where spikes are drawn independently with a probability p of having a spike, K is equal to $p^2(1 - p^2)$. Thus, fluctuations are Gaussian and their mean-square deviation decay with T as $\sqrt{\frac{K}{T}}$. As a consequence, even if neuron j and k are

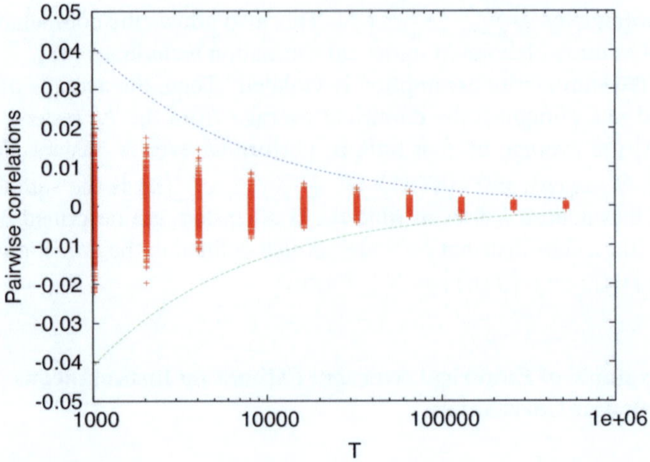

Fig. 8.6 Correlation (8.12) as a function of sample length T in a model where spikes are independent. For each T we have generated $1,000$ rasters of length T, with two independent neurons, drawn with a firing rate $p = \frac{1}{2}$. For each raster we have computed the pairwise correlation (8.12) and plotted it in log-scale for the abscissa (*red point*). In this way we have a view of the fluctuations of the empirical pairwise correlation about its (zero) expectation. The full lines represent respectively the curves $3\sqrt{\frac{p^2\,(1-p^2)}{T}}$ (*blue*) and $-3\sqrt{\frac{p^2\,(1-p^2)}{T}}$ (*green*) accounting for the Gaussian fluctuations of $C_\omega^{(T)}\,(k,j)$: 99 % of the $C_\omega^{(T)}\,(k,j)$'s values lie between these two curves (called "confidence bounds")

independent, the quantity $C_\omega^{(T)}\,(k,j)$ will in general not be 0: it has *fluctuations* around 0.

This can be seen by a short computer program drawing at random 0's and 1's independently, with the probability p to have a '1', and plotting $C_\omega^{(T)}\,(k,j)$ for different values of ω, while increasing T (Fig. 8.6).

As a consequence, it is stricto-sensu not possible to determine whether random variables are uncorrelated, by only computing the empirical correlation from samples of size T, since even if these variables are uncorrelated, the empirical correlation will never be zero. There exist statistical tests of independence from empirical data, beyond the scope of this chapter. A simple test consists of plotting the empirical correlation versus T and check whether it tends to zero as $\sqrt{\frac{K}{T}}$. Now, experiments affords only sample of limited size, where T rarely exceeds 10^6. So, fluctuations are of order $\sqrt{K} \times 10^{-3}$ and it makes a difference whether K is small or large.

It is therefore difficult to interpret *weak* empirical correlations. Are they sample fluctuations of a system where neurons are indeed independent, or are they really significant, although weak? This issue is further addressed in Sect. 8.4.2.

8.3.2.5 Matching Experimental Averages

Assume that an experimentalist observes $\mathcal{N}$ rasters, and assume that all those rasters are distributed according to an hidden probability distribution μ. Is to possible to determine or, at least, to approach μ from those rasters? One possibility relies on the maximal entropy principle described in the next sections. We assume for the moment that statistics is stationary.

Fix $\mathcal{K}$ observables $\mathcal{O}_k$, $k = 1, \ldots, \mathcal{K}$, and compute their empirical average $\pi_\omega^{(T)}[\mathcal{O}_k]$. The remarks of the previous sections hold: since all rasters are distributed according to μ, $\pi_\omega^{(T)}[\mathcal{O}_k]$ is a random variable with mean $\mu[\mathcal{O}_k]$ and Gaussian[1] fluctuations about its mean, of order $\frac{1}{\sqrt{T}}$. If there are $\mathcal{N} > 1$ rasters the experimentalist can estimate the order of magnitude of those fluctuations and also analyze the raster-length dependence. *In fine*, he obtains an empirical average value for each observable, $\pi_\omega^{(T)}[\mathcal{O}_k] = C_k, k = 1, \ldots, \mathcal{K}$. Now, to estimate the hidden probability μ, by some approximated probability μ_{ap}, he has to assume, as a minimal requirement, that:

$$\pi_\omega^{(T)}[\mathcal{O}_k] = C_k = \mu_{ap}[\mathcal{O}_k], \quad k = 1, \ldots, \mathcal{K}, \tag{8.13}$$

i.e., the expected average of each observable, computed with respect to μ_{ap} is equal to the average found in the experiment. This fixes a set of *constraints* to approach μ. We call μ_{ap} a *statistical model*.

Unfortunately, this set of conditions does not fix a unique solution but infinitely many! As an example if we have only one neuron whose firing rate is $\frac{1}{2}$, then a straightforward choice for μ_{ap} is the probability where successive spikes are independent ($P[\omega_k(n)\omega_k(n-1)] = P[\omega_k(n)]\,P[\omega_k(n-1)]$) and where the probability of a spike is $\frac{1}{2}$. However, one can also take a one-step memory model where transition probabilities obey $P[\omega_k(n) = 0 \mid \omega_k(n-1) = 0] = P[\omega_k(n) = 1 \mid \omega_k(n-1) = 1] = p$, $P[\omega_k(n) = 0 \mid \omega_k(n-1) = 1] = P[\omega_k(n) = 1 \mid \omega_k(n-1) = 0] = 1-p, p \in [0,1]$. In this case, indeed the invariant probability of the corresponding Markov chain is $\mu_{ap}[\omega_k(n) = 0, 1] = \frac{1}{2}$, since from Eq. (8.5),

$$\mu_{ap}[\omega_k(n) = 0] = \sum_{\omega_k(n-1)=0,1} P[\omega_k(n) = 0 \mid \omega_k(n-1)]\, \mu_{ap}[\omega_k(n-1)],$$

$$\left(\frac{p}{2} + \frac{1-p}{2}\right) = \frac{1}{2}.$$

The same holds for $\mu_{ap}[\omega_k(n) = 1]$. In this case, we match the constraint too but with a model where successive spikes are *not* independent. Now, since p takes values

[1]Fluctuations are not necessarily Gaussian, if the system undergoes a second order phase transition where the topological pressure introduced in Sect. 8.3.1.5 is not twice differentiable.

in the interval $[0, 1]$ there are uncountably many Markov chains with memory depth 1 matching the constraint. One could also likewise consider memory depth $D = 2, 3$ and so on.

Since transition probabilities reflect assumptions on the underlying (causal) mechanisms taking place in the observed neural network, the choice of the statistical model defined by those transition probabilities is not anecdotal. In the example above, that can be easily generalized, one model considers that spikes are emitted like a coin tossing, without memory, while other models involve a causal structure with a memory of the past. Even worse, there are infinitely many choices for μ_{ap} since (1) the memory depth can be arbitrary; (2) for a given memory depth there are (infinitely) many Markov chains whose Gibbs distribution matches the constraints (8.13). Is there a way to selecting, *in fine*, only one model from constraints (8.13), by adding some additional requirement? The answer is "yes".

8.3.2.6 Entropy

The entropy rate or Kolmogorov-Sinai entropy of a *stationary* probability distribution μ is:

$$h\left[\mu\right] = -\lim_{n\to\infty}\frac{1}{n}\sum_{\omega_1^n}\mu\left[\omega_1^n\right]\log\mu\left[\omega_1^n\right], \tag{8.14}$$

where the sum holds over all possible blocks ω_1^n. This definition holds for systems with finite or infinite memory. In the case of a Markov chain with memory depth $D > 0$, we have [12]

$$h\left[\mu\right] = -\sum_{\omega_1^{D+1}}\mu\left[\omega_1^D\right]P\left[\omega(D+1)\,|\,\omega_1^D\right]\log P\left[\omega(D+1)\,|\,\omega_1^D\right], \tag{8.15}$$

Note that, from time-translation invariance the block ω_1^D can be replaced by ω_n^{D+n-1}, for any integer n.

When $D = 0$, the entropy reduces to the classical definition:

$$h\left[\mu\right] = -\sum_{\omega(0)}\mu\left[\omega(0)\right]\log\mu\left[\omega(0)\right]. \tag{8.16}$$

8.3.2.7 Gibbs Distributions in the Stationary Case

In the stationary case Gibbs distributions obey the following variational principle [10, 28, 57]:

$$\mathcal{P}(\phi) = \sup_{\nu\in\mathcal{M}_{inv}}\left(h\left[\nu\right] + \nu\left[\phi\right)\right]\right) = h\left[\mu\right] + \mu\left[\phi\right], \tag{8.17}$$

where $\mathcal{M}_{inv}$ is the set of all possible stationary probabilities ν on the set of rasters with N neurons; $h[\nu]$ is the entropy of ν and $\nu[\phi]$ is the average value of ϕ with respect to the probability ν. Looking at the second equality, the variational principle (8.17) selects, among all possible probability ν, *one* probability which realizes the supremum, the Gibbs distribution μ.

The quantity $\mathcal{P}(\phi)$ is called the *topological pressure*. For a normalized potential it is equal to 0. However, the variational principle (8.17) holds for non-normalized potentials as well i.e., functions which are not the logarithm of a probability [10,28,57].

In particular, consider a function of the form:

$$\mathcal{H}_{\beta}(\omega_{-D}^{0}) = \sum_{k=1}^{\mathcal{K}} \beta_k \mathcal{O}_k(\omega), \tag{8.18}$$

where $\mathcal{O}_k$ are observables, β_k real numbers and β denotes the vector of β_k's, $k = 1, \ldots, \mathcal{K}$. We assume that each observable depends on spikes in a time interval $\{-D, \ldots, 0\}$.

To the non-normalized potential $\mathcal{H}_{\beta}(\omega_{-D}^{0})$ one can associate a normalized potential ϕ of the form:

$$\phi(\omega_{-D}^{0}) = \mathcal{H}_{\beta}(\omega_{-D}^{0}) - \log \zeta_{\beta}(\omega_{-D}^{0}), \tag{8.19}$$

where $\zeta(\omega_{-D}^{0})$ is a function that can explicitly computed. In short, one can associate to the potential $\mathcal{H}_{\beta}(\omega_{-D}^{0})$ a matrix with positive coefficient; $\zeta_{\beta}(\omega_{-D}^{0})$ is is a function of the (real positive) largest eigenvalue of this matrix as well as of the corresponding right eigenvector (see [75] for details). This function depends on the model-parameters β. The topological pressure is the logarithm of the largest eigenvalue.

In this way, $\mathcal{H}_{\beta}$ defines a stationary Markov chain with memory depth D, with transition probabilities:

$$P\left[\omega(0) \,\middle|\, \omega_{-D}^{-1}\right] = \frac{e^{\mathcal{H}_{\beta}(\omega_{-D}^{0})}}{\zeta_{\beta}(\omega_{-D}^{0})}. \tag{8.20}$$

Denote μ_{β} the Gibbs distribution of this Markov chain. The topological pressure $\mathcal{P}(\phi_{\beta})$ obeys:

$$\frac{\partial \mathcal{P}(\phi_{\beta})}{\partial \beta_k} = \mu_{\beta}[\mathcal{O}_k], \tag{8.21}$$

while its second derivative controls the covariance of the Gaussian matrix characterizing the fluctuations of empirical averages of observables about their mean. Note that those fluctuations are Gaussian if the second derivative of $\mathcal{P}$ is defined. This holds if all transitions probabilities are positive.

In the memory-less case $D = 0$ where only the statistics of instantaneous spiking patterns is considered, the Gibbs distribution reads:

$$\mu_\beta(\omega(0)) = \frac{e^{\mathcal{H}_\beta(\beta,\omega(0))}}{\sum_{\omega(0)} e^{\mathcal{H}_\beta(\beta,\omega(0))}}. \tag{8.22}$$

In this case,

$$\zeta_\beta = \sum_{\omega(0)} e^{\mathcal{H}_\beta(\beta,\omega(0))}. \tag{8.23}$$

This is a constant (it does not depend on the raster). It is called *partition function* in statistical physics.

8.3.2.8 The Maximal Entropy Principle

Assume now that we want to approximate the exact (unknown) probability μ by an approximated probability μ_{ap} that matches the constraints (8.13). The idea is to take as a statistical model μ_{ap} the Gibbs distribution of a function of the form (8.18), corresponding to a set of constraints attached to observables $\mathcal{O}_k$, where the β_k's are free parameters of the model. Thus, the statistical model is fixed by the set of observables and by the value of β. We write then, from now on, μ_β instead of μ_{ap}.

Looking at the variational principle (8.17), we have to take the supremum over all probabilities ν that matches (8.13), i.e., $\mu_\beta\left[\,\mathcal{O}_k\,\right] = C_k$ so that $\mu_\beta\left[\,\mathcal{H}\,\right]_\beta$ is a constant for fixed β. Therefore, in this case (8.13) reduces to *maximizing the entropy rate given the constraints* (8.13). This the classical way of introducing Gibbs distributions in physics courses. Then, the β_k's appear as Lagrange multipliers, that have to be tuned to match (8.13). This can been done thanks to (8.21). Note that the topological pressure is convex so that the solution of (8.21) is unique.

The important point is that procedure provides a *unique* statistical model defined by the transition probabilities (8.20). Thus, we have solved the degeneracy problem of Sect. 8.3.2.5 in the stationary case.

8.3.2.9 Range-1 Potentials

Let us now present a few examples used in the context of spike train analysis of MEA data, among others.

The easiest examples are potentials with a zero memory depth, in the stationary case, where therefore the spiking pattern $\omega(0)$ is independent of $\omega(-1)$. This corresponds to *range*-1 *potentials*.

Among them, the simplest potential has the form:

$$\phi_\beta(\omega(0)) = \sum_{k=1}^{N} \beta_k\, \omega_k(0) \;-\; \log\left(\,\zeta_\beta\,\right). \tag{8.24}$$

It corresponds to impose constraints only on firing rates of neurons. We have $\zeta_\beta = \prod_{k=1}^{N}(1+e^{\beta_k})$ and the corresponding Gibbs distribution is easy to compute:

$$\mu\left[\omega_m^n\right] = \prod_{l=m}^{n}\prod_{k=1}^{N}\frac{e^{\beta_k\,\omega_k(l)}}{1+e^{\beta_k}}. \tag{8.25}$$

Thus, the corresponding statistical model is such that spikes are independent. We call it a *Bernoulli model*. The parameter β_k is directly related to the firing rate r_k since $r_k = \mu\left(\omega_k(0)=1\right) = \frac{e^{\beta_k}}{1+e^{\beta_k}}$, so that we may rewrite (8.25) as:

$$\mu\left[\omega_m^n\right] = \prod_{l=m}^{n}\prod_{k=1}^{N} r_k^{\omega_k(l)}\left(1-r_k\right)^{1-\omega_k(l)},$$

the classical probability of coin tossing with independent probabilities.

Another prominent example of range-1 potential is inspired from statistical physics of magnetic systems and has been used by Schneidman and collaborators in [60] for the analysis of retina data (Sect. 8.4). It is called *Ising potential* and reads, with our notations:

$$\phi(\omega(0)) = \sum_{k=1}^{N}\beta_k\,\omega_k(0) + \sum_{1\le j<k\le N}\beta_{kj}\,\omega_k(0)\omega_j(0) - \log\zeta_\beta. \tag{8.26}$$

The corresponding Gibbs distribution provides a statistical model where synchronous pairwise synchronizations $\omega_k(0)\omega_j(0)$ between neurons are taken into account, but neither higher order spatial correlations nor other time correlations are considered. The function ζ_β is the classical partition function (8.23).

The Ising model is well known in statistical physics and the analysis of spike statistics with this type of potential benefits from a diversity of methods leading to really efficient algorithms to obtain the parameters β from data [11, 51, 55, 71].

8.3.2.10 Markovian Potentials

Let us now consider potentials of the form (8.7) allowing to consider spatial dependence as well as time dependence upon a past of depth D.

Consider first a stationary Markov chain with memory depth 1. The potential has the form:

$$\phi(\omega_{-1}^{0}) = \sum_{k=0}^{N-1}\beta_k\,\omega_k(0) + \sum_{k=0}^{N-1}\sum_{j=0}^{k-1}\sum_{\tau=-1}^{0}\beta_{kj\tau}\,\omega_k(0)\,\omega_j(\tau) - \log\zeta_\beta(\omega_{-1}^{0}). \tag{8.27}$$

This case has been investigated in [34] for spike train analysis of spike trains in the parietal cat cortex, assuming stationarity.

More general form of time-translation invariant potentials of the form:

$$\phi_n\left(\omega_{-D}^{0}\right) = \sum_{l=-D}^{0} \sum_{\mathcal{P}(N,D),} \beta_{i_1,n_1,\ldots,i_l,n_l}\, \omega_{i_1}(n_1)\ldots\omega_{i_l}(n_l), \tag{8.28}$$

where the sum $\sum_{\mathcal{P}(N,D)}$ holds on the set of non repeated pairs of integers (i,n) with $i \in \{1,\ldots,N\}$ and $n \in \{-D,\ldots,0\}$, have been considered in [75]

All examples treated above concerns stationary situations. In the non-stationary case, the entropy rate is not defined and the Gibbs distributions cannot be defined via the maximal entropy principle, while it is still possible to define them as done in Sect. 8.3.1.5. Now, the most general form for non-stationary Markov chain with memory depth D corresponds to potentials of the form:

$$\phi_n\left(\omega_{n-D}^{n}\right) = \sum_{l=-D}^{0} \sum_{\mathcal{P}(N,D),} \beta_{i_1,n_1,\ldots,i_l,n_l}(n)\, \omega_{i_1}(n+n_1)\ldots\omega_{i_l}(n+n_l), \tag{8.29}$$

In the non-stationary case the coefficients $\beta_{i_1,n_1,\ldots,i_l,n_l}(n)$ depend explicitly on n. They can be chosen so that (8.29) is normalized.

8.3.2.11 Non-Markovian Potentials

One can conceptually extend the definition of Markovian potentials to the case when $D \to -\infty$. This corresponds to a process with an infinite memory called "chain with complete connections" which is widely studied in the mathematical and mathematical physics literature (see [33] for a review). In this case the potential is a "function" $\phi_n\left(\omega_{-\infty}^{n}\right)$ depending on a infinite past $\omega_{-\infty}^{n-1}$. Although this case seems rather abstract, it turns out that the only known examples where spike train statistics can be exactly characterized in neural networks models are of this form. An example is given below. Moreover, this potential form allows to recover all the examples above.

8.3.2.12 How Good Is the Estimation?

Once we have chosen a set of constraints and once we have found the parameters β to match (8.13), how can we check the goodness of the model? Additionally, changing the set of constraints provides another model. How can we compare two statistical models?

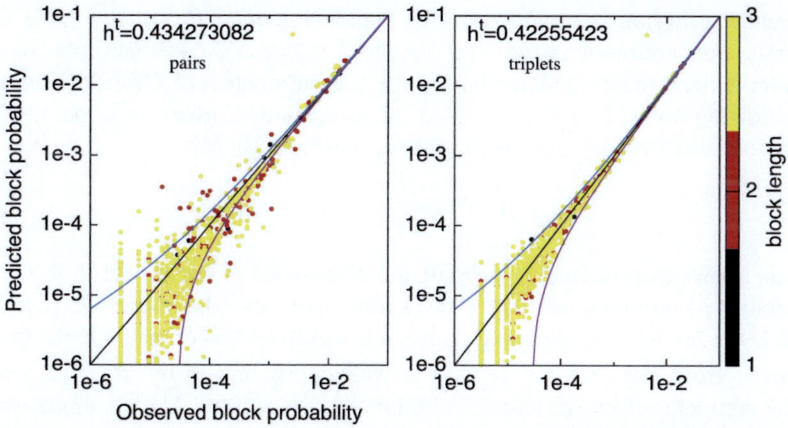

Fig. 8.7 Analysis of salamander retina data, from [75]. The estimated block probability versus the observed block probability for all blocks from range 1–4 (coded by *colors*), for $N = 4$ neurons with a model of range $R = 3$ for pairs and triplets. We include the equality line $y = x$ and the confidence bounds (*black lines*) for each model, corresponding to $\pi^{(T)}(w) \pm 3\sigma_w$ with σ_w being the standard deviation for each estimated probability given the total sample length $T \sim 3 \cdot 10^5$. In the figure, h^t corresponds to $\tilde{h}$, Eq. (8.31)

There is a wide literature in statistics dealing with the subject. In the realm of spike train analysis an important reference is [48] and references therein. Here, we point out two criteria for model comparison, used in this chapter as an illustration.

A first and straightforward criterion consists of computing the empirical probability of blocks of range $1, 2, \ldots$ and to compare it to the probability predicted by the model. Of course, the number of blocks of range R increases like 2^{NR}; moreover the probability of large blocks is expected to decrease fast with the block range. So, practically, one considers a subset of possible blocks. The challenge is in fact to predict the probability of events which have not been included as constraints in the Gibbs potential of the model. For example, does an Ising model well predict the probability of occurrence of triplets, quadruplets, of non simultaneous spikes?

The typical representation of this criterion is a graph, with, on abscissa, the observed probability of blocks and, on ordinate, the predicted probability. Thus, to each block corresponds a point in this two-dimensional graph. A "good" model is such that all points spread around the diagonal $y = x$. The distance to the diagonal is expected to increase as the probability of the block decreases thanks to the central limit theorem. Indeed, if the exact probability of a block is P, then the empirical estimation of this probability is a random variable with a Gaussian distribution, of mean P, and a variance that can be computed from the topological pressure. A usual approximation of this variance is $P(1-P)$. Thus, in a similar way as in Fig. 8.6, the set of points in the graph spreads around the diagonal in a region delimited by the curves $\pm 3\sqrt{\frac{P(1-P)}{T}}$ called "confidence bounds". An example is given in Fig. 8.7.

Another criterion is provided by the *Kullback-Leibler divergence (KL)* which provides some notion of asymmetric "distance" between two probabilities. Its computation is numerically delicate but, in the present context of Gibbs distributions, the following holds. If μ is the hidden (time-translation invariant probability) and μ_β a Gibbs distribution with a potential ϕ_β, one has, [10, 28]:

$$d(\mu, \mu_\beta) = \mathcal{P}(\phi_\beta) - \mu[\phi_\beta] - h(\mu). \tag{8.30}$$

This allows in principle to estimate the divergence of our model to the hidden probability μ, providing the exact spike train statistics. The smaller $d(\mu, \mu_\beta)$ the better is the model. Unfortunately, since μ is unknown this criterion looks useless. However, from Sect. 8.3.2.3, $\mu[\phi_\beta]$ is well approximated by $\pi_\omega^{(T)}[\phi_\beta]$ which can be computed from the raster. Additionally, the entropy $h(\mu)$ is unknown and its estimation by numerical algorithms for a large number of neurons is delicate [68]. However, when considering two statistical models $\mu_{\beta_1}, \mu_{\beta_2}$ with potentials $\phi_{\beta_1}, \phi_{\beta_2}$ to analyze the *same* data, $h(\mu)$ is a constant (it only depends on data). Thus, comparing these two models amounts to comparing $P[\phi_{\beta_1}] - \pi_\omega^{(T)}[\phi_{\beta_1}]$ and $P[\phi_{\beta_2}] - \pi_\omega^{(T)}[\phi_{\beta_2}]$. Thus, the quantity

$$\tilde{h}[\phi] = P[\phi] - \pi_\omega^{(T)}[\phi], \tag{8.31}$$

provides a relative criterion to compare models, i.e., determining if model ϕ_{β_2} is significantly "better" that model ϕ_{β_1}, reduces to the condition:

$$\tilde{h}[\phi_{\beta_2}] \ll \tilde{h}[\phi_{\beta_1}]. \tag{8.32}$$

Its computation is detailed in [74, 75].

8.4 Using Gibbs Distributions to Analysing Spike Trains Statistics

In this section we show how the statistical tools presented in this chapter can be used to analyze spike trains statistics. In the "challenge" section we mention the current controversy about the question: Are G cells sensors independent encoders or, on the opposite, are neural correlations important for coding? We present here recent works where Gibbs distributions has been be used to address this question with important implications on neural coding. However, as we show, those examples also raise additional and fundamental questions. Some of them that can be addressed on theoretical grounds, by studying neural networks models. A third section presents an example of such a model where spike trains is known to have a Gibbs statistics and where the potential is explicitly known. We compare those results to the current state of the art in spike train analysis with Ising distributions.

8.4.1 Are Ganglion Cells Independent Encoders?

This question can be now reformulated in the context of Gibbs distributions. Independence between neurons means that spike statistics is described by a potential, possibly non-stationary, of the form:

$$\phi_n(\omega) = \sum_{k=1}^{N} \phi_{n,k}(\omega_k). \tag{8.33}$$

This assumption can be stated independently on the memory depth, so we write here $\phi_n(\omega)$ instead of $\phi_n(\omega_n^{n-D})$ to alleviate notations and to be as generic as possible. In (8.33) ω_k is the spike train $\{\omega_k(l)\}_{l\leq n}$ produced by neuron k only. In this way the transition probabilities of the global network are products of transition probability for each neuron and the Gibbs distribution (8.5) is a product of marginal distributions for one neuron. On the opposite, if one believes that spike correlations play an important role in statistics one has to include them in the Gibbs potential. Typically, spike correlations are characterized by monomials and the potential takes the generic form (8.29). Obviously, there are many possible choices for this potential, depending on the set of observables assumed to be relevant.

To compare different models one can use the criteria described in Sect. 8.3.2.12. Does an independent model predicts correctly the spike blocks occurring in the observations? If not, which correlations has to be included? How evolves the Kullback-Leibler divergence as the type of correlations (monomials) taken into account growth?

However, the application of those criteria is delicate in experimental data, taking into account the large number of cells, their different types and the relatively small size of spike train samples obtained from experiments. For these reasons analysis of retina data has been performed either for memory-less models where the number of neurons can be up to 100 neurons, or to models with memory with small range and a small number of neurons. Let us present some of those works.

8.4.2 Weak-Pairwise Correlations Imply Strongly Correlated Network States in a Neural Population

A breakthrough in spike train analysis has been made by the seminal paper of E. Schneidman and collaborator [60]. They have considered carefully instantaneous pairwise spike correlations in experiments on the vertebrate retina. It appears that, mostly, those correlations are *weak*. But are they *significant*? In Sect. 8.3.2.4 we have shown that weak correlations can be hardly interpretable without further analysis. The maximal entropy principle provides a way to quantify the role of those correlations. Take a statistical model (a Gibbs potential) without constraint on instantaneous pairwise spike correlations, and fix only constraints on rates. This provides a Bernoulli model with a potential (8.24). Take another model where pairwise correlations are taken into account (but no higher-order interactions). This is an Ising model with potential (8.26). The authors compare those two models using the comparison between estimated block probability versus the observed block probability for *spatial* blocks, but also some other criteria such as the probability of having n spiking G cells in a time window of 20 ms.

Their work convincingly shows that although pairwise correlations are *weak*, they are nevertheless *necessary* to characterize spike trains statistics. The Ising model clearly does quite better than a Bernoulli model and the authors claim to "predict" about 90 % of the multi spiking structure of a large G cells population in salamander and guinea pig. Note however that they focus on spatial pattern. This shows that weak correlations between pairs of neurons coexist with strongly collective behavior in the responses of ten or more neurons. Similarly, Shlens et al. [65] predict 99 % of a complete ON and OFF parasol G cells population in primates with a Ising model . This suggests that the neural code is dominated by correlation effects.

On the other hand, this work does not directly unravel the importance of neural correlation for carrying information. Moreover, in [54], it has been shown that although Ising model is good for small populations, this is an artifact of the way data is binned and of the small size of the system. Additionally, it might be questionable whether more general forms of Gibbs distributions (e.g., involving more general monomials) could improve the estimation and account for deviations to Ising-model [42, 66, 71] and provide a better understanding of the neural code from the point of view of the maximal entropy principle [27]. Very recently, Ganmor and collaborators [19] have extended the maximal entropy principle introducing higher order *instantaneous* spikes correlations. Triplets and so on are considered although all spikes arise at the same time. This therefore still corresponds to a memory-less model. These authors have convincingly shown that such model describes more accurately retina responses to natural images than Ising model. In particular spatio temporal patterns where well predicted: binary words of 10 retinal G cells over 10 time steps. Note however that those data are binned (Sect. 8.5.1 for a definition and a discussion).

As a matter of fact, back to 1995, [36] already considered multi-unit synchronizations and proposed several tests to understand the statistical significance of spike synchronizations. A few years later, [35] generalized this approach to arbitrary spatio-temporal spike patterns and compared this method to other existing estimators of high-order correlations or Bayesian approaches. More recently, several papers have pointed out the importance of temporal patterns of activity at the network level [31,63,76]. As a consequence, a few authors [2,34,53] have attempted to define time-dependent Gibbs distributions on the base of a Markovian approach, but with one time step memory only, and under assumptions such as detailed balance, [34] or conditional independence between neurons, see Eq. (1) in [52]. A more general method relying on Gibbs distributions has been proposed in [75], and applied to the same data as [60], in [75]. The results shows a clear increase in the model accuracy (measured with the KL divergence (8.30)) when adding *spatio-temporal* constraints.

In some sense, it is clear that the more parameters (or constraints) the more the model fits the data, up to some limit, fixed by the data and especially the raster size, where models become indistinguishable. Is there a minimal statistical model? A related question is: what does a statistical model teach us about the underlying neural network (e.g., the retina) and about neural coding? Let us first present an important work addressing the second question, before addressing the first one.

8.4.3 The Architecture of Functional Interaction Networks in the Retina

A critical question to elucidating the neural code, at least from a theoretical point of view, is to confront multi-electrodes real data against sophisticated statistical models testing for the underlying neural structure involved. Since the connectivity of any neural network can in principle growth exponentially as a function of the number of neurons, classical numerical methods can rapidly become inefficient. The work of Schneidman and collaborators [60] has nevertheless suggested that although the number of possible activity patterns and underlying interactions is exponentially large in a neural network, a pairwise-based Ising model gives a surprisingly accurate description of neural population activity patterns. So, an economical assumption to reducing a putative network dimensionality, is to use a pairwise correlation model able to recover the main network structure involved in the encoding of a natural movie [60].

More recently, searching for further reduction on the dimensionality of the network structure, Ganmor et al. [18] have shown the presence of small groups of neurons having strong correlated activity. As an outcome of their MEA spike train analysis the authors introduce the notion of *functional connectivity*. This corresponds to associating with the parameters in the Gibbs potential a network of "effective" interactions (β_{ij} for spike pairs, β_{ijk} for triplets, ...).

The performance of their model is clearly higher than when assuming independence between neurons. It is able to predict the neural activity, including synchrony, for larger neural networks as in [60] and the contribution of small groups of neurons to the complete population activity. To further reduce the structure of the network to most critical neural interactions, the authors apply a nearest-neighbors paradigm. An interesting result is that the reduction of close to 50 % of the original pairwise interactions does not change the accuracy of predictions, and models most functional groups of nearest neurons with an accuracy >95 %. Additionally, small functional overlapping units (10–20 neurons) seem to be a critical structure for the encoding of natural movies stimulus.

This work, together with [19], shows that a Gibbs potential with a relatively small number of parameters, corresponding to effective interactions between neurons, is able to reproduce spike trains of populations of neurons in the retina submitted to natural images. The network of these effective interactions is organized in a hierarchical and modular manner so that large network models can be constructed from smaller sub-networks in a modular fashion. Thus, by a suitable scaling of parameters, one could be able to extrapolate the Gibbs potential of a small population of neurons to large populations. Moreover, in some sense, this effective network "underlies the code", from the terminology of the authors. This means that the spike generation, as a response to a stimulus (an image), results from a dynamical process which can be summarized by the Gibbs potential of the model.

This work raises however several questions. First, the potential considered by the authors is memory-less. No time dependent process takes place in the potential. In some sense, the time-causality expected in a neural network is hidden by the effective potential. Another critical aspect is the interpretation of the effective interaction. It is stated in [18] that "although the pairwise interactions in the model do not necessarily reflect a physical interaction between cells, they give a unique functional interaction map between the neurons in the network and represent statistical dependence between pairs of units." But, if they do not represent physical interactions (synapses), what do these functional interactions correspond to? To our knowledge this question has not been yet resolved.

8.4.4 Spike Train Analysis in a Neural Network Model

The maximal entropy principle relies on the assumption of stationarity as well as an a priori and somewhat ad hoc choice of observables. This choice severely constrains the form of the statistical model and the information that can be extracted about the underlying neuronal network producing the spike. In particular, the choice of observables determines the transition probabilities and implicitly fixes a causal structure to analyze spike events. Especially, memory-less models focuses on synchronous events, hiding somewhat temporal causality.

Obviously, it is extremely difficult to obtain a clear cut characterization of spike trains statistics from experimental data, taking into account the experimental set up, spike acquisition, spike sorting, but also the relatively small size of spike trains (typically, in retina experiments $T < 10^{5-6}$). So, a natural question is: "Can one have a reasonable idea of what spike statistics could be in neural network *model*"? The answer is "yes".

In neural networks spikes result from the collective and non linear dynamics of neurons coupled by synapses (electrical or chemical) and submitted to "external" stimuli. As a consequence statistics of spike train is closely related to this network structure (neurons and synapses characteristics) and to the stimulus. The idea is to show here the relationships between the neural network structure and the form of transition probabilities in an explicit example, a conductance-based neural network model of Integrate-and-Fire (IF) type called "generalized Integrate-and-Fire" (gIF) and introduced by Rudolph and Destexhe [56]. This section summarizes the paper [7].

8.4.4.1 The gIF Model

We consider the evolution of a set of N neurons. Here, neurons are considered as "points" instead of spatially extended and structured objects. As a consequence, we define, for each neuron $k \in \{1 \dots N\}$, a variable $V_k(t)$ called the "membrane potential of neuron k at time t" without specification of which part of a real neuron (axon, soma, dendritic spine, ...) it corresponds to.

Fix a firing threshold θ. The sub-threshold dynamics of the model is:

$$C_k \frac{dV_k}{dt} + g_k\,(t,\omega\,)\,V_k = i_k(t,\omega). \tag{8.34}$$

C_k is the membrane capacity. $g_k\,(t,\omega\,)$ is the integrated conductance at time t given the past spike activity encoded in the raster ω. It is defined in the following way. Denote $\alpha_{kj}\,(t-\tau\,)$ the synaptic response of neuron k, at time t, to a pre-synaptic spike from neuron j that aroused at time τ. Classical examples of synaptic responses are $\alpha_{kj}\,(t\,) = e^{-\frac{t}{\tau_{kj}}} H(t)$ or $\alpha_{kj}\,(t\,) = \frac{t}{\tau_{kj}} e^{-\frac{t}{\tau_{kj}}} H(t)$, where H the Heaviside function (that mimics causality) and τ_{kj} is the characteristic decay times of the synaptic response. In gIF model the conductance $g_k\,(t,\omega\,)$ integrates the synaptic responses of neuron k to spikes arising in the past. Call $t_j^{(r)}(\omega)$ the r-th spike-time emitted by neuron j in the raster ω, namely $\omega_j(n) = 1$ if and only if $n = t_j^{(r)}(\omega)$ for some r. Then

$$g_k\,(t,\omega\,) = g_{L,k} + \sum_{j=1}^{N} G_{kj}\alpha_{kj}\,(t,\omega\,), \tag{8.35}$$

where $\alpha_{kj}\,(t,\omega\,) = \sum_{t_j^{(r)}(\omega)<t} \alpha_{kj}\left(t - t_j^{(r)}(\omega)\right)$ sums up the spike responses of post-synaptic neuron k to spikes emitted by the pre-synaptic neuron j at times $t_j^{(r)}(\omega) < t$. $g_{L,k}$ is the leak conductance of neuron k.

Returning to Eq. (8.34) the term $i_k(t,\omega)$ is a current given by $i_k(t,\omega) = g_{L,k}\,E_L + \sum_{j=1}^{N} W_{kj}\,\alpha_{kj}\,(t,\omega\,) + i_k^{(ext)}(t) + \sigma_B \frac{dB_k}{dt}$, where E_L is the Nernst leak potential, W_{kj} is the synaptic weight from j to k, $i_k^{(ext)}(t)$ the (time-dependent) external stimulus received by neuron k, and dB_k is a Brownian noise whose variance is controlled by σ_B.

As in all IF models, when V_k reaches the firing threshold θ, it is reset. Here, however, it is not instantaneously reset to a constant value, but to a Gaussian random variable with mean zero and variance σ_R^2, after a time delay $\delta > 0$ including the depolarization-repolarization and refractory period. We call $\tau_k\,(t,\omega\,)$ the *last time before t where neuron k has been reset.*

8.4.4.2 Membrane Potential Decomposition

Given the spike history of the network it is easy to integrate (8.34) and to find an explicit expression for the membrane potential at time t. It depends on the past spike history ω. Denote $V_k(t,\omega)$ the membrane potential at time t given the spike history before t. Set:

$$\Gamma_k(t_1,t_2,\omega) = e^{-\frac{1}{C_k}\int_{t_1}^{t_2} g_k(\,u,\omega\,)\,du}, \tag{8.36}$$

corresponding to the flow of (8.34). We have $V_k(t,\omega) = V_k^{(det)}(t,\omega) + V_k^{(noise)}(t,\omega)$,

$$V_k^{(det)}(t,\omega) = V_k^{(syn)}(t,\omega) + V_k^{(ext)}(t,\omega) \tag{8.37}$$

where

$$V_k^{(syn)}(t,\omega) = \frac{1}{C_k}\sum_{j=1}^{N} W_{kj}\int_{\tau_k(\,t,\omega\,)}^{t} \Gamma_k(t_1,t,\omega)\,\alpha_{kj}\,(\,t_1,\omega\,)\,dt_1, \tag{8.38}$$

is the synaptic interaction term which integrates the pre-synaptic spikes from the last time where neuron k has been reset. Additionally,

$$V_k^{(ext)}(t,\omega) = \frac{1}{C_k}\left[g_{L,k}\,E_L + \int_{\tau_k(\,t,\omega\,)}^{t} i_k^{(ext)}(t_1)\,\Gamma_k(t_1,t,\omega)dt_1\right], \tag{8.39}$$

contains the stimulus $i_k^{(ext)}(t)$ influence. Thus $V_k^{(det)}(t,\omega)$ contains the deterministic part of the membrane potential. On the opposite, $V_k^{(noise)}(t,\omega)$ is the stochastic part of the membrane potential. This is a Gaussian variable with mean zero and variance

$$\sigma_k^2\,(\,t,\omega\,) = \Gamma_k^2(\tau_k\,(\,t,\omega\,)\,,t,\omega)\,\sigma_R^2 + \left(\frac{\sigma_B}{C_k}\right)^2\int_{\tau_k(\,t,\omega\,)}^{t} \Gamma_k^2(t_1,t,\omega)\,dt_1. \tag{8.40}$$

The first term in the right-hand side comes from the reset of the membrane potential to a random value after resetting. The second one is due the integration of synaptic noise from $\tau_k\,(\,t,\omega\,)$ to t.

8.4.4.3 Statistics of Raster Plots

It has been shown in [7] that the gIF model (8.34) has a unique Gibbs distribution with a *non-stationary* potential:

$$\phi_n\,(\,\omega\,) = \sum_{k=1}^{N}\phi_{n,k}\,(\,\omega\,), \tag{8.41}$$

with[2],

$$\phi_{n,k}(\omega) \stackrel{\mathrm{def}}{=} \omega_k(n)\log\pi\left(X_k(n-1,\omega)\right) + \left(1-\omega_k(n)\right)\log\left(1-\pi\left(X_k(n-1,\omega)\right)\right). \qquad (8.42)$$

Here

$$X_k(n-1,\omega) = \frac{\theta - V_k^{(det)}(n-1,\omega)}{\sigma_k(n-1,\omega)}, \qquad (8.43)$$

and

$$\pi(x) = \frac{1}{\sqrt{2\pi}} \int_x^{+\infty} e^{-\frac{u^2}{2}}\, du. \qquad (8.44)$$

As announced in the beginning of this section, the Gibbs potential that constraints spike statistics summarizes several contributions. The effect of the synaptic network structure, integrated over the past spike history as well as over the contribution of the stimulus (external current), also integrated over the past spike history, appears in the term $V_k^{(det)}(n-1,\omega)$. The synaptic noise and the reset to a random value, integrated over the history, depends on the term $\sigma_k(n-1,\omega)$. We insist that the result holds for a *time-dependent external current, i.e., a non-stationary dynamics.*

The potential exhibits therefore clearly the causal structure of spikes generation. The probability to have a spike at time n depends *explicitly* on synaptic weights, defining the neural network structure, and on the stimulus, via the external current. Moreover, it introduces a clear history dependence. Now, there are several important remarks.

1. The number of parameters on which spike train statistics depend is relatively small. It increases like a polynomial in the number on neurons, e.g., at most N^2 synaptic weights.
2. Those parameters are physical parameters: the synaptic weights between neurons, the stimulus, plus physical characteristics of the neuron such as leak Nernst potential. Thus, the potential outlines a network which is not *effective*, as in Sect. 8.4.3, but is the *real* underlying network. Additionally, this potential is *non-linear* function of the synaptic weights and stimulus while Ising-like models are *linear* in the interactions.
3. The memory is mathematically infinite. In other words, to handle the spike statistics *exactly*, one has to consider a non-Markovian potential with an infinite history ($D \to -\infty$).

[2] This is an approximation of the exact potential holding when the noise variance is small.

8.4.4.4 From Non-Markovian to Markovian Potentials

Since the dependence on the past decays exponentially fast, thanks to the exponential decay of synaptic response, it is possible to provide Markovian approximations of the potential (8.41). Basically, one truncates the synaptic response after a characteristic time D and performs a series expansion of the function (8.44), using the fact that the power of a monomial is the same monomial. So, the series becomes a polynomial, which provides a Markovian potential of the form (8.29). Here, the coefficients $\beta_{i_1,n_1,\dots,i_l,n_l}(n)$'s depend explicitly on the synaptic weights (network structure) as well as on the stimulus. Now, for N neurons and a memory depth D, the truncated potential contains 2ND coefficients $\beta_{i_1,n_1,\dots,i_l,n_l}(n)$, while the exact ($D \to -\infty$) potential depends only on a polynomial number of parameters. This shows that, in this model, a potential of the form (8.29) induces a strong, and somehow pathological, redundancy.

Additionally, the truncated potential is *far from Ising, or more elaborated potentials used in experimental spike train analysis*. As we have seen most of these models are memory-less and non causal. Now, the best approximation of the potential (8.41) by a memory-less potential is... Bernoulli. This is because of the specific form of ϕ: a term $\omega_k(0)$ multiplied by a function of $\omega_{-\infty}^{-1}$. To have a memory-less potential one has to replace this function by a constant, giving therefore a Bernoulli potential. So, Ising model as well as memory less models are rather poor in describing the statistics of model (8.34). But, then, how can we explain the success of Ising model to analyze retina data? We return to this point in the conclusion section.

8.4.4.5 Are Neurons Independent?

For this model we can answer the question of neurons independence. The potential (8.41) is a sum over neurons, similarly to (8.33), but it has not the same form as (8.33). The difference is subtle but essential. While in (8.33) the potential of neuron k, ϕ_k depends upon the past via the spike-history ω_k of *neuron k only*, in (8.41) ϕ_k depends upon the past via the spike-history ω of the *entire network*. The factorization in (8.41) reflects only a conditional independence: if the past spike history of the network is fixed then the only source of randomness is the noise, which is statistically independent by hypothesis, for all neurons. So, there is nothing deep in the factorization (8.41). On the opposite, a factorization like (8.33) would reflect a deeper property. Neurons would somewhat be able to produce responses which are well approximated by a function of their own history only, although each neuron receives inputs from many other neurons. Considering the form of the potential ϕ, given by (8.42) there is no chance to obtain the strong factorization property (8.33) unless neurons are disconnected. This property could however arise if the model obeys a mean-field theory as the number of neurons tends to infinity. This requires, in general, strong constraints on the synaptic weights (vanishing correlations), not necessarily realistic.

8.4.4.6 Links with the Retina

This model is not sufficient to describe the retina since it neglects the specific types of bipolar, horizontal and some amacrine cells that do not "fire". Additionally, it neglects electric synapses (gap junctions) playing an important role in the connectivity as shown in Fig. 8.1b. Recent investigations show that the conditional factorization property (8.41) *disappears* in the presence of gap junctions, so that statistics is expected to be even more complex, with no independence at all (in preparation).

8.5 Conclusion

In this chapter we have attempted to give a short overview of recent questions related with the concept of spike train analysis. Taking as an example the case of the retina we have presented a summary of recent experimental progresses from MEA recording to spike train analysis. On the theoretical side, we have introduced a general formalism connecting spike train statistics to Markov chains and Gibbs distributions. This formalism looks appropriate since, on one hand it allows to recover the Gibbs distributions forms used currently in the literature of spike train analysis, and on the other hand it affords analytical developments to characterize spike train probabilities in neural networks models. Finally, we have presented three examples of recent successes in spike trains analysis. These examples are encouraging but raise salient questions that we would like now to address.

8.5.1 Ising or Not Ising?

In Sects. 8.4.2 and 8.4.3 we have outlined the relative success of Ising model to analyze retina data, while in Sect. 8.4.4.4 we have computed explicitly the potential and concluded that it is quite far from Ising. What is the reason of this discrepancy? A first explanation, exposed in Sect. 8.4.4.6, is that the model (8.34) is not a good model for the retina. Another possible reason is the difference of *time scales* considered in both approaches. While the theoretical results of Sect. 8.4.4 consider neurons dynamics at the time scale of a spike (about 1 ms), statistical analysis of experimental data use, in all the examples we know, *data binning*. From preliminary analyzes of spike train (correllograms), one extracts a characteristic time scale τ (about 10–20 ms) from which spike trains are binned. Recall that a binned spike train is a raster Ω, obtained by cutting the original raster ω into time-slices of size τ and setting $\Omega_k(n) = 1$ in the slice n if and only if neuron k as fired at least once in this slice. In this way, one smooths out the dynamical interactions occurring at a time scale smaller than τ (especially synaptic interactions). So the coefficient β_{kj}'s in a binned-Ising model with a binning of 10–20 ms somewhat integrate the synaptic transmission effects and neurons pairwise interactions appear as instantaneous. In this way, one looses an important part of the dynamics and of the network structure. The "functional interactions" evoked in Sect. 8.4.3 corresponds to an integration of physical interactions over the binning time scale. For example, in the Ising model, the pairwise coefficient β_{kj} integrates the effect of several circuits connecting neuron j to neuron k, as well as dynamic-dependent effects. As a matter of fact its interpretation is rather delicate.

This is however certainly not the end of the story and this aspect has to be still investigated on the theoretical and experimental side.

8.5.2 Linear Potentials Versus Combinatorial Explosion

Experimental attempts to go "beyond Ising" [34, 73] suggest that Markovian models with increasing range should describe better and better the spike statistics. This is also expected from the theoretical analysis summarized in Sect. 8.4.4. However, this raises several remarks and questions. First, it is evident that the more parameters, the best is the model, but what do we learn from this plethora of parameters? Second, one has to face the critical issue of an exponential increase of parameters, with the potential range and with the number of neurons, so that numerical methods can rapidly become inefficient. Moreover, the sample size required to determine those coefficients is expected to increase also exponentially, ruining any hope to extract reliable coefficients from empirical data. Finally, as outlined in the previous section, the interpretation of the coefficients is difficult even for the simplest pairwise interactions β_{kj}.

Our point of view is that the linear potential approach, based on the maximal entropy principle, is maybe inappropriate. On the opposite, non linear potentials of the form (8.41), truncated to a finite memory depend on a number of parameters, the physical parameters of the network, which increases only polynomially with the number of neurons. Although, the number of blocks determining the potential increases exponentially fast with the memory depth, it could well be that only a small proportion of blocks are sufficient to extract most of the information about the hidden parameters. Finally, the interpretation of parameters is here straightforward and such a model allows to treat the non-stationary case. This may provide an alternative to Ising-like models to study spike train statistics in experiments.

8.6 Outlook

In this section we would like to point out a few challenges for the next future, on the theoretical and experimental sides.

8.6.1 Gibbs Distributions and the Neural Code

Although interesting results have come out from the Gibbs analysis of retina spike trains, the link between spike statistics and their modeling with Gibbs distribution on one hand, and the way how a visual scene is encoded by G cells spikes emission on the other hand, remains rather tight. Defining a priori a Gibbs potential from a set of constraints superimposes upon spike trains a causal structure, purely spatial or spatio-temporal, associated with "effective interactions", e.g., the coefficients $\beta_{i_1,n_1,\ldots,i_l,n_l}(n)$ in the polynomial expansion (8.29). What do these interactions teach us about the neural code? How are they related to a visual scene? Given a Gibbs distribution that fits well retina spike statistics is it possible to infer information about the visual scene perceived by this retina? Is it possible to build retina "decoders" from Gibbs statistics? If yes, does a "correlated decoder", with correlated G cells perform better than a "rate decoder" where G cells outputs are independent? Although interesting advances has been done on these questions (see, e.g., [59]) we believe that they are far from being solved, and that they constitute a challenge for the next years.

A related question concerns the concept of receptive field. We have presented in Sect. 8.2.1.2 the classical notion of RF which is associated with an isolated G cell, *independently of what the neighboring G cells* are perceiving. Additionally, e.g., Fig. 8.3, describes well the response of a G cell in terms of firing rates without need of considering higher order statistics. It is a current approach to model RF as filters, typically linear, followed by a non-linear correction [46]. How does the paradigm of linear-non linear filter connects with the paradigm of Gibbs distribution? Can we infer the shape of the RF filter from the Gibbs potential? Classical RF filters are based on firing rates for individual G cells; on the opposite Gibbs statistics deals with spike events (monomials) for collective behaviors. Are these two visions coherent? Is there a link e.g., between effective interactions and RF?

To our best knowledge those questions have not been resolved. On the theoretical side they can be addressed in the context of Gibbs distributions. Given a Gibbs potential modeling retina response it could be possible to compute the linear response to a stimulus considered as a weak perturbation of dynamics. This linear response is characterized by a convolution kernel which could be compared to the models of receptive field filters used in the literature. This work remains still to be done though.

8.6.2 Experimental Limits

To our opinion the current experimental set up is faced with the following limits.

- Natural stimulus must reproduce ecological environment, where animals lives, including the way how animals explore it, how they are in action, moving their head and eyes. As a consequence the captured images used in experiments need to be dynamically displayed to the retina, reproducing natural motions.

- Sophisticated experiment require MEA devices recording from a large numbers of cells (e.g., >100). For example a new MEA technology using 4,096 electrodes matrix makes it possible to recording most of the neurons in a single small surface, but the recording data string would take several gigabytes of saving space and requires fast technologies to access data. Moreover, their numerical analysis, spike sorting + spike train analysis, in any computer or even a cluster will take a very long time. Actually,
- The MEA outputs need adequate spike sorting algorithms able to deal with larger and larger numbers of cells. Current algorithms allow to treat about 200 electrodes signals.
- Spike train analysis necessitates adequate statistical tools applying to a large number of interacting cells to evaluate different possible models for neural encoding (e.g., population coding). Current algorithms allow to treat less than 100 neurons for an Ising model.
- A validation of any neural coding model is required. This is done by contrasting its performance against real behavioral sensory results for the animals under study [26]. Additionally, without the precise quantification of the animal performance for a particular behavioral task, responding to natural stimulus, it will not be possible to access the extended validity of any proposed model. Thus, both the animal capacity and the theoretical model need to be contrasted.

Clearly, those constraints constitute high level challenges for the scientific community. Probably, this is only the beginning of a long story.

8.7 Online Resources

8.7.1 Database

Webvision. The Organization of the Retina and Visual System

http://webvision.med.utah.edu/

This site summarizes recent advances in knowledge and understanding of the visual system through dedicated chapters and evolving discussion to serve as a clearing house for all things related to retina and vision science.

The brain from top to bottom: the retina

http://thebrain.mcgill.ca/flash/d/d_02/d_02_cl/d_02_cl_vis/d_02_cl_vis.html

This web site contains a series of topics dealing with the brain: "memory and the brain", "evolution and the brain" and so on. Each topic is developed at different levels: beginner, intermediate, advanced. This is a very useful and didactic reference.

Information Processing in the Retina

http://www.sumanasinc.com/webcontent/animations/content/receptivefields.html

This page contains an animation illustrating the functioning of On-Off Receptive Fields. This is part of the web site http://www.sumanasinc.com/webcontent/animations.html which contains nice animations on different topics, including neuroscience.

Multi-Electrode arrays

http://en.wikipedia.org/wiki/Multielectrode_array

From the famous web site wikipedia.

Paul Avery's Home Page. Image Gallery: Vision and the Eye

http://www.phys.ufl.edu/~avery/course/3400/gallery/gallery_vision.html

This page contains a series of nice pictures illustrating the functioning of eyes and vision.

8.7.2 Software

Event neural assembly Simulation

http://enas.gforge.inria.fr/v2/

EnaS is a library providing numerical tools for the mesoscopic simulation of neural networks (temporal computations in micro-columns models or spiking neural networks) and the analysis of spike trains either coming from neural simulators or from biological experiments.

Virtual Retina

http://www-sop.inria.fr/neuromathcomp/public/software/virtualretina/

Virtual Retina is a simulation software developed at Inria Sophia Antipolis Méditerannée by Adrien Wohrer during his Ph.D. (2005–2008) supervised by Pierre Kornprobst and Thierry Viéville.

Virtual Retina allows large-scale simulations of biologically-plausible retinas, with customizable parameters, and different possible biological features

Acknowledgements This work has been supported by ERC grant Nervi 227747 (BC), European grant BrainScales (BC), ANR-CONICYT grant (KEOPS), FONDECYT 1110292 (AP) and ICM-IC09-022-P (AP).

List of Acronyms

P	Photoreceptor
B cell	Bipolar Cell
H cell	Horizontal Cell
A cell	Amacrine Cell
RF	Receptive Field
IF	Integrate and Fire

References

1. *Principles of Neural Science. 4th edition*. McGraw-Hill, 2000.
2. S.-I. Amari. Information geometry of multiple spike trains. In Sonja Grün and Stefan Rotter, editors, *Analysis of Parallel Spike trains*, volume 7 of *Springer Series in Computational Neuroscience*, part 11, pages 221–253. Springer, 2010. DOI: 10.1007/978-1-4419-5675.
3. J.J. Atick. Could information theory provide an ecological theory of sensory processing? *Network: Computation in Neural Systems*, 3(2):213–251, 1992.
4. B.B. Averbeck, P.E. Latham, and A. Pouget. Neural correlations, population coding and computation. *Nat Rev Neurosci*, 7(5):358–66, 2006.
5. V. Balasubramanian and P. Sterling. Receptive fields and functional architecture in the retina. *J Physiol (Lond)*, 587(12):2753–67, 2009.
6. I.H. Brivanlou, D.K. Warland, and M. Meister. Mechanisms of concerted firing among retinal ganglion cells. *Neuron*, 20(3):527–39, 1998.
7. B. Cessac. Statistics of spike trains in conductance-based neural networks: Rigorous results. *Journal of Computational Neuroscience*, 1(8), 2011.
8. B. Cessac, H. Paugam-Moisy, and T. Viéville. Overview of facts and issues about neural coding by spikes. *J. Physiol. Paris*, 104(1–2):5–18, February 2010.
9. B. Cessac and T. Viéville. On dynamics of integrate-and-fire neural networks with adaptive conductances. *Frontiers in neuroscience*, 2(2), July 2008.
10. J.R. Chazottes and G. Keller. *Pressure and Equilibrium States in Ergodic Theory*, chapter Ergodic Theory. Encyclopedia of Complexity and System Science, Springer, 2009.
11. S. Cocco, S. Leibler, and R. Monasson. Neuronal couplings between retinal ganglion cells inferred by efficient inverse statistical physics methods. *PNAS*, 106(33):14058–14062, 2009.
12. I.P. Cornfeld, S.V. Fomin, and Y.G. Sinai. *Ergodic Theory*. Springer, Berlin, Heidelberg, New York, 1982.
13. J.B. Demb, K. Zaghloul, and P. Sterling. Cellular basis for the response to second-order motion cues in y retinal ganglion cells. *Neuron*, 32:711–721, 2001.
14. S.H. DeVries. Correlated firing in rabbit retinal ganglion cells. *Journal of Neurophysiology*, 81(2):908–920, 1999.
15. J.E. Dowling. *The retina: an approachable part of the brain*. Harvard University Press, Cambridge, Mass. (USA)., 1987.
16. R. Fernandez and G. Maillard. Chains with complete connections : General theory, uniqueness, loss of memory and mixing properties. *J. Stat. Phys.*, 118(3–4):555–588, 2005.
17. G.D. Field and E.J. Chichilnisky. Information processing in the primate retina: circuitry and coding. *Annu Rev Neurosci*, 30:1–30, 2007.
18. E. Ganmor, R. Segev, and E. Schneidman. The architecture of functional interaction networks in the retina. *The journal of neuroscience*, 31(8):3044–3054, 2011.
19. E. Ganmor, R. Segev, and E. Schneidman. Sparse low-order interaction network underlies a highly correlated and learnable neural population code. *PNAS*, 108(23):9679–9684, 2011.
20. W.S. Geisler. Visual perception and the statistical properties of natural scenes. *Annu. Rev. Psychol.*, 59:167–192, 2008.
21. W.S. Geisler, J.S. Perry, and A.D. Ing. Natural systems analysis. *Human Vision and Electronic Imaging XIII*, 6806:8060–M8060, 2008.
22. H.-O. Georgii. *Gibbs measures and phase transitions*. De Gruyter Studies in Mathematics:9. Berlin; New York, 1988.
23. I.I. Gikhman and A.V. Skorokhod. *The Theory of Stochastic Processes*. Springer, 1979.
24. T. Gollisch and M. Meister. Eye smarter than scientists believed: neural computations in circuits of the retina. *Neuron*, 65(2):150–164, January 2010.
25. J. M. Hammersley and P. Clifford. Markov fields on finite graphs and lattices. *unpublished*, 1971.
26. A.L. Jacobs, G. Fridman, R.M. Douglas, N.M. Alam, P.E. Latham, G.T. Prusky, and S. Nirenberg. Ruling out and ruling in neural codes. *Proc Natl Acad Sci U S A*, 106(14):5936–41, 2009.

27. E.T. Jaynes. Information theory and statistical mechanics. *Phys. Rev.*, 106:620, 1957.
28. G. Keller. *Equilibrium States in Ergodic Theory*. Cambridge University Press, 1998.
29. K. Koch, J. McLean, M. Berry II, P. Sterling, V. Balasubramanian, and M.A. Freed. Efficiency of information transmission by retinal ganglion cells. *Curr Biol*, 14(17):1523–30, 2004.
30. K. Koch, J. McLean, R. Segev, M.A. Freed, M.J. Berry II, V. Balasubramanian, and P. Sterling. How much the eye tells the brain. *Curr Biol*, 16(14):1428–34, 2006.
31. B.G. Lindsey, K.F. Morris, R. Shannon, and G.L. Gerstein. Repeated patterns of distributed synchrony in neuronal assemblies. *Journal of Neurophysiology*, 78:1714–1719, 1997.
32. N.K Logothetis. Vision: A window on consciousness. *Scientific American*, 281:44–51, 1999.
33. G. Maillard. *Introduction to chains with complete connections*. Ecole Federale Polytechnique de Lausanne, winter 2007.
34. O. Marre, S. El Boustani, Y. Frégnac, and A. Destexhe. Prediction of spatiotemporal patterns of neural activity from pairwise correlations. *Phys. rev. Let.*, 102:138101, 2009.
35. L. Martignon, G. Deco, K. Laskey, M. Diamond, W. Freiwald, and E. Vaadia. Neural coding: Higher-order temporal patterns in the neurostatistics of cell assemblies. *Neural Computation*, 12(11):2621–2653, November 2000.
36. L. Martignon, H. von Hasseln, S. Grün, A. Aertsen, and G. Palm. Detecting higher-order interactions among the spiking events in a group of neurons. *Biological Cybernetics*, 73(1):69–81, July 1995.
37. R. Masland. The fundamental plan of the retina. *Nature neuroscience*, 4(9), September 2001.
38. R.H. Masland and P.R. Martin. The unsolved mystery of vision. *Curr Biol*, 17(15):R577–82, 2007.
39. D.N. Mastronarde. Correlated firing of cat retinal ganglion cells. I. Spontaneously active inputs to X-and Y-cells. *Journal of Neurophysiology*, 49(2):303–324, 1983.
40. M. Meister, J. Pine, and D.A. Baylor. Multi-neuronal signals from the retina: acquisition and analysis. *J Neurosci Methods*, 51(1):95–106, 1994.
41. S. Nirenberg, S. M. Carcieri, A. L. Jacobs, and P. E. Latham. Retinal ganglion cells act largely as independent encoders. *Nature*, 411(6838):698–701, 2001.
42. I.E. Ohiorhenuan, F. Mechler, K.P. Purpura, A.M. Schmid, Q. Hu, and J.D. Victor. Sparse coding and high-order correlations in fine-scale cortical networks. *Nature*, 466(7):617–621, 2010.
43. B.A. Olshausen and D.J. Field. Natural image statistics and efficient coding. *Network*, 7(2):333–9, 1996.
44. S. Panzeri and S.R. Schultz. A unified approach to the study of temporal, correlational, and rate coding. *Neural Comput*, 13:1311–1349, 2001.
45. A. Petrusca, D. Grivich, M. Sher, A. Field, G. Gauthier, J. Greschner, M. Shlens, J. Chichilnisky, and E. Litke. Identification and characterization of a y-like primate retinal ganglion cell type. *J Neuros*, 27(41):11019–27, 2007.
46. J.W. Pillow, L. Paninski, V.J. Uzzell, E.P. Simoncelli, and E.J. Chichilnisky. Prediction and decoding of retinal ganglion cell responses with a probabilistic spiking model. *J. Neurosci*, 25:11003–11013, 2005.
47. A. Pouget, P. Dayan, and R. Zemel. Information processing with population codes. *Nat Rev Neurosci*, 1(2):125–32, 2000.
48. F. Rieke, D. Warland, R. de Ruyter van Steveninck, and W. Bialek. *Spikes: Exploring the Neural Code*. Bradford Books, 1997.
49. R. L. Rockhill, F. J. Daly, M. A. MacNeil, S. P. Brown, and R. H. Masland. The diversity of ganglion cells in a mammalian retina. *J Neurosci*, 22(9):3831–43, 2002.
50. R. W. Rodieck. Maintained activity of cat retinal ganglion cells. *J Neurophysiol*, 30(5):1043–71, 1967.
51. Y. Roudi, E. Aurell, and J.A. Hertz. Statistical physics of pairwise probability models. *Frontiers in Computational Neuroscience*, page 15, 2009.
52. Y. Roudi and J. Hertz. Mean field theory for non-equilibrium network reconstruction. *Phys. Rev. Lett.*, 106(048702), 2011.

53. Y. Roudi and J.A. Hertz. Mean field theory for non-equilibrium network reconstruction. *arXiv*, page 11, Sept 2010.
54. Y. Roudi, S. Nirenberg, and P.E. Latham. Pairwise maximum entropy models for studying large biological systems: when they can work and when they can't. *PLOS Computational Biology*, 5(5), 2009.
55. Y. Roudi, J. Tyrcha, and J.A. Hertz. Ising model for neural data: Model quality and approximate methods for extracting functional connectivity. *Physical Review E*, page 051915, 2009.
56. M. Rudolph and A. Destexhe. Analytical integrate and fire neuron models with conductance-based dynamics for event driven simulation strategies. *Neural Computation*, 18:2146–2210, 2006.
57. D. Ruelle. *Thermodynamic formalism*. Addison-Wesley, Reading, Massachusetts, 1978.
58. R. Sarpeshkar. *Ultra Low Power Bioelectronics: Fundamentals, Biomedical Applications, and Bio-Inspired Systems*. Cambridge University Press, 2010.
59. M.T. Schaub and S.R. Schultz. The ising decoder: reading out the activity of large neural ensembles. *arXiv:1009.1828*, 2010.
60. E. Schneidman, M.J. Berry II, R. Segev, and W. Bialek. Weak pairwise correlations imply strongly correlated network states in a neural population. *Nature*, 440(7087):1007–1012, 2006.
61. E. Schneidman, W. Bialek, and M.J. Berry II. Synergy, redundancy, and independence in population codes. *J Neurosci*, 23(37):11539–53, 2003.
62. G. Schwartz and M.J. Berry II. Sophisticated temporal pattern recognition in retinal ganglion cells. *J Neurophysiol*, 99(4):1787–98, 2008.
63. R. Segev, I. Baruchi, E. Hulata, and E. Ben-Jacob. Hidden neuronal correlations in cultured networks. *Physical Review Letters*, 92:118102, 2004.
64. E. Seneta. *Non-negative Matrices and Markov Chains*. Springer, 2006.
65. J. Shlens, G.D. Field, J. L. Gauthier, M.I. Grivich, D. Petrusca, A. Sher, A. M. Litke, and E.J. Chichilnisky. The structure of multi-neuron firing patterns in primate retina. *J Neurosci*, 26(32):8254–66, 2006.
66. J. Shlens, G.D. Field, J.L. Gauthier, M. Greschner, A. Sher, A.M. Litke, and E.J. Chichilnisky. The structure of large-scale synchronized firing in primate retina. *The Journal of Neuroscience*, 29(15):5022–5031, April 2009.
67. E.P. Simoncelli and B.A. Olshausen. Natural image statistics and neural representation. *Annu Rev Neurosci*, 24:1193–216, 2001.
68. S.P. Strong, R. Koberle, R.R. de Ruyter van Steveninck, and W. Bialek. Entropy and information in neural spike trains. *Phys. Rev. Let*, 80(1):197–200, 1998.
69. M. Taketani and M. Baudry. *Advances in Network Electrophysiology: Using Multi-Electrode Arrays*. Springer, 2006.
70. E. Thompson, A. Palacios, and F. Varela. Ways of coloring: Comparative color vision as case study for cognitive science. *Behavioral and Brain Sciences*, 15:1–75, 1992.
71. G. Tkačik, E. Schneidman, M.J. Berry II, and W. Bialek. Spin glass models for a network of real neurons. *arXiv*, page 15, 2009.
72. J. L. Van Hemmen and T.J. Sejnowski. 23 problems in systems neuroscience. *Oxford University Press, Inc.*, 2006.
73. J.-C. Vasquez, H. Nasser, A. Palacios, B. Cessac, T. Viéville, and H. Rostro-Gonzalez. Parametric estimation of spike train statistics by gibbs distributions : an application to bio-inspired and experimental data. In *Proceedings of Neurocomp 2010 (Lyon)*, 2010.
74. J.-C. Vasquez, A.G. Palacios, O. Marre, M.J. Berry II, and B. Cessac. Gibbs distribution analysis of temporal correlation structure on multicell spike trains from retina ganglion cells. *J. Physiol. Paris*, 2011. submitted.
75. J.-C. Vasquez, T. Viéville, and B. Cessac. Entropy-based parametric estimation of spike train statistics. *Inria Research Report*, 2010.
76. A.E.P. Villa, I.V. Tetko, B. Hyland, and A. Najem. Spatiotemporal activity patterns of rat cortical neurons predict responses in a conditioned task. *Proc Natl Acad Sci USA*, 96(3):1106–1111, 1999.
77. A. Wohrer and P. Kornprobst. Virtual retina: a biological retina model and simulator, with contrast gain control. *Journal of Computational Neuroscience*, 26(2):219–249, 2009.

Biology, Medicine and Biophysics Index

F. Cazals and P. Kornprobst (eds.), *Modeling in Computational Biology and Biomedicine*,
DOI 10.1007/978-3-642-31208-3,

Mathematics and Computer Science Index

F. Cazals and P. Kornprobst (eds.), *Modeling in Computational Biology and Biomedicine*,
DOI 10.1007/978-3-642-31208-3,

Overall Index

F. Cazals and P. Kornprobst (eds.), *Modeling in Computational Biology and Biomedicine*,
DOI 10.1007/978-3-642-31208-3, © Springer-Verlag Berlin Heidelberg 2013

MIX
Papier aus verantwortungsvollen Quellen
Paper from responsible sources
FSC® C105338

If you have any concerns about our products,
you can contact us on
ProductSafety@springernature.com

In case Publisher is established outside the EU,
the EU authorized representative is:
Springer Nature Customer Service Center GmbH
Europaplatz 3, 69115 Heidelberg, Germany

Printed by Libri Plureos GmbH
in Hamburg, Germany